"For sixty years, *Parabola* has entertained its readers with ~~articles, puzzles,~~ surveys, and problems aimed at encouraging students ~~of mathematics~~ at ~~second-~~ ary school level... The key word here is not 'mathema~~tics' but 'entertained.'~~

—Ian Stewart, from the foreword

Parabolic Problems

Parabola is a mathematics magazine published by UNSW, Sydney. Among other things, each issue of *Parabola* has contained a collection of puzzles/problems, on various mathematical topics and at a suitable level for younger (but mathematically sophisticated) readers.

Parabolic Problems: 60 Years of Mathematical Puzzles in Parabola collects the very best of almost 1800 problems and puzzles into a single volume. Many of the problems have been re-mastered, and new illustrations have been added. Topics covered range across geometry, number theory, combinatorics, logic, and algebra. Solutions are provided to all problems, and a chapter has been included detailing some frequently useful problem-solving techniques, making this a fabulous resource for education and, most importantly, fun!

Features

- Hundreds of diverting and mathematically interesting problems and puzzles.
- Accessible for anyone with a high school-level mathematics education.
- Wonderful resource for teachers and students of mathematics from high school to undergraduate level, and beyond.

AK Peters/CRC Recreational Mathematics Series

Series Editors

Robert Fathauer
Snezana Lawrence
Jun Mitani
Colm Mulcahy
Peter Winkler
Carolyn Yackel

Luck, Logic, and White Lies
The Mathematics of Games, Second Edition
Jörg Bewersdorff

Mathematics of The Big Four Casino Table Games
Blackjack, Baccarat, Craps, & Roulette
Mark Bollman

Star Origami
The Starrygami™ Galaxy of Modular Origami Stars, Rings and Wreaths
Tung Ken Lam

Mathematical Recreations from the Tournament of the Towns
Andy Liu, Peter Taylor

The Baseball Mysteries
Challenging Puzzles for Logical Detectives
Jerry Butters, Jim Henle

Mathematical Conundrums
Barry R. Clarke

Lateral Solutions to Mathematical Problems
Des MacHale

Basic Gambling Mathematics
The Numbers Behind the Neon, Second Edition
Mark Bollman

Design Techniques for Origami Tessellations
Yohei Yamamoto, Jun Mitani

Mathematicians Playing Games
Jon-Lark Kim

Electronic String Art
Rhythmic Mathematics
Steve Erfle

Playing with Infinity
Turtles, Patterns, and Pictures
Hans Zantema

Parabolic Problems
60 Years of Mathematical Puzzles in Parabola
David Angell and Thomas Britz

For more information about this series please visit: https://www.routledge.com/AK-PetersCRC-Recreational-Mathematics-Series/book-series/RECMATH?pd=published,forthcoming&pg=2&pp=12&so=pub&view=list

Parabolic Problems
60 Years of Mathematical Puzzles in Parabola

David Angell and Thomas Britz
UNSW Sydney, Australia

CRC Press
Taylor & Francis Group
Boca Raton London New York

CRC Press is an imprint of the
Taylor & Francis Group, an **informa** business

AN A K PETERS BOOK

Designed cover image: Helena Brusic The Imagination Agency

First edition published 2024
by CRC Press
2385 NW Executive Center Drive, Suite 320, Boca Raton FL 33431

and by CRC Press
4 Park Square, Milton Park, Abingdon, Oxon, OX14 4RN

CRC Press is an imprint of Taylor & Francis Group, LLC

© 2024 David Angell and Thomas Britz

Library of Congress Cataloging-in-Publication Data

Names: Angell, David (Mathematician), author. | Britz, Thomas
(Mathematician), author.
Title: Parabolic problems : 60 years of mathematical puzzles in Parabola /
David Angell and Thomas Britz.
Description: First edition. | Boca Raton : AK Peters ; CRC Press, 2024. |
Series: AK Peters/CRC recreational mathematics series | Includes
bibliographical references and index.
Identifiers: LCCN 2023057449 (print) | LCCN 2023057450 (ebook) | ISBN
9781032499987 (hardback) | ISBN 9781032483191 (paperback) | ISBN
9781003396413 (ebook)
Subjects: LCSH: Mathematical recreations. | Mathematics--Problems,
exercises, etc.
Classification: LCC QA95 .A545 2024 (print) | LCC QA95 (ebook) | DDC
793.74--dc23/eng/20240310
LC record available at https://lccn.loc.gov/2023057449
LC ebook record available at https://lccn.loc.gov/2023057450

ISBN: 978-1-032-49998-7 (hbk)
ISBN: 978-1-032-48319-1 (pbk)
ISBN: 978-1-003-39641-3 (ebk)

DOI: 10.1201/9781003396413

Typeset in Latin Modern font
by KnowledgeWorks Global Ltd.

Publisher's note: This book has been prepared from camera-ready copy provided by the authors.

In memory of
Charles Cox and George Szekeres,
with gratitude and admiration.

Contents

Foreword I

by Ian Stewart

"I FOUND A STONE but did not weigh it. After I added one seventh and added one eleventh, I weighed it: result, 1 [unit of weight]. What was the original weight of the stone?"

Apart from the slightly cryptic statement – added one seventh of what? One eleventh of what? – this could be a problem from a modern school text. It actually comes from a Babylonian clay tablet, date around 1800–1600 B.C. The unit of weight was the *mina*, and the answer is given as $\frac{2}{3}$ *mina*, 8 *sheqels* and $22\frac{1}{2}$ *se*. (1 *mina* = 60 *sheqels*, 1 *sheqel* = 180 *se*.) You can now reverse–engineer the question and figure out the cryptic bits.

The tablet appears to have been inscribed by a student at the tablet house, the Babylonian term for "school". Education consisted of copying texts, and probably learning them by heart. This presumably worked, given the high quality of Babylonian mathematics and astronomy, but even so, I don't recommend it. Education has moved on, and it's easier to *understand* than it is to memorise.

Another clay tablet records the process:

Son of the tablet house, where did you go in your early days?
I went to the tablet house.
I read out my tablet, ate my lunch.
Prepared my [new] tablet, inscribed it, finished it.
When the tablet house was dismissed, I went home.
I entered my house. My father was sitting there.
I read my tablet to him, and he was pleased.

As this document shows, little has changed in four thousand years. In particular, from time immemorial, maths teachers have sought to make their subject more palatable for students by phrasing exercises in the form of stories, puzzles, or intriguingly offbeat problems. I'm not thinking of questions like, "If a dog and a half digs a hole and a half in an hour and a half, how many holes will 17 dogs dig in 365 days?". Such efforts, mercifully defunct, tried to present matters of arithmetical technique in "practical" situations but fell flat because there was nothing remotely practical about the resulting problem.

This is not to say that you can't amuse yourself solving such problems. Mathematicians can find entertainment in the driest areas of their subject. (When I first typed that sentence it read "direst areas". Well, those too. Perhaps my computer was trying to tell me something.) I'll prove it. One dog, in one hour, still digs a hole and a half. Whatever half a hole may be. So 17 dogs, in one hour, dig $17 \times 1.5 = 25.5$ holes. In 365 days, they will dig $25.5 \times 24 \times 365 = 223,380$ holes. Assuming they keep assiduously to their task for an entire year, day and night, at a uniform rate.

It wasn't just the Babylonians. Alcuin of York was a polymath: scholar, poet, clergyman, and teacher. He was born in Northumbria, England, around 735, and died in 804. He taught at the Court of Charlemagne, and was described by a contemporary as "the most learned man anywhere to be found". Many scholars believe he was the author of a curious book, *Propositiones ad Acuendos Juvenes* (Problems to Sharpen Youths), and I'll assume they know what they're talking about. The evidence is a letter from Alcuin to Charlemagne of 799 referring to sending "certain figures of arithmetic for the joy of cleverness". *Whose* cleverness he didn't say: possibly that of the reader, of Charlemagne, or of Alcuin himself. The *Propositiones* is a collection of 53 puzzles in mathematics and logic, stated as "word problems" (a phrase that today strikes terror in the hearts of innumerable schoolchildren and university students, especially in America). He does give the answer, immediately after each puzzle. Another edition, known as the Bede text but probably not edited by the Venerable Bede, has three more problems, but only $34\frac{1}{2}$ answers. This time they are in their time–honoured place at the back of the book.

Alcuin's best-known puzzle is probably the parable of the wolf, the goat, and the cabbage. I'm sure you've come across this one, but bear with me.

A farmer is taking his goods to market: one of each of the aforementioned. They come to a river, with a boat that can hold the farmer plus *one* of the three items. (It is, presumably, a gigantic lettuce, and a sizeable goat also seems in order). The wolf wants to eat the goat and the goat is desperate to pig out on cabbage, but when the farmer is present, he can prevent nature from taking its course and ruining his market prospects. However, should he leave the wolf alone with the goat, on his return he will find that the goat has been wolfed. Ditto goat/cabbage, *mutatis mutandis*. Problem: cross the river without letting the goods consume each other.

It can be done in seven crossings, but no less, and in two symmetrically related ways. (You *have* to start by taking the goat over, and there's no point in bringing it back immediately, so you return without it. That's the first two crossings. Now take wolf or cabbage – either will do. *Bring the goat back...*) The lovely feature of this puzzle is that it leads to some beautiful, general mathematical methods. For example, there's a neat graphical solution using the eight vertices of a cube to represent which of the goods is on which side of the river. The state of play is a 3-vector (w, g, c), where an entry is 0 for one side of the river and 1 for the other. So, for instance, $(0, 1, 0)$ means the goat is on the far side while the wolf and cabbage are on the near side. The farmer is always assumed to be in the boat. Moves across the river correspond to edges of the cube; delete any edges that result in devoured produce. Now all you have to do is find a connected path from one corner to the diametrically opposite one.

Years ago I analysed the puzzle that way for the French magazine *Pour la Science*, having disguised it as a highly original conundrum featuring a lion, a llama, and a lettuce. Transported by farmer Al Quinn. I also discussed a beautiful link to graph–tracing algorithms and depth–first search, which have numerous practical uses. I also extended the problem to one with four items: leviathan, lion, llama, and lettuce. Now the leviathan will eat the lion *unless* a lettuce is also present, "for leviathans become docile when subjected to the smell of fresh lettuce". The transition graph becomes a four–dimensional hypercube with some edges deleted.

The original version with three items turns up all over the place. One is a Chinese legend involving a tiger mother with three cubs. One of these, following a traditional belief, is a leopard, much fiercer than the two tiger cubs, and prone to the odd tigerburger. The puzzle was one of Lewis Carroll's favourites, and it has even appeared on *The Simpsons*.

This example shows that stories, puzzles, and intriguingly offbeat problems can lead, in a painless and fun manner, to significant mathematics with important applications. Do not make the mistake of judging the mathematics by the manner of its presentation.

All of which, in an indirect manner, explains why I'm writing this Foreword.

For sixty years, *Parabola* has entertained its readers with mathematical articles, surveys, and problems aimed at encouraging students of maths at senior secondary school level (or equivalent, such as high school). The key word here is not "mathematics", but "entertained".

The late Graham Hoare, a much–loved maths teacher who was Head of Dr Challoner's Grammar School in England, often said that "Mathematics is not a spectator sport". The way to enjoy maths is to *engage* with it. (Warning: it will fight back. But the feeling of joy when you win makes the struggle worthwhile.) To aid this process, *Parabola* has generally included numerous puzzles and problems. To celebrate its 60th anniversary, the editors have issued this collection, thereby following the trail blazed by unknown Babylonian scribes, Alcuin of York, and *The Simpsons*. It's a *fantastic* collection of problems, from every area of senior–school mathematics. Some of them look like ordinary questions from a textbook, but those are deceptive: there's always an intriguing twist, a clever insight, that takes the problem out of the realm of the routine exercise. (Which is why my dog and a half won't cut it.)

The 329 stories/puzzles/offbeat problems in this book – more than six times as many as Alcuin managed – will help you hone your mathematical skills and stretch your ingenuity to the limit. You can dip in anywhere. Let me emphasise that although nothing is *easy*, every question is *fair*, in the sense that however weird it looks, you can get the answer in a reasonable period of time using standard senior school maths. Just not using it in the routine way you've probably been taught.

I'm sticking my neck out here, but it seems to me that it's not just kids who can benefit from something challenging on which to hone their mathematical skills. It's all of us. Me included. Being a retired Maths Professor is no excuse.

Maths isn't just something you do at school. It's what makes today's world function. There's more maths in your mobile phone than there is in any textbook on the subject. You can *use* hi–tech gadgets without knowing much maths, but no one could invent them that way. So we have to pass on mathematical skills to the next generation, and for self–protection it pays to keep our own up to scratch as well. As Hoare said, it's not a spectator sport.

Have fun.

Ian Stewart
Coventry
August 2023

Foreword II

by Adelle Coster and Bruce Henry

F UN AND CHALLENGING mathematical problems have been a primary focus of *Parabola* since it was established as a magazine for secondary school students in 1964. In the first issue, the editor, Charles Cox, stated that: "One of the aims of this magazine is to provide problems and puzzles that will test your insight, ingenuity and determination to the limit". The early issues of *Parabola* drew attention to famous mathematical problems including the Isoperimetric Problem: to find the curve of a given length that encloses the largest area; the Four Colour Problem: to show that no more than four colours are required to distinguish different regions on a map; and Fermat's Last Theorem: if n is an integer greater than 2, then there are no positive integers x, y, z for which it is true that $x^n + y^n = z^n$.

At the time of publication in *Parabola*, the solution to the Isoperimetric Problem was known, but solutions to the other two problems were not known. It was conjectured by many that the problems might never be solved. The Four Colour Problem was solved by Kenneth Appel and Wolfgang Haken in 1976; and Andrew Wiles obtained a proof for Fermat's Last Theorem in 1994. A simply stated, but famous and unsolved, mathematics problem of our time is to prove the Collatz Conjecture: if a sequence of positive integers is constructed in which each integer is determined from the previous integer in the sequence by dividing by two if it is even, or multiplying by three and then adding one if it is odd, then regardless of the integer chosen to begin the sequence, it must sooner or later reach the number one.

This book brings together the collection of problems from *Parabola*, curated by current *Parabola* Problems Editor David Angell and Editor Thomas Britz.

The creation of challenging and intriguing mathematical problems is done with ingenuity and the ability to stand on the shoulders of giants. The problems presented come from the great efforts of Charles Cox, David Angell and numerous other people over the years, often involving heroic efforts and sleepless nights formulating problems with exceptionally talented students in mind. The establishment of *Parabola*, with problems, as an enrichment resource for secondary school students was largely instigated by George Szekeres who was recruited as the first Chair in Pure Mathematics at the University of New South Wales, in 1963. George, who was born in Budapest, Hungary, had himself been inspired by the enrichment journal *Kőzépiskolai Matematikaiés Fizikai Lapok*. While studying at university he met with fellow students Paul Erdős, Paul Turán and Esther Klein, who all shared a passion for posing and solving mathematical problems. One of their shared problems, the so–called Happy Ending Problem, which led to the marriage of George and Esther, was featured in Volume 38 Issue 1 of *Parabola*, which celebrated George's 90th Birthday.

A few years before the establishment of *Parabola*, the School of Mathematics at the University of New South Wales had established a Mathematics Competition for Secondary School Students. The *Parabola* problems with solutions became an invaluable aid to students preparing for the competition, and, in their turn, the competition problems and solutions were published in *Parabola*. This has added to the abundance and depth of problems in

Parabola, and still does: the UNSW School Mathematics Competition is the longest-running competition of its type in Australia.

To this day, *Parabola* is proudly supported by the School of Mathematics and Statistics at UNSW Sydney. Via its online access at www.parabola.unsw.edu.au, it is now the foremost mathematical journal of its kind in the world, providing a valuable resource of articles and fun and fascinating problems to an ever–growing international readership. It also provides an opportunity for students and others to publish their contributed articles, problems and solutions.

The *Parabola* problems are an important and ongoing legacy that has helped to shape some of the brightest minds in Australia and now the world. We congratulate David Angell and Thomas Britz for their outstanding efforts in maintaining and strengthening this invaluable resource.

Adelle Coster (Head of School)
Bruce Henry (previous *Parabola* Editor)
Sydney
December 2023

Preface

IT IS NOW 60 YEARS since the School of Mathematics at the University of New South Wales published the first issue of *Parabola*, a magazine for secondary mathematics students. In furtherance of its principal aim – that of inspiring students with the highest level of enthusiasm for mathematics – *Parabola* has featured articles on accessible topics, news from the world of mathematics, cartoons, and reports on student mathematical competitions from the local to the worldwide. And puzzles. There have always been puzzles, problems, brainteasers – call them what you will – aimed at diverting readers. "Diverting" in both senses of the word: while we hope that readers will find *Parabola* problems entertaining, we also strive unceasingly to *divert* readers away from mere reading, and into an active participation in the mathematical endeavour. The suggestion one sometimes hears, that mathematical education from the first steps through to the end of an undergraduate degree consists essentially of "solving problems someone else has already solved", is a thoroughly misguided criticism. Ideas discovered for oneself are not the less valuable, educationally and psychologically, for being already known to others; following unseen footsteps is not all that different from exploring an untracked world.

PARABOLA

A Mathematics Magazine
for Secondary School Students

PUBLISHED BY THE UNIVERSITY OF NEW SOUTH WALES

Vol. 1, No. 1   July, 1964

Since its inception, *Parabola* has published (mostly) three issues per year, with (mostly) ten problems per issue. In this celebration of sixty years of *Parabola*, we present a selection of the almost 1800 puzzles that have appeared. Our first criterion is, very simply, that the problems chosen should be *fun!* While this, inevitably, really means "fun to us", we feel confident that readers will share our delight in the unexpected questions that can be asked, and the unexpected solutions that can be found through the application of generally school–level mathematics.

Problems in this book are not presented thematically, as is often the case in puzzle books, but in chronological order as they appeared in *Parabola*. We feel that one of the great attractions of the *Parabola* problems section has always been that it consists of a variety of separate problems in unrelated areas. Readers who wish to attempt problems in a specific area such as algebra, geometry or logic may consult the classification provided in Chapter 5. Some problems in this list are assigned to more than one category, as many of the best and most enlightening solutions employ techniques from varied and unexpected areas of mathematics. Assigning a problem to a specific area inevitably entails giving some kind of "spoiler" for the solution; therefore, readers who wish to solve the problems by their own unaided efforts should read this list only with caution.

Parabolic Problems is intended not only as a compilation of attractive puzzles, but also as a celebration of 60 years of *Parabola*. For this reason, problems have been taken verbatim, or nearly so, from the original publication. This means that some problems, particularly the earlier ones, have a rather dated feel to them. The mathematical issues addressed, however, are eternal, and, we trust, will not be diminished by the narrative in which they are embedded. Sometimes a diagram has been added to a problem statement: our rationalisation for this is that including diagrams in published material was an onerous task in the 1960s and 1970s, and we like to think that the original problem authors and editors would certainly have included more diagrams had it been practicable. In early issues (up to 1974), problems were classified as "junior", "intermediate" or "open": this distinction has been removed.

Solutions in this book have been closely modelled upon those originally published in *Parabola*. However, we have allowed ourselves a little more freedom in the solutions than we have in the problems, and have improved upon the originals wherever we felt it possible and appropriate to do so. *Parabola* was originally intended as a magazine for secondary school mathematics students in Australia. The problems published were meant to be such as could be solved by means of a simple yet insightful idea, requiring little or no mathematical technique beyond that taught up to year 10, or with methods that could easily be explained to thoughtful students at upper secondary level. Since 2014, *Parabola* has been freely available online, and its readership has expanded globally to students and adults with enthusiasm for recreational mathematics. In view of this, we have begun to allow ourselves some more advanced techniques in our solutions. For instance, problems involving calculus were rarely, if ever, set in the early years; this policy has now been relaxed. Nevertheless, it is of interest how frequently "calculus" problems can be solved by "non–calculus" methods, and we occasionally suggest that readers do both. There is a "no calculus" category in the topic list in Chapter 5.

In order to bring the problems and their solutions within the range of an even wider readership, *Parabolic Problems* includes a chapter on miscellaneous problem–solving techniques. Schoolwork has been avoided in this chapter: our intention is to present topics that are unlikely to be universally included in school syllabi (though, naturally, syllabus content will vary around the world), but which should be within the grasp of a diligent and thoughtful secondary–age student. It must be understood that we have had no aim to produce a "textbook" treatment of the topics addressed; we present a minimum of theory, together with useful illustrative examples, which we hope will assist readers to apply the ideas described to specific problems. Content in Chapter 3 has not been chosen at random: every topic is relevant to the solution of at least one puzzle in this book.

It has become something of a tradition in recreational mathematics to set puzzles involving the number of the present year. Readers will find many examples of this in *Parabolic Problems*. We could, of course, have updated the years to match the publication date of this book; in many cases, the year number could have been replaced by an arbitrary positive integer. For historical reasons, we have chosen to do neither of these, but to leave such puzzles in their original form.

It is often difficult to identify the original source of a problem. While it has always been the case that a large proportion of the puzzles in *Parabola* have been invented by the editorial staff, some have occasionally been culled from other publications, or contributed by readers. Many well–known problems circulate widely: this was already true when *Parabola* was founded in 1964, and is much more so in the age of the internet. In compiling our selection of *Parabolic Problems*, we have tried largely to avoid questions which are too well known. A list of sources – necessarily, for the reason just given, incomplete – has been compiled and is given at the end of the book.

Why should anyone seek to solve problems such as those presented in *Parabola*, and in this book? No doubt there are important mathematical techniques which can be learned from doing so: the pigeonhole principle, ingenious counting techniques and proof by induction come immediately to mind, and the attentive reader will note many more. But in the end, the most important reasons are those we stated earlier: to be inspired with enthusiasm for mathematics, and to have fun! We hope that the reader has fun.

We are boundlessly grateful for all the assistance we have received in compiling our selection of *Parabolic Problems*. First and foremost, of course, to those who provided the collection of problems upon which we have drawn, and in particular, two of our predecessors at *Parabola*, Charles Cox and George Szekeres. Among more recent contributors, a primary place must be accorded to Sin Keong Tong. *Parabola* would never have been started, and if started, would most likely not have continued, without the financial support of the University of New South Wales. Our thanks are also due to Taylor and Francis editorial staff Mansi Kabra and Callum Fraser; and to Helena Brusic for her spectacular cover image. Our gratitude to Ian Stewart, Adelle Coster and Bruce Henry for their engaging introductions to the book. Finally, our love and gratitude to our respective partners: to Suzanne; and to Ania.

David Angell and Thomas Britz
School of Mathematics and Statistics
UNSW Sydney
November 2023

Author Biographies

David Angell studied mathematics at Monash University and the University of New South Wales, Australia, earning a Ph.D. from the latter institution with a thesis on Mahler's method in transcendence theory. He has been a member of the academic staff in the School of Mathematics at UNSW since 1989, and has consistently received glowing evaluations of his teaching both from colleagues and from students. David has taught a wide variety of mathematics subjects, but his favourites have always been number theory and discrete mathematics. He is particularly interested in teaching students to produce proofs and other mathematical writing which are clearly expressed, logically impeccable and engaging for the reader. David is strongly committed to extension activities for secondary school students. He has for many years been the problems editor for *Parabola*, the online mathematics magazine produced by UNSW, as well as contributing a number of articles to the magazine. He has also given talks on a wide variety of topics to final-year secondary students. His text *Irrationality and Transcendence in Number Theory* was published by CRC Press, Taylor & Francis in 2022.

Beyond mathematics, David is an enthusiast for wilderness activities and has undertaken expeditions in Australia, Greenland, Nepal, Morocco and many other areas. He is a keen amateur musician, and is the founding conductor of the Bourbaki Ensemble, a chamber string orchestra based in Sydney, Australia.

Thomas Britz studied mathematics and physics at Aarhus University, M.I.T. and Queen Mary University of London. After receiving a Ph.D. in mathematics from Aarhus University, Thomas completed postdoctoral fellowships at Victoria University, BC, the Technical University of Denmark and the University of New South Wales. Thomas has been a member of the School of Mathematics and Statistics at UNSW Sydney since 2010 and is currently the president of the Combinatorial Mathematics Society of Australasia (CMSA).

In addition to his research on combinatorics and its applications, Thomas spends great effort to support and care for students and their education, and he strongly promotes the use of caring and emotions in education among fellow teachers. As chief editor of Parabola since 2014, Thomas has invited readers and contributors from all around the world to enjoy and create its mathematical content, so that it is now the largest international maths journal for students and anyone else interested in maths. Thomas is also a father, and he enjoys a busy life with his awesome partner and lovely children. He loves to solve puzzles and think about things, often more than is good for him.

CHAPTER 1

Problems

Parabola *was founded in 1964, a publication (as it remains) of the School of Mathematics at the University of New South Wales. It has always had a strong tradition of posing engaging mathematical problems and puzzles for its readers to solve and enjoy. The magazine's founding editor, Charles Cox, expressed himself as follows in an editorial in the first issue.*

"One of the aims of this magazine is to provide problems and puzzles which will test your insight, ingenuity and determination to the limit. Most people enjoy solving even simple puzzles. But if the problem at first appears quite intractable, if a neat solution is effected only by a flash of inspiration after complete concentration, the sense of achievement increases enormously. It is this creative experience which transforms the career of a professional mathematician from merely a day job. . . to a continual source of adventure and delight".

We begin with two problems from Parabola *Volume 1 Issue 1, July 1964.*

Q1 The very first Parabola problem!

(a) Two numbers a and b are such that a is smaller and b is greater than 1. If S is the sum of a and b and P is their product, prove that S and P differ by more than 1.

(b) Hence show that if the product of two positive numbers is 1, their sum cannot be less than 2.

(c) Using this result, or otherwise, prove that amongst all right–angled triangles of equal area, the isosceles triangle has the shortest hypotenuse.

Q2 One morning after church, the verger, pointing to three departing parishioners, asked the bishop, "How old are those three people?"

The bishop replied, "The product of their ages is 2450, and the sum of their ages is twice your age".

The verger thought for some moments and said, "I'm afraid I still don't know".

The bishop answered, "I am older than any of them".

"Aha!" said the verger, "Now I know".

Problem: How old was the bishop?

(Ages are in whole numbers of years and no one is over 100.)

Q3 Show that 49 is not a divisor of $n^2 + n + 2$ for any integer n.

DOI: 10.1201/9781003396413-1

Q4 The map to the right is obtained by drawing five circles. The resulting regions into which the plane has been divided are coloured in two colours, green and white, in such a way that regions with a positive length of boundary in common have different colours.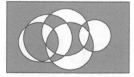

Show that two colours are sufficient for every map obtained in this way, irrespective of the number of circles it contains.

Q5 Four brothers A, B, C and D sold a herd of cattle, each beast fetching as many pounds as there were animals in the herd. The brothers shared the money in the following way. First, A took £10, then B took £10, then C, then D, and A again, and so on. Eventually, it being D's turn, there was less than £10 remaining. However, A, B and C each returned a sum of money to D so that the division was perfectly fair.

How much did each return?

Q6 (a) $\angle AOB$ is an acute angle and OX is a variable ray dividing it into two smaller angles. Show that $\sin \angle AOX + \sin \angle BOX$ is greatest when OX bisects $\angle AOB$.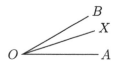

(b) A polygon with n sides is inscribed in a circle. Show that the area of the polygon is greatest if it is regular.

Q7 The diagram on the right shows a rectangular floor that is x long and y wide, which is covered with $1 \times a$ tiles, where a, x and y are all positive integers.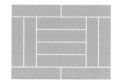

Prove that such a covering is possible (without cutting any tiles) if and only if a divides either x or y exactly.

Q8 A lives in a street in which there are houses numbered from 1 to 30. B wants to know the number of A's house.

B asks A: "Is it greater or less than 15?" A answers but lies.

B asks A: "Is it a perfect square?" A answers truthfully.

B asks A: "Is it odd or even? That will determine it definitely". "Even", says A truthfully.

B tells A the number but is wrong.

What was the number of A's house?

Q9 A man lost in the desert hears a train whistle due west of him. Although the track is too far away to be seen, he knows that it is straight and runs in a direction somewhere between south and east; however, he does not know its exact course. He realises that his only chance to avoid perishing from thirst is to reach the track before the train has passed. In which direction should he travel to give himself the greatest possible chance of survival? (The man and train both move at constant speeds, the former more slowly than the latter.)

Q10 Suppose that the polynomial

$$P(x) = a_0x^7 + a_1x^6 + a_2x^5 + a_3x^4 + a_4x^3 + a_5x^2 + a_6x + a_7,$$

whose coefficients are all integers, has for seven different integral values of x the value 1 or the value -1. Prove that $P(x)$ cannot be factorised as the product of two polynomials with integral coefficients.

Q11 Show how to dissect a regular hexagon into five pieces which can be rearranged to make a square.

Q12 Show that every triangle can be dissected into seven acute–angled triangles.

Q13 A number of towns $T_1, T_2, \ldots, T_n$, no three collinear, are connected by a system of one–directional roads with the following properties. Given any pair of towns, there is a straight road joining the two towns. There are no road junctions except at the towns. Roads may cross each other at other points, but flyovers are used at those points so that it is impossible to leave one road and enter the other. Also, it is possible to reach any town from any other. Prove that it is possible to leave T_1 and drive in a circuit so that, on returning to T_1, every other town has been visited exactly once.

Q14 In a certain community, there are a number of clubs, none of which includes the entire community in its memberships. For every pair of distinct clubs, there are at least two people who are members of each. No two clubs have exactly the same members; if any three people are considered, then there is one and only one club having all three as members. Also, there are at least two clubs. Show that

(a) for any two clubs, exactly two people are members of both;

(b) the community contains at least four people;

(c) every club has a membership of at least three people.

Q15 C is a circle of diameter D, touching a line L. A circle C_0 is drawn that touches C and L. Circles $C_1, C_2, \ldots, C_k$ are constructed in succession, each circle C_{i+1} touching circles C and C_i as well as the line L, as shown in the diagram. Let $D_0, D_1, \ldots, D_k$ be the diameters of the circles $C_0, C_1, \ldots, C_k$ respectively. Find a formula for D_k in terms of D and D_0.

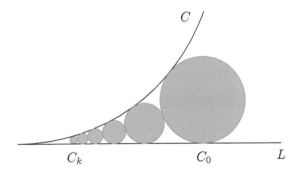

Q16 Three motorists A, B and C often travel on a certain highway, and each motorist always travels at a constant speed. A is the fastest of the three and C is the slowest. One day when the three travel in the same direction, B overtakes C; five minutes later, A overtakes C; and in another three minutes, A overtakes B. On another occasion when they again travel in the same direction, A overtakes B first; then, nine minutes later, overtakes C.

When will B overtake C?

Q17 A rectangular base has a transparent semi–cylindrical cover. A thin elastic thread joining two of the diametrically opposite vertices is stretched over the cover. It may be assumed that the thread takes up a position such that its length is minimised. Vertical light rays cast a shadow of the thread upon the base. Determine the shape of the shadow.

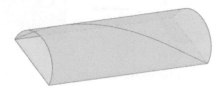

Q18 There are n people at a party, and n handshakes are made, no pair of people shaking hands twice.

Prove that, for some $r \in \{3, \ldots, n\}$, it is possible to find people $P_1, \ldots, P_r$ such that P_1 has shaken hands with P_2; P_2 has shaken hands with P_3; and so on, and P_r has shaken hands with P_1.

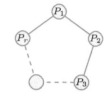

Q19 Prove that if the lowest common multiple of two numbers is equal to the square of their difference, then their greatest common divisor is the product of two consecutive integers.

Q20 A, B and C are the angles of triangle ABC, and a, b and c are the lengths of the corresponding sides. Prove that if

$$\frac{1}{\tan A}, \quad \frac{1}{\tan B} \quad \text{and} \quad \frac{1}{\tan C}$$

are in arithmetic progression, then so are a^2, b^2 and c^2. Is the converse true?

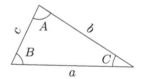

Q21 A triangle is cut into two smaller shapes by a straight line through the centroid, the point of intersection between the lines from each triangle corner to the midpoints of the opposing sides. Show that the ratio of the area of the smaller piece to that of the larger may take any value between 4/5 and 1.

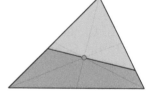

Q22 There are two prime numbers p such that $(2^{p-1} - 1)/p$ is a perfect square. Find them, and prove they are the only ones.

Q23 You can't construct $\sqrt[3]{2}$, at any rate not with a pair of compasses and an unmarked ruler. Many tried; it took over 2000 years to prove the task impossible. There is, however, a simple way of doing it if one does a bit of cheating!

Draw the equilateral triangle $\triangle ABC$ with side lengths 1. Extend the line segment AB to a point D so that $|BD| = 1$. Join DC. Extend the line segments DC and BC as far as necessary to let you to draw a line from A that intersects both of the extended lines DC and BC, say in points E and F, so that the distance $|EF|$ is exactly 1. (You would need to first mark this unit distance on the straight edge, which is where you would cheat.)

Prove that $|AE|$ is the required distance $\sqrt[3]{2}$.

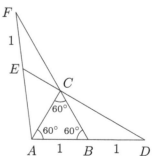

The next problem was contributed by Sin Keong Tong when he was 16 years old. Sin Keong enjoyed Parabola problems from the very first issue in 1964 and has been enjoying them ever since. He is now inventing and contributing problems as one of the three Problem Editors for Parabola.

Q24 In the gambling game *two–up*, a player pays an amount of money as bet and then plays the game. If the player wins, then they win twice the amount of money paid as bet.

In this problem, Joe brought $2500 with him to a two-up club and employed the following system of betting. He bet $1 and tripled the bet after each loss. After each win, Joe took all of the won money and put it into a bag, and then bet $1 once more. After having won his last bet and gone through the $2500 exactly, Joe went to celebrate and mislaid the bag. It was found by an honest club manager who announced that the bag would be given to the person who knew the exact amount in the bag. All Joe could remember was that he had won exactly 100 bets but, after consulting a mathematician friend, the amount was identified and the bag returned to Joe.

How much money was there in the bag?

Q25 How many solutions are there to the equation

$$\cos x = \frac{x}{50} \ ?$$

Here, the cosine function is in radians; for instance, $\cos \pi = -1$.

Q26 Four buttons labelled A, B, C and D are placed on top of the numbers 1, 2, 3 and 4 of a clock face. A button may be moved in either a clockwise or an anticlockwise direction over four other numbers to a fifth number (for instance, A can move from 1 to 6 or 8), provided that its destination is not already occupied by another button. After a certain number of such moves, the buttons again cover the numbers 1, 2, 3 and 4. How many different rearrangements of the four buttons are possible as a result of this process?

Q27 Does there exist a natural number n such that the fractional part of the number $(2 + \sqrt{2})^n$ exceeds 0.999999?

Q28 An ordinary deck of cards is thoroughly shuffled and cut at random. Player A then exposes the cards from the top, one at a time, and for each card that is not an ace, B pays 10 cents to A. However, as soon as an ace turns up, A pays $1 to B and the game is finished. Is this game perfectly fair? If not, then which player does it favour?

Q29 By inserting brackets in $1 \div 2 \div 3 \div 4 \div 5 \div 6 \div 7 \div 8 \div 9$, the value of the expression can be made to equal 7/10. How? Also, find the largest value and the smallest value that can be obtained by insertion of brackets.

Q30 Find all positive integers n such that $(n - 1)! = 1 \times 2 \times \cdots \times (n - 1)$ is not divisible by n^2.

Q31 Let m and n be two positive integers with no common factor. Prove that if each of the $m + n - 2$ fractions

$$\frac{m + n}{m}, \quad \frac{2(m + n)}{m}, \quad \cdots, \quad \frac{(m - 1)(m + n)}{m},$$
$$\frac{m + n}{n}, \quad \frac{2(m + n)}{n}, \quad \cdots, \quad \frac{(n - 1)(m + n)}{n}$$

are plotted as points on the real number line, then exactly one of these fractions lies inside each of the unit intervals

$$(1,2), \quad (2,3), \quad (3,4), \quad \ldots, \quad (m+n-2, m+n-1).$$

For instance, if $m = 3$ and $n = 4$, then $\frac{7}{4}$ lies inside $(1,2)$; $\frac{7}{3}$ lies inside $(2,3)$; $\frac{14}{4}$ lies inside $(3,4)$; $\frac{14}{3}$ lies inside $(4,5)$; and $\frac{21}{4}$ lies inside $(5,6)$.

Q32 On each side of a convex quadrilateral, a circle is drawn having that side as diameter. Prove that no point of the quadrilateral lies outside all four circles.

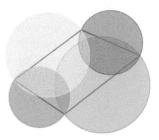

Q33 Find all solutions to the simultaneous equations:

$$y = x + \sqrt{x + \sqrt{x + \cdots + \sqrt{x + \sqrt{y}}}} \quad \text{and} \quad x + y = 6,$$

where there are 1975 square roots in the first equation.

Q34 A king moves on an 8×8 chessboard so that, in 64 moves, it visits all squares and returns to its original position on the last move. Furthermore, if the king's movements are drawn by joining the centre points of consecutive positions by straight line segments, the circuit obtained does not cross itself. Prove that at least 28 of the moves have been either horizontal or vertical.

Q35 You are given 50 intervals of finite length on the real number line. Prove that at least one of the following statements about those intervals is true:

(a) There are 8 intervals all of which have at least one point in common.

(b) There are 8 intervals so that no two of them have a common point.

Q36 Seven towns $T_1, T_2, \ldots, T_7$ are connected by a network of 21 one–way roads for which exactly one road runs directly between any two towns. Given any pair of towns T_i and T_j, there is a third town T_k that can be reached by a direct route from both T_i and T_j.

(a) Prove that each town T_i has at least three roads directed away from it. Hence, prove that exactly three roads are directed away from T_i.

(b) Suppose that the towns which can be reached directly from T_1 are T_2, T_3 and T_4. Show that the roads between T_2, T_3 and T_4 form a circuit.

(c) Draw a possible orientation of traffic on the 21 roads.

Q37 Show how to cut up and reassemble five squares of side length 1 into a single square.

Q38 Given a positive integer n, find the largest integer N for which the set of integers $S = \{n, n+1, \ldots, N\}$ can be split into two subsets A and B such that $A \cup B = S$ and that the difference $x - y$ between any two elements x and y in one of the sets A and B is never in that same set.

Q39 Suppose that $a_1, a_2, \ldots, a_7$ are any seven integers and that $b_1, b_2, \ldots, b_7$ are the same integers rearranged in some order. Show that the following integer is even:

$$(a_1 - b_1)(a_2 - b_2) \cdots (a_7 - b_7).$$

Q40 The King's men have captured a band of outlaws with an odd number of men. The rangers demand to know which ones shot the King's deer. The outlaws in panic each point to the nearest man. Prove that at least one man will not be accused. (No two pairs of outlaws are the same distance apart.)

Q41 Show how to place non–overlapping squares having side lengths $\frac{1}{2}, \frac{1}{3}, \frac{1}{4}, \frac{1}{5}, \ldots$ inside a square with side length 1.

Q42 Six circular areas are lying in the plane so that no one of them covers the centre of another. Show that there is no point in common to the six circular areas.

Q43 Consider a circle of radius 1. Using six regularly spaced points on the circle as centres, draw arcs of radius 1 to form the six petals of the "flower" inside the circle, as shown in the figure below. Next, draw the largest circles that will fit between the petals; then draw the next largest circles that will now fit between the petals, and so on. What are the radii of these circles?

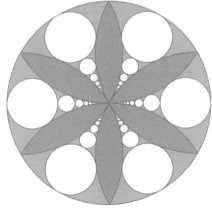

Q44 Certain convex polygons can be dissected into squares and equilateral triangles all having the same length of side. For example, the illustration below shows a hexagon dissected in such a way. If a convex polygon can be dissected in this way, then how many sides did it have originally? Prove your answer.

Q45 Given that $x+y = 1$ and $x^4+y^4 = 7$ for real numbers x and y, find x^2+y^2 and x^3+y^3.

Q46 Find a set A of 2^n different integers in the range $0, 1, 2, \ldots, 3^n-1$, having the property that no number a in A is the arithmetic mean of any other two different numbers b, c in A.

Q47 Equilateral triangles ABK, BCL, CDM, DAN are constructed inside the square $ABCD$. Prove that the midpoints of the four segments KL, LM, MN, NK and the midpoints of the eight segments $AK, BK, BL, CL, CM, DM, DN, AN$ are the twelve vertices of a regular dodecagon.

Q48 When the fire alarm went off, the six patrons in the restaurant all hurriedly exited, seizing a coat on the way.

(1) Safely outside, they discovered that no–one had their own coat.

(2) The coat held by A belonged to the person who had seized B's coat.

(3) The owner of the coat grabbed by C held a coat owned by the person holding D's coat.

(4) The person who seized E's coat was not the owner of that grabbed by F.

Who borrowed A's coat? Whose coat did A seize?

Q49 Let p be a prime number, let $\lfloor n/p \rfloor$ be the largest integer not greater than n/p and consider the binomial coefficient

$$\binom{n}{p} = \frac{n(n-1)\cdots(n-p+1)}{1 \times 2 \times \cdots \times p}.$$

Prove that $\binom{n}{p} - \lfloor \frac{n}{p} \rfloor$ is divisible by p.

Q50 On the sides of an arbitrary triangle $\triangle ABC$, three triangles $\triangle ABR$, $\triangle BCP$ and $\triangle CAQ$ are constructed with angles $\angle PBC = \angle CAQ = 45°$ and $\angle BCP = \angle QCA = 30°$ and $\angle ABR = \angle RAB = 15°$.

Prove that $\triangle QRP$ is a right–angled triangle with side lengths $|RP| = |RQ|$.

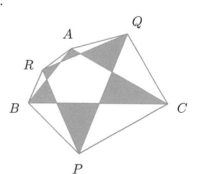

Q51 Each of three classes has n students. Each student knows altogether $n + 1$ students in the other two classes. Prove that it is possible to select one student from each class so that all three know one another. Here, acquaintances are always mutual.

Q52 Pegs are arranged in a rectangular grid on a board, and rubber bands can be placed around the pegs to form geometrical shapes. The figure on the right shows how to construct squares of areas 1 and 5 using pegs and rubber bands.

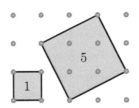

(a) Show how to construct squares of areas 8 and 10.

(b) Prove that it is not possible to construct a square of area $4n + 3$ where n is an integer.

Q53 Show that the diameter d of the inscribed circle of a right–angled triangle with legs a and b and hypotenuse c satisfies the equation $d = a + b - c$.

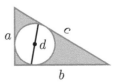

Q54 Given any ten distinct positive integers each less than 100, show that there are two subsets of this set having no elements in common such that the sums of the numbers in the two subsets are equal.

Q55 Mount Zircon is shaped like a perfect cone whose base is a circle of radius 2 miles, and the straight line paths up to the top are all 3 miles long. From the point A at the southernmost point of the base, a path leads to B, a point on the northern slope $\frac{2}{5}$ of the way to the top. If AB is the shortest path on the mountainside joining A to B, find

(i) the length of the whole path AB, and

(ii) the length along the path between P and B, where P is a point on the path at which it is horizontal.

Q56 Prove that $|BC| = |BD| + |DA|$ in the figure below.

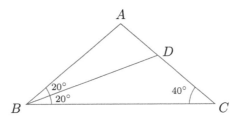

Q57 Thirty–two counters are placed on a (normal 8×8) chessboard in such a way that there are four in every row and four in every column. Show that it is always possible to select eight of them so that there is one of the eight in each row of the chessboard and one in each column.

Q58 Given two intersecting straight lines a and b and a point P on b, show how to construct a circle whose centre is on b, which passes through P and has a as a tangent.

Q59 Let p and q be natural numbers such that

$$\frac{p}{q} = 1 - \frac{1}{2} + \frac{1}{3} - \frac{1}{4} + \cdots - \frac{1}{1318} + \frac{1}{1319}.$$

Prove that p is divisible by 1979.

Q60 At a party, there are $2n$ people, each one of whom is acquainted with at least n others. Prove that it is possible to seat them at a circular table so that each is placed between two acquaintances.

Q61 Given a point P and three distances x, y, z, it is desired to construct an equilateral triangle $\triangle ABC$ with P inside it, such that $|PA| = x$, $|PB| = y$ and $|PC| = z$.

(a) Find conditions on x, y, z such that this is possible.

(b) When the conditions in (a) are satisfied, show how to achieve the construction.

Q62 Find a set S of seven consecutive positive integers such that there exists a polynomial $P(x)$ of the fifth degree with the following properties:

- all coefficients of $P(x)$ are integers, and the coefficient of x^5 is 1;
- $P(n) = n$ for five elements n of S, including the least and the greatest;
- $P(n) = 0$ for one element n of S.

Q63 In a party of 9 persons, among any three there are at least two who do not know each other. Prove that in the party there are four persons, none of whom knows another of the four. (Assume that "knowing another" is *symmetric*: if A knows B, then also B knows A.)

Q64 A pirate ship sees another ship 20 miles due east sailing due north. The pirate ship gives chase by always sailing directly towards its quarry. However, both ships have the same speed, so that the pirate ship eventually is also sailing just about northwards, some distance behind the other. How far behind?

Q65 Determine the maximum value of $m^3 + n^3$, where m and n are integers satisfying $m, n \in \{1, 2, 3, \ldots, 1981\}$ and $(n^2 - mn - m^2)^2 = 1$.

Q66 I have two different integers greater than 1. I inform Sam and Pam of this fact, and I tell Sam the sum of my two numbers and I tell Pam their product. The following dialogue then occurs.

> Pam: "I can't determine the numbers".
>
> Sam: "The sum is less than 23".
>
> Pam: "Now I know the numbers".
>
> Sam: "Now I know the numbers too".

What are the numbers?

Q67 We colour each point of a 1×1 square (including the boundary) with one of three colours. Prove that there will always be two points of the same colour which have a distance at least $\sqrt{65}/8$ from each other.

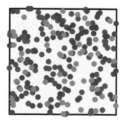

Q68 A number of circles are drawn in a plane in such a manner that there are exactly 12 points in the plane that lie on more than one of the circles. What is the smallest number of regions into which the plane can be subdivided by the circles?

Q69 Points A and B lie on opposite sides of a line ℓ. How can one construct the circle passing through A and B having the shortest possible chord on the line ℓ?

Q70 The function $f(n)$ is defined for all positive integers n and takes on non–negative integer values. Also, $f(2) = 0$, $f(3) > 0$ and $f(9999) = 3333$, and for all m, n,

$$f(m+n) - f(m) - f(n) \in \{0, 1\}.$$

Determine $f(1982)$.

Q71 Prove that if n is a positive integer such that the equation

$$x^3 - 3xy^2 + y^3 = n$$

has a solution in integers (x, y), then it has at least three such solutions. Show that the equation has no solutions in integers when $n = 2891$.

Q72 Let S be a square with sides of length 100, and let L be a path within S which does not meet itself and which is composed of line segments A_0A_1, $A_1A_2, \ldots$, $A_{n-1}A_n$ with $A_0 \neq A_n$. Suppose that for every point P of the boundary of S there is a point of L at a distance from P not greater than $1/2$. Prove that there are two points X and Y in L such that the distance between X and Y is not greater than 1, and the length of that part of L which lies between X and Y is not smaller than 198.

Q73 A convex polygon is dissected into triangles by non–intersecting diagonals. Every vertex of the polygon is a vertex of an odd number of these triangles. Prove that the number of edges of the polygon is a multiple of 3.

Q74 The *harmonic mean* of two positive numbers x and y is the reciprocal of the arithmetic mean of their reciprocals,

$$\frac{1}{h} = \frac{1}{2}\left(\frac{1}{x} + \frac{1}{y}\right).$$

We seek finite or infinite sequences of positive integers in which no integer appears more than once, and every term except for the first and last is the harmonic mean of its neighbours. (Of course, if the sequence is infinite, then there is no last term.)

(a) Prove that such finite sequences exist, and that they can have any number of terms.

(b) Prove that there are no such infinite sequences.

Q75

(a) If a_n is the nearest integer to $\sqrt{n}$, then calculate

$$\frac{1}{a_1} + \frac{1}{a_2} + \frac{1}{a_3} + \cdots + \frac{1}{a_{1984}}.$$

(b) If b_n is the nearest integer to $\sqrt[3]{n}$, then calculate

$$\frac{1}{b_1^2} + \frac{1}{b_2^2} + \frac{1}{b_3^2} + \cdots + \frac{1}{b_{1984}^2}.$$

Q76 Find all functions f defined on the set of positive real numbers which take positive real values and satisfy the following conditions:

(i) $f(xf(y)) = yf(x)$ for all positive x, y;

(ii) $f(x) \to 0$ as $x \to \infty$.

Q77 Find the smallest positive multiple of 99999 with no 9 among its digits.

Q78 Using a calculator, you will be able to check that

$$\sqrt{5 + \sqrt{21}} + \sqrt{8 + \sqrt{55}} \quad \text{and} \quad \sqrt{7 + \sqrt{33}} + \sqrt{6 + \sqrt{35}}$$

are approximately equal. Either prove that they are exactly equal, or decide (with proof) which is larger.

Q79 $ABCD$ is any rectangle, and $\triangle ABX$, $\triangle BCY$, $\triangle CDV$ and $\triangle DAW$ are outward-drawn equilateral triangles. Prove that the sum of the areas of the triangles $\triangle AXW$, $\triangle BYX$, $\triangle CVY$ and $\triangle DWV$ equals the area of the rectangle.

Q80 Show that there is no party of ten people in which the people have $9, 9, 9, 8, 8, 8, 7, 6, 4, 4$ acquaintances among themselves. Acquaintances may be assumed to be mutual.

Q81 If a, b, c, d are four positive integers such that $ab = cd$, then prove that $a + b + c + d$ is not a prime number.

Q82 From a point P inside a cube, line segments are drawn to each of the eight vertices of the cube, forming six pyramids each having P as the apex and a face of the cube as the base. Is it possible to place P in such a position that the volumes of these pyramids are in the ratios $1 : 2 : 3 : 4 : 5 : 6$?

Q83

1	2	3	4	5	$\cdots$
2	4	6	8	10	$\cdots$
3	6	9	12	15	$\cdots$
4	8	12	16	20	$\cdots$
5	10	15	20	25	$\cdots$
$\vdots$	$\vdots$	$\vdots$	$\vdots$	$\vdots$	$\ddots$

In the above array, the entries in the nth row are successive multiples of n. Find a formula for S_n, the sum of the numbers in the nth diagonal. For example, if $n = 5$, then we want the sum of the numbers marked in red, $S_5 = 35$.

Q84 Andy, Betty and Colin, having tied for first place in their chess club tournament, are to play off for the championship. Each is to play one game with each of the others, scoring 1 point for a win, $\frac{1}{2}$ for a draw and 0 for a loss. If their scores are still all level, then this will be repeated; however, if two of them are level ahead of the third, then those two will continue to play until one of them scores a win.

Andy plays sound but cautious chess. He never loses against either of the others, but has a 10% probability of beating Betty in any given game, and 20% probability of winning against Colin. Games between Betty and Colin are swashbuckling affairs which never result in draws: Betty wins 60% of the time and Colin 40%. Compare the three players' chances of emerging as club champion.

Q85

(a) Prove that there exist two powers of 3 having the same first five digits.

(b) Show that there exists a power of 3 whose first four digits are 1111.

Although Parabola *has no pretensions to being a research publication, it can happen that a problem which is quite accessible to readers may serve as the beginning of a research project. The following problem, published in* Parabola *in 1985, had been the inspiration for an investigation some twenty years earlier which led to a published research paper by Charles Cox (founding editor of* Parabola*) and a colleague.*

Q86 A list of prime numbers $p_1, p_2, p_3, \ldots$ is generated as follows. We take $p_1 = 2$, and if $n > 1$, then p_n is the largest prime factor of $p_1 p_2 \cdots p_{n-1} + 1$. Thus, p_2 is 3, the largest prime factor of $2 + 1$, and we find $p_3 = 7$, $p_4 = 43$, $p_5 = 139$: this last because $2 \times 3 \times 7 \times 43 + 1 = 1807 = 13 \times 139$. It does not follow from this rule that p_n is always larger than p_{n-1}; however, prove that the prime number 5 never occurs in the list.

Comment. The sequence $p_1, p_2, p_3, \ldots$ soon reaches rather large numbers: in 1967, when the article mentioned above appeared, the question of whether or not p_n is always larger than p_{n-1} was far too difficult to answer by direct calculation. Over fifty years later, with increased computing power readily available, it is not difficult to confirm that

$$p_9 = 4680225641471129, \quad p_{10} = 1368845206580129,$$

and p_{10} is smaller than p_9. The paper cited is C.D. Cox and A.J. Van der Poorten, *On a sequence of prime numbers*, and can be found online; more information about the sequence may be found through the On–Line Encyclopedia of Integer Sequences at `https://oeis.org/A000946`.

Q87 The numbers $1, 2, \ldots, 64$ are written onto the squares of an 8×8 chessboard, one to a square. Prove that there is a pair of adjacent squares which contain numbers differing by 16 or less. (Squares are considered to be adjacent if they share a side or a corner.)

Q88 The triangle $\triangle ABC$ is right–angled at A, and the lengths of AB and AC are 3 and 4 respectively. It is possible to draw two squares inside the triangle all of whose vertices lie on the sides of the triangle. Find the area of the overlap region common to both squares.

Q89 Let $\square$ be an operation which combines two integers in some way to give a new integer. Suppose it is known that

$$x \square (y + z) = (y \square x) + (z \square x)$$

for all x, y, z. Prove that this operation is *commutative*:

$$u \square v = v \square u \quad \text{for all integers } u, v.$$

Q90 Let $f(k)$ denote the number of zeros in the decimal representation of the positive integer k. Compute

$$S_n = \sum_{k=1}^{n} 2^{f(k)}$$

for $n = 9999999999$: that is, the sum of all values of $2^{f(k)}$ for k from 1 to 9999999999.

Q91 The bisectors of the angles C and D in a convex quadrilateral $ABCD$ meet at a point P on AB such that $\angle CPD = \angle DAB$. Prove that P is the midpoint of AB.

Q92 The sum of two positive integers is 10000000000. The two integers have the same digits, but not in the same order. Prove that both numbers are multiples of 5.

Q93 Let N be the number of strings of 20 letters, all either O or X, which contain the word OX precisely 7 times: for example, such a string as XXOXOOXOXXXOXOXOXOXXOX, where the pairs OX are shown in red. Find an argument to show that

$$N = C(7,7)C(13,7) + C(8,7)C(12,7) + C(9,7)C(11,7) + C(10,7)C(10,7)$$
$$+ C(11,7)C(9,7) + C(12,7)C(8,7) + C(13,7)C(7,7),$$

and a different argument to show that

$$N = C(21,15).$$

The terms $C(n,r)$ in these expressions are *binomial coefficients* – see page 250.

Q94 In the 3×4 array of equally spaced dots to the right, there are 10 sets of 4 dots that lie at the vertices of a square; one such set is shown by the square in the figure. There are six small squares, two larger ones and two at an angle. Find for each $m \times n$ array with $m \le n$ a formula for the number of squares all of whose vertices are in the array.

Q95 We are given a rectangular array of numbers arranged in m rows and n columns. We are allowed to change the sign of every number in any row, or to change the sign of every number in any column. Prove that, if we perform these admissible operations repeatedly, then we can eventually obtain an array in which the sums of the numbers along every row and down every column are all non–negative.

Q96 Show that if p is any odd prime number except 5, then there is at least one integer k for which p divides the number $11 \cdots 11$ composed of k ones.

Q97 Let $x_1, x_2, \ldots$ be an infinite list of digits not containing the digit 9. Consider the list of numbers $y_1, y_2, \ldots, y_n, \ldots$ where y_n has decimal representation $x_1 x_2 \cdots x_n$; that is, $y_n = 10^{n-1} x_1 + \cdots + 10 x_{n-1} + x_n$. Prove that this list contains infinitely many composite numbers.

Q98 Let $P(x) = a_0 + a_1 x + a_2 x^2 + \cdots + a_n x^n$ be a polynomial with integer coefficients such that $P(2) = 3$ and $P(5) = 7$. Prove that $P(x)$ has no integer roots.

Q99 A regular polygon with n vertices is inscribed in a circle of radius 1. Let L be the set of all *mutually distinct* lengths of all line segments joining the vertices of the polygon. For instance, L contains three elements for the regular polygon below.

What is the sum of the squares of the elements of L?

Q100 Let n be a power of 2, say $n = 2^m$, and let $(a_1, a_2, \ldots, a_n)$ be a sequence of n numbers each equal to either $+1$ or -1. Another sequence $(b_1, b_2, \ldots, b_n)$ is obtained by the following operation:

$$b_i = a_i a_{i+1} \quad \text{for} \quad i = 1, 2, \ldots, n-1 \quad \text{and} \quad b_1 = a_1 a_n .$$

This operation is performed repeatedly. Prove that, after enough repetitions of the operation, all terms of the resulting sequence are equal to $+1$.

Q101 A number x is divisible by 9 but not by 10. It has the property that, when a certain one of its digits is deleted, the result is the whole number $y = x/9$ which is also a multiple of 9. Prove that y also has the property that, when one of its digits is deleted, it decreases by 9 times. Find all possible values of x.

Q102 Let x and y be positive integers with $y > 1$. Prove that

$$\underbrace{\sqrt{x + \sqrt{x + \sqrt{x + \cdots + \sqrt{x}}}}}_{y \text{ square roots}}$$

cannot be a whole number.

Q103 You are given 100 points in a plane, no three of which are collinear. Of these points, 50 are chosen and labelled A and the rest are labelled B. Show that it is possible to draw 50 straight line segments each linking a point labelled A to a point labelled B so that

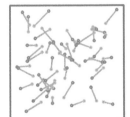

(i) each of the 100 points is an endpoint of one line;

(ii) no two line segments intersect.

Q104 I have four infinite sets of non–negative numbers X_1, X_2, X_3 and X_4. Every non–negative integer can be expressed uniquely in the form

$$m = x_1 + x_2 + x_3 + x_4$$

where $x_1 \in X_1$, $x_2 \in X_2$, $x_3 \in X_3$ and $x_4 \in X_4$.

(i) Show that 0 is in each set but that no other number is in two or more sets.

(ii) For any $N > 0$, let $n(N)$ be the number of elements of $X_1 \cup X_2 \cup X_3 \cup X_4$ that are less than N. Show that $n(N) \geq 4\sqrt[4]{N} - 3$ for any N.

(iii) Find four sets having the stated properties and, if possible, a number N for which $n(N) = 4\sqrt[4]{N} - 3$.

Q105 A set of cups is arranged in a rectangular array of m rows and n columns, and a random number of beans is placed in each cup, with no cup being left empty. The following operations are permitted.

(1) One bean is taken from every cup in a row.
(This is not possible if some cup is already empty.)

(2) The number of beans in every cup in any column is doubled.

Is it always possible to perform these operations repeatedly so that all cups are eventually empty?

Q106 Let P be a point on a circle that touches the vertices of a regular n–sided polygon. Label the n vertices clockwise as $A_1, \ldots, A_n$, so that P lies between A_1 and A_n; for each $i = 1, \ldots, n$, let x_i be the distance between P and A_i.

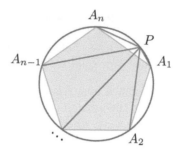

Prove that $\dfrac{x_2 + \cdots + x_{n-1}}{x_1 + x_2 + \cdots + x_{n-1} + x_n} = \cos\left(\dfrac{\pi}{n}\right)$.

Q107 Let L be a set of n line segments with the property that any three of them can be assembled to form a triangle. A pair of line segments is "exceptional" if one is more than twice as long as the other. What is the maximum possible number of exceptional pairs in L?

Q108 Let S be a set of numbers with the property that the product of every two distinct elements of S is an integer.

(a) Show that if every number in S is rational, then the product of any three or more distinct elements of S is an integer.

(b) Show that if not every number in S is rational, then this need not be true.

Q109 Suppose that v, w, x, y, z are real numbers such that

$$v + w + x + y + z = 11 \quad \text{and} \quad v^2 + w^2 + x^2 + y^2 + z^2 = 25.$$

Find the smallest and largest possible values of z.

Q110 Given two half–lines extending from a point O and a point P lying between these half–lines, construct points A and B on the half–lines so that the line segment AB passes through P and the triangle $\triangle AOB$ has minimal area.

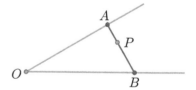

Q111 Define

$$f(x) = \sum_{k=114}^{184} \frac{k}{x - k}.$$

Describe the graph of $f(x)$ and observe that, if c is any positive number, then the set $\{x : f(x) > c\}$ is the union of 71 intervals. Let $L(c)$ denote the sum of the lengths of these intervals. Show that

$$L\left(\frac{1988}{336}\right) = 1788 \quad \text{and} \quad L\left(\frac{1788}{336}\right) = 1988.$$

Q112 B. Rainy, the school genius, is using a calculator to complete a maths exam but accidentally knocks it off the desk. The calculator is broken and only the algebraic operations $+, \times, -, \div$ are still operational. The remaining exam question requires a calculation involving $\ln 11$, the natural logarithm of 11. Our protagonist recollects seeing in a calculus text that, whenever $x > 1$,

$$\ln \frac{x+1}{x-1} = 2\left(\frac{1}{x} + \frac{1}{3x^3} + \frac{1}{5x^5} + \cdots\right).$$

After a little calculating, B. Rainy finds whole numbers A, B, C such that $\ln 11$ is given to eight decimal places by the expression

$$A\left(\frac{1}{23} + \frac{1}{3 \times 23^3} + \frac{1}{5 \times 23^5}\right) + B\left(\frac{1}{65} + \frac{1}{3 \times 65^3}\right) - C\left(\frac{1}{485} + \frac{1}{3 \times 485^3}\right);$$

and evaluates this on the calculator, completing the examination and obtaining 100% as always. Find A, B, C and calculate $\ln 11$.

Q113 Simplify the sum

$$\sec 0 \sec 1 + \sec 1 \sec 2 + \sec 2 \sec 3 + \cdots + \sec(n-1)\sec n.$$

Q114 How many of the numbers $0, 1, \ldots, 9999$ have digit sum less than 20?

Q115 A set S consists of 100 different positive whole numbers, the largest of which is x. The sum of three different numbers chosen from S is never equal to a fourth number in S. Find the smallest possible value of x. Exhibit a set S having this value of x, and prove that no smaller value is possible.

Q116 Let $a_1 < a_2 < \cdots$ be an increasing list of positive integers and define the number $b_n = |\{i \; : \; a_i < n\}|$ for each $n \geq 1$. Show that each positive integer N can be expressed either as $N = a_k + k$ for some k or as $N = b_\ell + \ell$ for some ℓ but cannot be expressed both as $N - a_k + k$ and as $N = b_\ell + \ell$.

Q117 Two bodies travel in opposite directions around a circular track, one at constant speed v (in m/s), the other at a speed increasing at constant rate a (in m/s²). At time $t = 0$, the bodies are at the same point A and the second body is at rest.

After how many seconds did their first meeting occur, if their second meeting is again at the point A?

Q118 In the figure, $ABCRSA$ is a semicircle, $PQRS$ is a square, and the area of $\triangle ABC$ is equal to the area of $PQRS$. Prove that X, the point of intersection of BS and QR, is the incentre of $\triangle ABC$.

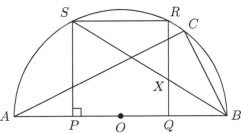

Q119 The isosceles triangle $\triangle ABC$ is obtuse–angled at A, and P is a point on its circumcircle (other than A) such that $|PB| + |PC| = 2|AB|$. If X is the point of intersection of AP and BC, then prove that $|BC| = 2|AX|$.

Q120 Let $P_1, P_2, \ldots, P_n$ be any n distinct points in the plane. Show that it is always possible to dissect the plane into n connected regions $R_1, R_2, \ldots, R_n$, all congruent to one another, such that P_i, lies inside R_i for $i = 1, 2, \ldots, n$.

Q121 If a, b, c are the side lengths of a triangle, T its area, and R the radius of its circumcircle, then prove that $abc = 4TR$.

Q122 The convex quadrilateral $ABCD$ is not cyclic, and no two sides are parallel. How many circles can be drawn which are equidistant from all four vertices?

Q123 Every point on the circumference of a circle is coloured red, blue or green. Show that however this is done, there will be three points A, B, C on the circle, all of the same colour, such that $\triangle ABC$ is isosceles.

Q124 A 1–metre square masonry slab which formed part of a 1 metre–wide path has become displaced as shown in the figure, so that a triangular piece projects beyond each side of the path. Find the sum of the perimeters of these two triangles.

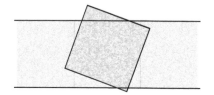

Q125 In a club with 36 members, each two members are either friends or enemies; each member has exactly 13 enemies. In how many different ways can one select three members so that they are either all friends or all enemies?

Q126 The polynomial $p(x)$ of degree n has the values

$$p(k) = 3^k \quad \text{for} \quad k = 0, 1, 2, \ldots, n.$$

Find the value of $p(n + 1)$.

Q127 Prove (without extensive calculations!) that

$$\frac{31}{45} < \frac{1000}{1001} \frac{1002}{1003} \frac{1004}{1005} \cdots \frac{1992}{1993} < \frac{23}{31}.$$

> *The next problem was published in* Parabola *in 1993... but that information couldn't possibly be of any relevance... could it...?*

Q128 Last year was a busy one for my family, with six of my brothers and sisters having children. Writing myself a timetable for buying niecely and nephewly birthday presents, I noticed some odd facts about the six dates. The difference (in days) between successive birthdays was always the same, and this difference was a prime number. No two children were born in consecutive months, and no two were born on the same date in different months. One of the birthdays was on August 8. When were the others?

Q129 Alexander, David, Esther, Jacinda and Simon all received different marks in the maths test which was held unexpectedly last week. In the following, those students who made correct statements invariably had obtained higher marks than those who made incorrect statements.

Simon:	Alexander and Esther gained the top two places.
Jacinda:	No, what Simon just said is wrong.
David:	I was ranked in between Simon and Jacinda.
Alexander:	Jacinda came second.
Jacinda:	I scored fewer marks than Esther.
Esther:	Exactly three of the previous five statements are correct.

Find the order in which the students finished.

Q130 Find all positive integers m such that if

$$(1993 + m)^{1993}$$

is expanded by the Binomial Theorem, then two adjacent terms are equal.

Q131 A certain race of beings lives on two planets. Each nation owns one connected piece of territory on each planet. It is desired to colour a pair of maps of the two planets according to two rules. First, two countries with a common border must bear different colours; second, each nation must have its two territories (one on each planet) coloured with the same colour. Find an example of a pair of maps which requires eight colours.

Q132 Using pencil, paper and old–fashioned long division, a six–digit number was divided by a three–digit number, giving a three–digit quotient with no remainder. In the working, every even digit $(0, 2, 4, 6, 8)$ was replaced by an E, while every odd digit $(1, 3, 5, 7, 9)$ was replaced by an O. Given the result shown below, reconstruct the working. No number begins with a zero.

```
            O O E
OEO  ┃ E E O E E
       E E E O
       ─────────
         E O E
         O E O
         ─────────
           E O E
           E O E
           ─────────
```

Q133 Take a normal 8×8 chessboard and remove from it as few as possible individual squares, in such a way that no 3×1 rectangle can be placed so as to completely cover three squares on the remaining part of the board.

Q134 Take a square–based pyramid whose triangular faces are all equilateral, and a regular tetrahedron whose faces are all the same size as the triangular faces of the pyramid. Join these two solids along a pair of triangular faces. In the combined solid you will see two triangles which appear to lie very nearly in the same plane. Do they in fact lie exactly in the same plane? Prove your answer.

Q135

(a) We have seven coins, apparently identical, of which two are heavier than the other five (and the two heavy coins weigh the same as each other). With three weighings on a beam balance, find the heavy coins.

(b) Show that the above problem cannot be solved with certainty if we have eight coins, two being heavier than the other six.

Q136 A pentagon with all its diagonals drawn in has a coin placed on each intersection of lines, with either heads or tails facing up (see the diagram).

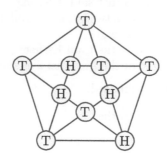

It is permitted to select any one of the ten lines in the figure, and turn over all the coins lying on that line; such a move may be performed as many times as you like.

(a) Find a sequence of moves starting with the above position and finishing with all coins heads up.

(b) Find a rule by which it is possible to tell merely by studying any initial position whether or not the task in (a) can be accomplished.

Q137 In the binomial expansion of $(x + 1)^n$, three consecutive coefficients are such that the second and third are respectively a times and 23 times the first, where a is a positive integer. Find a and n.

Just in case you are concerned... we can assure readers that nothing has been accidentally omitted from the next problem!

Q138

(a) All of the following.

(b) None of the following.

(c) Some of the following.

(d) All of the above.

(e) None of the above.

Q139 In the middle of a large field, a patch is fenced off in the shape of a regular hexagon. A goat outside the hexagon is tethered by a rope attached to one corner of the hexagon and having length $2 + \sqrt{3}$ times the side of the hexagon. Find the total area within which the goat can graze.

Q140 Given three non–collinear points X, Y, Z such that the angle $\angle XYZ$ is obtuse, show how to construct a triangle $\triangle ABC$ such that the median of the triangle through A intersects the circumcircle at X, the angle bisector through A intersects the circumcircle at Y and the altitude through A intersects the circumcircle at Z.

The topical references which lay behind the phrasing of the next problem are now, happily, beyond recall. The mathematics, on the other hand, needs no alteration!

Q141 The Australian No–Hopers Party (ANHP) has 25 members in Parliament, including exactly two who live in Canberra, ride bicycles and own whiteboards. It is known that

(a) if the number of ANHP members who live in Canberra and own whiteboards is greater than 4 or the number who ride bicycles but do not live in Canberra is greater than 3, then the total number who own whiteboards is 15;

(b) if the number who ride bicycles is at most 8 or the number who do not own whiteboards is greater than 7, then the number who both ride bicycles and own whiteboards is 8;

(c) if the number who live in Canberra but do not ride bicycles is not equal to 12 or the number who ride bicycles and own whiteboards is less than 8, then the total number who ride bicycles is 4.

If you knew the total number of ANHP members who ride bicycles, then you could calculate how many live in Canberra. Determine how many members of the ANHP own whiteboards, how many live in Canberra, and how many ride bicycles.

Q142 A supermarket runs a free contest for its customers, who have to collect numbered tickets adding up to 1994 or 1995. For 1994, the customer wins \$100; for 1995, \$1000. The company can print any number of tickets, with any integers on them, but in order to create public interest, they wish to print a large variety of different tickets. However, they also wish to be *certain* that at most ten \$1000 prizes and fifty \$100 prizes can be won. Devise a scheme to achieve this.

Q143 Show how to cut a square of area 12 into seven pieces which can be rearranged to form either a 3×4 or a 2×6 rectangle.

Q144 On an island there are 50 brown, 57 green, 62 yellow and 68 red frogs. Whenever three frogs of three different colours meet, they change immediately into two frogs of the fourth colour. Later on, it is observed that all frogs on the island have the same colour. Which colour, and what is the maximal possible number of frogs on the island at this stage?

Q145 Show that, in order to write an odd number as a sum of two or more of its factors (without repetitions), at least 9 factors must be used.

Q146 Five line segments are given, with the property that any three of them will form a triangle. Show that at least one of these triangles is acute–angled.

Q147 Six points are located inside, or on the boundary of, a 3×4 rectangle. Show that two of them are separated by a distance of $\sqrt{5}$ or less.

Q148 In a large flat area of bushland, there are two fire–spotting towers, one exactly 20km east of the other. A bushfire is reported as being due north–east of the western tower, and simultaneously due north–west of the eastern tower. However, each of these directions could be in error by up to 1° either way. Find the total area within which the fire might be located

(a) by a simple approximate argument;

(b) exactly.

Q149 Let α be a constant, not a multiple of $\pi/2$, and consider the curve

$$y = x - \sin x - (1 - \cos x) \tan \alpha.$$

(a) Show that the x–axis is tangent to the curve at the origin.

(b) Show that the x–axis is also tangent to the curve at some point other than the origin if and only if $\tan \alpha - \alpha$ is a multiple of π.

Q150 Which of the statements in the following list are true?

1. At least one odd–numbered statement in this list is false.
2. Either the second or the third statement in this list is true.
3. This list does not contain two consecutive false statements.
4. There are at least two false statements in this list.
5. If the first statement in this list is deleted, then the number of true statements will decrease.

Q151 Show how to dissect any triangle of the following types into three isosceles triangles:

(a) acute–angled triangles;
(b) triangles with at least one 45° angle;
(c) triangles with one angle five times another;
(d) triangles with one angle six times another;
(e) triangles with one angle seven times another.

Q152 Let $a_1, a_2, a_3, \ldots$ be a sequence of positive integers which is *non–decreasing*, that is, $a_1 \leq a_2 \leq a_3 \leq \cdots$. Define a sequence $b_1, b_2, b_3, \ldots$ of non–negative integers, where b_k is the number of terms among $a_1, a_2, a_3, \ldots$ which are less than k. Prove that, for any positive integer m, if $n = a_m$, then

$$(a_1 + a_2 + \cdots + a_m) + (b_1 + b_2 + \cdots + b_n) = mn.$$

Q153 Starting from home, I drive some distance due east. Then I turn 90° left and drive another distance, then turn left again, and so on until I have driven $4n$ segments. Suppose that the lengths of these segments are $1, 2, \ldots, 4n$ kilometres but not necessarily in that order.

(a) What is the greatest distance I can eventually be from home?
(b) Can I end up at home?

Q154 Prove that

$$\binom{n}{1} - \frac{1}{2}\binom{n}{2} - \cdots - (-1)^n \frac{1}{n}\binom{n}{n} = 1 + \frac{1}{2} + \cdots + \frac{1}{n}.$$

Q155 The set of numbers $\{21, 24, 25, 29\}$ is given. It is permitted to multiply any two numbers from the set (multiplying one number by itself is also allowed) and place in the set the last two digits of the product. For example, since $21 \times 29 = 609$, we may put the number $09 = 9$ into the set. By repeating this operation as often as desired, is it possible to obtain a set including the number 99?

Q156 Semicircles are drawn internally on the hypotenuse of a right–angled triangle and externally on the other two sides.

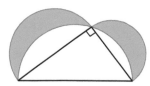

Find, without using algebra or calculus, the total area of the shaded crescents.

Q157 Show that there are integers a, b, c, not all zero, between -10^6 and 10^6, such that

$$-10^{-11} < a + b\sqrt{2} + c\sqrt{3} < 10^{-11}.$$

Q158 How many positive real numbers x are there for which $x \leq 1997$ and

$$\sqrt{x+1} + \sqrt{x}$$

is an integer?

Q159 Find the smallest positive integer n such that the sum of the digits of the product of the digits of the sum of the digits of n is 10 or more.

The 1000th problem to appear in Parabola *was introduced as follows.*

"To mark our thousandth problem we present a question with three unusual features. First, it is a rather well–known type of problem – we hope that this is an unusual feature for Parabola *problems! Second, we won't be giving the solution next issue, for reasons which will become obvious. Third, there will be a prize for the best solution sent in by our readers. Here is the problem".*

Q160 How many different ways can you find of using the digits $1, 2, \ldots, 9$, just once each, to make a total of 1000? Here are some examples to start you off:

$$1000 = 987 + 6 + 5 - 4 + 3 + 2 + 1;$$
$$1000 = 1978/2 + 46 - 35;$$
$$1000 = (56 \times 98) - (12 \times 374).$$

You may string numbers together to make multi–digit numbers, and you may use addition, subtraction, multiplication, division and brackets – but that's all. Variations which are nothing but a reordering of terms will not be counted – for example, $1000 = 1 + 2 + 3 - (4 - 5 - 6 - 987)$ is the same as the first expression above.

Q161 Without any calculations, can you determine whether it is it possible to circumscribe two non–congruent triangles of equal area around the same circle?

Q162 Let $ABCD$ be a rectangle and let X be the point of intersection between line segments AC and DM, where M is the midpoint of AB. Find the ratio of the area of $\triangle AMX$ to the area of the rectangle $ABCD$.

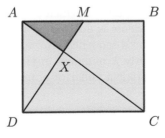

Q163 In a tennis tournament, every player plays every other player exactly once. We say that player X has *directly beaten* player Y if X plays against Y and wins; we say that player X has *indirectly beaten* player Y if X has directly beaten a player Z who has directly beaten Y. If a player has beaten all other players in the tournament, directly or indirectly, then that player wins a prize.

A tennis player (let's call him Pete) enters the tournament and is the only player to receive a prize. Show that Pete has directly beaten everyone else in the tournament. (In tennis it is not possible to draw).

Q164 Melbourne tram tickets have six–digit numbers, from 000000 to 999999. A ticket is called *lucky* if the sum of its first three digits is equal to the sum of its last three digits. How many consecutively numbered tickets should one buy to be sure of getting at least one lucky ticket, assuming that one does not know where the sequence will start?

Q165 Two players play the following game on a 9×9 chessboard. Playing alternately, the first player places a black counter in a square and then the second player places a white counter in a square. The game finishes when all 81 squares are filled with counters. There are 9 rows and 9 columns, so there are either more white counters or more black counters in each row and column. Each player gets a point if they have more counters in a row or a column, so the total number of points allocated is 18.

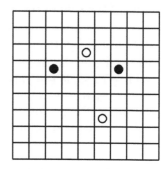

What is the highest number of points the first player can gain, if both players play as well as possible?

Q166 There are three types of line segments that connect the vertices of a cube: edges of the cube, face diagonals and long diagonals which connect opposite vertices.

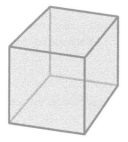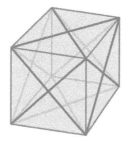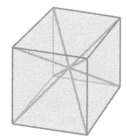

Consider a closed path along some of these line segments which passes through the 8 vertices without intersecting itself. Prove that the path must include at least one edge of the cube.

Q167 A group of 31 students plays a game in which each student writes the numbers from 1 to 30 in any order. One of the students is designated the "leader". The leader's sequence is compared with every other student's sequence, and each student earns k points if their sequence and the leader's have exactly k numbers in the same position. It turns out that each of the 30 players earned a different number of points. Prove that one of the students' sequences was the same as the leader's.

Q168 There are 25 people sitting around a table and they are playing a game with a deck of 50 cards. Each of the numbers $1, 2, \ldots, 25$ is written on two of the cards. Each player has exactly two cards. At a signal, each player passes one of their cards – the one with the smaller number – to their right–hand neighbour. This move is repeated. Prove that, sooner or later, one of the players will have two cards with the same numbers.

Q169 A total of 119 residents live in an apartment block consisting of 120 apartments. We call an apartment *intolerable* if there are 15 or more people living there. Every day, the inhabitants of an intolerable apartment have a quarrel and all go off to other apartments in the same block, no two to the same apartment. Is it certain that eventually there will be no more intolerable apartments?

Q170 Is the large pentagon twice, more than twice or less than twice the area of the star inside it? Try to answer without actually calculating the respective areas!

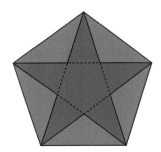

Q171 Let f be a function such that $f(n)$ is a positive integer whenever n is a positive integer. Suppose that

$$f(n+1) > f(n) \quad \text{and} \quad f(f(n)) = 3n$$

for all positive integers n. Determine the value of $f(2001)$.

Q172 Simplify the expression

$$\frac{2^3 - 1}{2^3 + 1} \frac{3^3 - 1}{3^3 + 1} \frac{4^3 - 1}{4^3 + 1} \cdots \frac{2002^3 - 1}{2002^3 + 1}.$$

Q173 Show that for any real numbers x, y we have

$$\frac{1}{x} + \frac{1}{y} \geq \frac{4}{x+y}.$$

Use this to prove that if $a_1, a_2, \ldots, a_n$ are positive reals, then

$$\frac{1}{a_1} + \frac{4}{a_2} + \frac{4^2}{a_3} + \cdots + \frac{4^{n-1}}{a_n} \geq \frac{2 \times 4^{n-1} - 1}{a_1 + a_2 + a_3 + \cdots + a_n}.$$

Q174 Fewer than 100 players entered a tournament in which each match has *three* contestants and one winner. Before each round, the names of all the survivors were drawn to provide as many matches as possible, with any competitors not drawn proceeding automatically to the next round. It turned out that the tournament was won outright by a player who contested only one match. What was the greatest possible number who took part?

Q175 Let

$$S = \frac{1}{1 \times 2 \times 3} + \frac{1}{4 \times 5 \times 6} + \cdots + \frac{1}{2002 \times 2003 \times 2004}.$$

It is clear that if we multiply S by $2004!$, then all the denominators will cancel and we shall obtain an integer. Prove that this integer is a multiple of 2005.

Q176 Find all real values of x satisfying the equation

$$\frac{x-1}{2004} + \frac{x-3}{2002} + \frac{x-5}{2000} + \cdots + \frac{x-2003}{2} = \frac{x-2}{2003} + \frac{x-4}{2001} + \frac{x-6}{1999} + \cdots + \frac{x-2004}{1}.$$

Q177 Annie and Brian, sitting beside the sea on the equator, watch the sun setting due west over the ocean on a calm day. Annie is 10 metres above sea level and Brian is on a cliff top 30 metres directly above Annie. Taking the circumference of the earth to be 40000 kilometres, determine how long after Annie Brian observes the instant of sunset

(a) by an approximate method without using trigonometry;

(b) exactly.

Q178 Metal bars of length 740 centimeters each are sold to a builder who wants to cut them into small pieces of two different lengths, 70cm and 50cm. He needs 1000 pieces of the first size and 2000 pieces of the second, and does not want to spend more money than necessary. Can you advise on the least number of bars he needs to buy, and show him how to perform the cutting?

Q179 Let D, E and F be three points on three sides of a triangle ABC, as in the figure. Prove that at least one of the triangles $\triangle AFE$, $\triangle BDF$ and $\triangle CED$ has an area not greater than one quarter that of $\triangle ABC$.

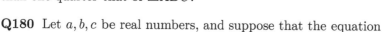

Q180 Let a, b, c be real numbers, and suppose that the equation

$$x^4 + ax^3 + bx^2 + cx + 1 = 0$$

has at least one real solution. Prove that $a^2 + b^2 + c^2 \geq \frac{4}{3}$.

Q181 Find all real numbers a, b, c in arithmetic progression such that $\sin^2 a$, $\sin^2 b$, $\sin^2 c$ are also in arithmetic progression.

Q182 Let $f(x) = ax^2 + bx + c$ be a quadratic with real coefficients. Prove that if $f(x)$ is an integer when $x = 0$, $x = 1$ and $x = 2$, then $f(x)$ is an integer whenever x is an integer.

Q183 An aeroplane leaves a town at latitude 1° south, flies x kilometres due south, then x kilometres due east, and x kilometres due north. At this point, the plane is $3x$ kilometres due east of its departure site. Find x.

Q184 Find all functions f satisfying the following functional equation in which $x \neq \pm 1$:

$$f\left(\frac{x - 3}{1 + x}\right) + f\left(\frac{x + 3}{1 - x}\right) = x.$$

Q185 Let ABC be a triangle such that sides AB and AC are fixed, but the angle $\angle BAC$ may vary. On the exterior of $\triangle ABC$, construct three squares $ABDE$, $ACGF$ and $BCHK$. Find the angle $\angle BAC$ such that the area of the hexagon $DEFGHK$ is maximal.

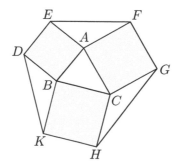

Q186 Prove that if A, B and C are the three angles of a triangle, then

$$\cos^2 A + \cos^2 B + \cos^2 C \geq \frac{3}{4}.$$

Q187 Let a and b be two unequal sides of a triangle, and let α and β be the angles opposite these sides, respectively. Prove that

$$\frac{a+b}{a-b} = \frac{\tan \frac{\alpha+\beta}{2}}{\tan \frac{\alpha-\beta}{2}}.$$

Q188 Let a, b, c and d be positive real numbers satisfying

$$\frac{1}{1+a^4} + \frac{1}{1+b^4} + \frac{1}{1+c^4} + \frac{1}{1+d^4} = 1.$$

Prove that $abcd \geq 3$.

Q189 Find the side lengths of the triangle with a given area S and one given angle γ, such that the length of the side opposite γ is minimal.

Q190 Given five different numbers a_1, a_2, a_3, a_4, a_5 satisfying

$$a_1^2 + a_2^2 + a_3^2 + a_4^2 + a_5^2 = 1,$$

prove that there are two unequal numbers a_i, a_j among these five such that

$$(a_j - a_i)^2 \leq \frac{1}{10}.$$

Q191 Alan, Betty, Chris and Debbie are of different ages. They all speak the truth whenever they talk to an older person, but they all lie when talking to a younger person. Alan says to Betty, "I am older than Chris". Debbie says to Chris, "You are the youngest of us all". List the four people in order from youngest to oldest.

Q192 Find infinitely many triangles which have integer side lengths with no common factor, and which contain an angle of $120°$.

Q193 A sequence $a_1, a_2, a_3, \ldots$ of positive integers has the properties

$$a_n^2 - a_{n-1}a_{n+1} = 1$$

for all $n \geq 2$, and $a_1 = 1$.

  (a) Prove that a_2 cannot equal 1.
  (b) Prove that if $a_2 = 2$, then $a_n = n$ for all n.
  (c) Prove that if $a_2 \geq 3$, then $a_n > a_{n-1} + 1$ for all $n \geq 2$.
  (d) Find all values of a_2 and all values of n such that $a_n = 2011$.

Q194 Let n be a positive integer and suppose that 2^n and 5^n begin with the same digit. Prove that there is only one possibility for this digit.

Q195 Show that 99999999999999999999999999999991 is not prime. No computing assistance!

Q196 Postman Pat has an infinite number of letters to deliver in Integer Avenue, which contains an infinite number of houses. Each letter is labelled with an integer (positive, negative or zero), and different letters have different labels. The letter with label n is to be delivered to house number $(4n+1)^2$. Prove that no house receives more than one letter.

Q197 Find a triangle with integer side lengths, all different, such that the longest side is 2011 and one of the angles is $\cos^{-1}\frac{1102}{2011}$.

Q198 Suppose that x satisfies the equation

$$x^{10^{10^{10}}} = 10^{10^{10^{11}}}.$$

We would like to know how many decimal digits x has.

  (a) What would you guess? Ten digits? A hundred? A million maybe?

  (b) Calculate the actual answer.

Q199 If

$$
\begin{aligned}
123x + 456y + 789z &= a \\
456x + 789y + 123z &= b \\
789x + 123y + 456z &= c,
\end{aligned}
$$

find *with least possible hard work* x, y, z in terms of a, b, c.

Q200 For any real number x, we write $\lfloor x \rfloor$ for x rounded to the nearest integer downwards, and $\lceil x \rceil$ for x rounded to the nearest integer upwards. For example,

$$\lfloor \pi \rfloor = 3 \quad \text{and} \quad \lceil \pi \rceil = 4 \quad \text{and} \quad \lfloor 5 \rfloor = \lceil 5 \rceil = 5.$$

Find all real numbers x which satisfy the equation

$$\lfloor 6x \rfloor + 7x - \lceil 8x \rceil = 9.$$

Q201 For this question, an n–digit integer is a string of n decimal digits with no prohibition on leading zeros: for example, 005105 is a valid six–digit integer. How many triples (x, y, z) of 10-digit integers are there which satisfy the conditions that x, y and z contain the digits $0, 1$ and 5 only, and that $x + y = z$?

Q202 Find all solutions of the simultaneous equations in $2n$ variables

$$
\begin{array}{ll}
x_1^2 + x_2 = 1, & 2x_2 + x_3 = 1 \\
x_3^2 + x_4 = 1, & 2x_4 + x_5 = 1 \\
\quad \vdots & \quad \vdots \\
x_{2n-1}^2 + x_{2n} = 1, & 2x_{2n} + x_1 = 1.
\end{array}
$$

> *Many, many puzzles and games are available online nowadays. They can frequently be solved with the aid of a little mathematical thinking; conversely, games will often give rise to challenging mathematical problems. A recently popular game goes under the name of KenKen (www.kenken.com). We flatter ourselves that possibly the most difficult KenKen puzzle ever was especially devised for the pages of Parabola.*

Q203 Fill in one of the numbers $1, 2, 3, 4, 5, 6$ in each empty square of the diagram given below, in accordance with the following rules.

(a) Each horizontal row must contain the numbers 1 to 6, once each.

(b) Each vertical column must contain the numbers 1 to 6, once each.

(c) The numbers in the "inner" region consisting of ten squares must have a sum of 26.

(d) The numbers in the "outer" region of twenty–five squares must have a product of 24186470400000.

Q204 Jack looked at the clock next to his front door as he left home one afternoon to visit Jill and watch a TV programme. Arriving exactly as the programme started, he set out for home again when it finished one hour later. As he did so he looked at her clock and noticed that it showed the same time as his had when he left home. Puzzling over how Jill's clock could be so wrong, Jack travelled home at half the speed of his earlier journey. When he arrived home, he saw from his clock that the whole expedition had taken two hours and fifteen minutes. He still hadn't worked out about Jill's clock and so he called her up on the phone. Jill explained that her clock was actually correct (as was Jack's), but it was an "anticlockwise clock" on which the hands travel in the opposite direction from usual. Jack had been in such a hurry to leave that he hadn't noticed the numbers on the clock face going the "wrong" way around the dial. At what time did Jack leave home?

Q205 Find all real numbers x which satisfy the equation

$$\lfloor x \rfloor - [2x] + \lceil 3x \rceil = 5,$$

where we write $\lfloor x \rfloor$ for x rounded to the integer below, $\lceil x \rceil$ for x rounded to the integer above, and $[x]$ for x rounded to the nearest integer, with halves rounding upwards.

Q206 There are n double seats in a railway carriage, all occupied. People get up and leave, one at a time, in random order. Find the probability that, after the last pair is broken, there remain exactly k people in the carriage.

Q207 In the game of poker, a pack of cards (consisting of the usual 52 cards) is shuffled and five cards are dealt to each player. A hand is referred to as "four of a kind" if it contains four cards of the same value and one other card. For example, ♠7, ♡7, ♢7, ♣7, ♢J constitutes four of a kind. Suppose that I deal two five–card hands from the same pack, one to myself and one to my opponent.

(a) What is the probability that my opponent has "four of a kind"?

(b) Suppose that I pick up my cards and discover that I have four of a kind. Is the probability that my opponent has four of a kind now greater or less than it was?

(c) Suppose on the other hand that I had picked up my cards and found that I *did not* have four of a kind. Is the probability that my opponent has four of a kind now greater or less than it was initially?

Q208 Seven different real numbers are given. Prove that there are two of them, say x and y, for which

$$\frac{1 + xy}{x - y}$$

is greater than $\sqrt{3}$.

Q209 A sheet of A4 paper has proportions $1 : \sqrt{2}$. Suppose such a sheet is white on one side and blue on the other, and label the corners A, B, C, D. The sheet is laid white side up and corner A is folded over, as shown in the diagram, in such a way that A, E, C are collinear and the length of the crease MN is equal to that of the side AD. What proportion of the visible figure $MNBCD$ is blue?

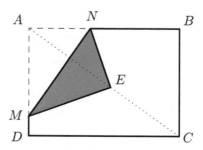

(And another question: can you think of a good reason *why* a sheet of A4 paper should have proportions $1 : \sqrt{2}$?)

Q210 In the game of chess, a knight moves two squares horizontally or vertically and then one square in a perpendicular direction. So, as long as the edges of the board do not get

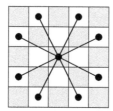 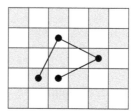

in the way, the knight has at any stage eight possible moves, as shown in the first diagram. From the second diagram, we can see that it is possible for a knight (making more than one move) to move from one square to that immediately to the right. Now, suppose that a "superknight" moves in a similar fashion but with three steps in one direction and eight in a perpendicular direction. Also, suppose that the board is infinitely large, so that there are no edges to get in the way. Can a superknight move from one square to that immediately to the right?

Q211 Find all solutions of the equation

$$x^4 + 4x^3 - 6x^2 - 20x + 13 = 0.$$

Q212 Let a and b be positive integers, and consider a "knight–like" piece which moves on a chessboard a squares up, down, left or right and then b squares in a perpendicular direction. We shall refer to such a piece as an (a, b)–*superknight*: for example, the piece described in Problem 210 is a $(3, 8)$–superknight, while a $(2, 1)$–superknight is just an ordinary knight.

Find all values of a and b for which an (a, b)–superknight can move from one square to that immediately to the right. As in Problem 210, we assume that the board is infinitely large, so that there are no edges to get in the way.

Q213 How often is the sun directly overhead at the equator? Once a day? Twice a day? Something else? Explain your answer!

Q214 Given an integer $n \geq 2$, what is the greatest number that can be obtained by writing n as a sum of positive integers and multiplying those integers? For example, if $n = 2013$, then we could write $n = 1006 + 1007$ and obtain the product $1006 \times 1007 = 1013042$, or $n = 1000 + 1000 + 13$ giving $1000 \times 1000 \times 13 = 13000000$, but neither of these is the maximum possible.

Q215 Let n be a positive integer. Show that if n is odd, then it is *not possible* to find polynomials $f(x)$ and $g(x)$ so that the only coefficients of these polynomials are 1 and -1, and

$$\frac{f(x)}{g(x)} = x^n - x^{n-1} + 1.$$

Note that in particular, $f(x)$ and $g(x)$ may not have any zero coefficients: for example $g(x) = x^3 - x + 1$ does not meet the requirements of the question.

Q216 As in Problem 212, an (a, b)–superknight is a knight–like chess piece which moves a squares up, down, left or right and then b squares in a perpendicular direction. In the solution to Problem 210, we saw that a $(3, 8)$–superknight can move from a square to the square immediately to its right in thirteen moves.

Next question: is it possible to do this in fewer than 13 moves?

Q217 As in Problem 214, we wish to write a given number n as a sum of positive integers in such a way that the product of the summands is as large as possible. For this question, however, the summands must be 4 or more. For example, $14 = 4 + 5 + 5$ is allowed but $14 = 2 + 3 + 4 + 5$ is not. If $n = 2013$, then what is the maximum product we can obtain in this way?

Q218 For what values of the coefficients a, b, c, d can the quartic equation

$$x^4 + ax^3 + bx^2 + cx + d = 0$$

be solved by using the method of Problem 211?

Q219 Find all solutions of the equation

$$x^2 - 12\lfloor x \rfloor + 23 = 0,$$

where $\lfloor x \rfloor$ denotes the integer part of x, that is, x rounded *down* to the nearest integer.

Q220 We saw in Problem 219 that the equation $x^2 - 12\lfloor x \rfloor + 23 = 0$ has four solutions, where the notation $\lfloor x \rfloor$ denotes x rounded to the nearest integer downwards. Show how to find positive integers a and b for which the equation

$$x^2 - 2a\lfloor x \rfloor + b = 0$$

has as many solutions as desired. In particular, find an equation of this type which has more than 2013 solutions.

Q221 Let $f(x)$ be a polynomial with degree 2012, such that

$$f(1) = 1, \quad f(2) = \frac{1}{2}, \quad f(3) = \frac{1}{3}, \quad \ldots, \quad f(2013) = \frac{1}{2013}.$$

Find the value of $f(2014)$.

Q222

(a) Factorise the quartic $x^4 + 6x^2 - 16x + 9$.

(b) Use (a) to find, without calculus, the maximum value of $\dfrac{x}{(x^2 + 3)^2}$.

(c) Find the maximum area of a triangle inscribed within the ellipse

$$\frac{x^2}{a^2} + \frac{y^2}{b^2} = 1$$

as shown: one vertex of the triangle is to be at the point $(0, -b)$ and the opposite side is to be horizontal.

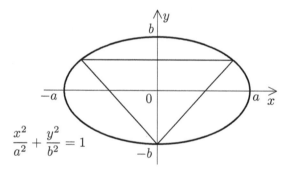

Q223 In a certain town, each of the inhabitants is either a truth–teller or a liar; however, this does not mean that everyone is actually able to answer every question they are asked. If a truth–teller is absolutely certain of the answer to a question, then she will give that answer; if not, she will say, "I don't know". On the other hand, a liar will never admit to not knowing something: he will give an answer that he knows is false, if any, but if there is nothing that he is *certain* is false, then he will give a randomly chosen answer (possibly even, by accident, the true answer). Moreover, everyone in this town can instantly deduce the logical consequences of any facts they know.

I meet four inhabitants of this town and ask them, "How many of you are truth–tellers?"

Kevin says, "I don't know"; then Laura says, "One"; then Mike says, "None". Noela, however, is asleep. Fortunately, I don't need to wake her up, since I can already tell whether she is a truth–teller or a liar. Which?

Q224 Use the ideas of the solution to Problem 222 to find *without calculus* the maximum value of

$$\frac{x}{(x^2 + a^2)^2},$$

where a is a positive real number.

Q225 In the town of the truth–tellers and liars from Problem 223, I meet four more people. I ask each of them, "How many of *the other three* are liars?"

George says "One". Helen says, "I don't know". Ian says, "Three". Jacqui says, "Two". Are these people truth–tellers or liars?

Q226 The diagram below shows two squares inscribed in a 60° sector of a circle; the points A, B and E are on the circular arc, and each square is symmetric about the angle–bisector OE. Prove that the squares have the same size.

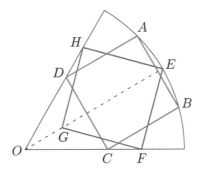

Q227 Mitchell and Dale are playing a game with dice. Mitchell has a die with five sides (each equally likely to show up) and Dale has a normal six–sided die. The two throw their dice alternately, with Mitchell going first. The first to throw a 1 wins. What are the winning chances of the two players?

Q228 As expected, Mitchell won his dice game with Dale (see Problem 227). Dale says, "Well, of course, you had the advantage: you had to get a 1 from a five–sided die and I had to do it from a six–sided die. Not only that, but you got to throw first. Let's play again, with me going first. Then we'll have equal chances". Mitchell replies, "OK, fine, but to make it more interesting we'll both double the number of sides on our dice: I'll use a 10–sided die and you'll use one with 12 sides. You can start, and the first to throw a 1 will win. Obviously, you still have an even chance of winning".

(a) Show that Dale was right in saying that the two had even chances in the game he proposed.

(b) Should he accept Mitchell's alternative offer?

Q229 The diagrams show a circle of radius a with a circumscribed square, and a circle of radius b with an inscribed square.

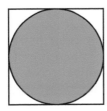 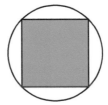

(a) Show that the ratio of the shaded area to the whole area for the first diagram is greater than $\frac{2}{3}$, and for the second diagram less than $\frac{2}{3}$.

(b) If the shaded area for both parts of the diagram together is exactly $\frac{2}{3}$ of the total area, find the ratio a/b.

(c) If the circles are the same size (that is, $a = b$) and we take m copies of the first diagram and n copies of the second diagram, then is it possible to get the shaded area being $\frac{2}{3}$ of the total area?

Q230 A triangle has sides a, b, c; the angles opposite these sides are A, B, C, respectively. Prove that
$$a^3 \sin(B - C) + b^3 \sin(C - A) + c^3 \sin(A - B) = 0.$$

Q231 Mitchell and Dale (see Problems 227 and 228) are now playing a game with ordinary (six–sided) dice. The rules are as follows: if a player throws a 6, then he wins; if he throws a 1 or a 2 then he loses; if he throws a 3, 4 or 5, then the other player has a turn. What is the first player's chance of winning?

Q232 Consider the number
$$S = \frac{2}{101} + \frac{4}{10001} + \frac{8}{100000001} + \cdots + \frac{1024}{1000\cdots0001},$$
where the numerators are powers of 2, and the number of zeros in each denominator is always one less than the corresponding numerator. Write S as an infinite recurring decimal.

Q233 How many 10–digit numbers x are there such that x ends with the digits 2015 and x^2 begins with the digits 2015?

Q234 Determine how many values of x satisfy the conditions
$$x^2 - x\lfloor x \rfloor = 20.15 \qquad \text{and} \qquad x < 2015.$$
Here, $\lfloor x \rfloor$ denotes x rounded to the nearest integer downwards; for example, $\lfloor \pi \rfloor = 3$.

Q235 Let m be a integer with $m \geq 2$. Prove that there is a cubic polynomial
$$p(x) = x^3 + ax^2 + bx + c$$
with integer coefficients, such that when x is an integer, $p(x)$ is never a multiple of m.

Q236 The point (s, t) is the centre of a square. Three vertices of the square lie on the parabola $y = x^2$. For $s = \frac{3}{2}$, find the coordinates of all four vertices of the square.

Q237 Find the 400th digit after the decimal point in the expansion of
$$\left(\sqrt{20} + \sqrt{15}\right)^{2016}.$$

Q238 A certain country has very unusual laws regarding the construction of highways: between every pair of towns, there must be a highway going in one direction but not in the other direction. (Except at towns, highways never intersect: overpasses are constructed where necessary.) Prove that there is a town which can be reached from every other town either directly, or with just one intermediate town.

Q239 Find the minimum value of
$$\sqrt{1 + x_1^2} + \sqrt{4 + x_2^2} + \sqrt{9 + x_3^2},$$
given that x_1, x_2, x_3 are real numbers with $x_1 + x_2 + x_3 = 8$.

Q240 Recall the country described in Problem 238: between every pair of towns there is a highway going in one direction but not in the other direction. We shall call a town "central" if it can be reached from every other town either directly, or with just one intermediate town. If the number of towns in the country is odd, show how to arrange the directions of the highways in such a way that *every* town is central.

Q241 Two more problems in the spirit of Problem 239.

(a) Find the maximum value of

$$\sqrt{1 - x_1^2} + \sqrt{4 - x_2^2} + \sqrt{9 - x_3^2},$$

given that x_1, x_2, x_3 are positive real numbers with $x_1 + x_2 + x_3 = 1$.

(b) Find the minimum value of

$$\sqrt{(x - 1)^2 + (x^2 - 2)^2} + \sqrt{(x - 3)^2 + (x^2 - 4)^2},$$

where x is a real number. Also, find the value of x which gives this minimum.

Q242 A sequence of numbers $a_1, a_2, a_3, \ldots$ is defined by the properties

$$a_1 = 3, \qquad a_{n+1} = a_n^2 - 2 \text{ for } n \geq 1.$$

Find the limit of the expression

$$\frac{a_n}{a_1 a_2 a_3 \cdots a_{n-1}}$$

as $n \to \infty$.

Q243 It's easy to see that we can start at the point A in the following diagram, then travel along the lines in such a way as to visit every one of the labelled points without repeating any points: for example, $ABCDEFGHJKL$. But is it possible to do the same thing starting at B?

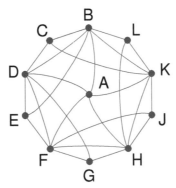

Q244 In a certain country (see Problems 238 and 240), between every pair of towns there is a highway going in one direction but not in the other direction. A town is called "central" if it can be reached from every other town either directly, or with just one intermediate town.

(a) Show that if there are 8 towns in this country, then it is possible for *every* town to be central.

(b) Show that the same is true for any number of towns except 2 or 4.

Q245 Find the greatest possible area of a quadrilateral having sides $2, 3, 4, 5$, in that order.

Q246 Divide the following array of numbers

2	2	1	2	1	1	1
2	4	3	1	1	2	2
1	1	2	1	1	2	1
5	1	3	1	6	1	1
1	3	3	1	1	1	4

into 11 connected regions, each containing numbers adding up to 6. A "connected region" means a set of squares in which every square is joined to some other square along an edge (not just a corner). No combination of numbers in a region may be used more than once. For example, you might use one of the following regions,

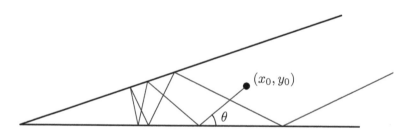

but you may not use more than one of them because they all contain the same numbers.

Q247 A ball (which can be thought of as a point of zero dimension) is projected into a "wedge–shaped billiard table" and continues to bounce off the sides as shown.

The ball starts at a distance x_0 to the right and y_0 above the vertex of the wedge, and the angle between its initial trajectory and the horizontal is θ. Find the closest distance the ball attains to the vertex.

Q248 The inspiration for this puzzle came from the "Plumber Game", which can be found at www.mathsisfun.com/games/plumber-game.html.

A game is played on a 4 by 8 grid of squares. The aim is to create a path from START to FINISH by placing in some or all of the squares either a quarter–circle connection or a straight connection. An example of a successful path is shown.

Prove that a successful path must contain an odd number of straight connections.

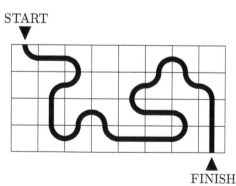

Q249 Solve the equation

$$\sqrt{x + 20} + \sqrt{x} = 17.$$

Q250 Consider a circle with centre $O = (a, b)$ and radius r, and a point $P = (p, q)$ which lies outside the circle. If Q and R are the two points on the circle such that PQ and PR are tangent to the circle, find the equation of the line QR.

Q251 In Problem 247, we considered a billiard ball being projected into a "wedge–shaped table" as shown in the diagram.

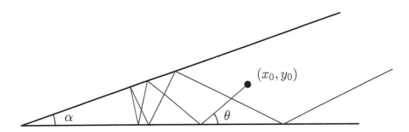

If the angle between the ball's initial trajectory and the horizontal is θ, and the angle at the vertex of the wedge is α, how many times does the ball hit the wedge?

Q252 Find all real numbers a, b, c such that $a < b < c$ and

$$a + b + c = 5, \quad a^2 + b^2 + c^2 = 15, \quad a^3 + b^3 + c^3 = 53.$$

Q253 A right circular cone is circumscribed about a sphere of radius r in such a way that the cone has minimal volume. Prove that the cone has exactly double the volume of the sphere.

Q254 To reach an exit of the MessConnex tollway, drivers have to get through a system of three roundabouts, shown as A, B and C in the diagram. Each roundabout has three

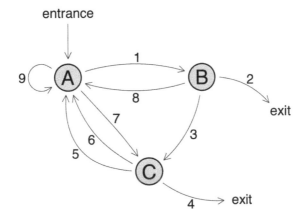

outgoing roads; two of the nine roads lead out of the MessConnex; the others stay within the system, one of them even returning to the very same roundabout. There are no signs to indicate the correct exit, so the drivers just have to guess; and all the exits look identical, so if drivers return to the same roundabout, then they just have to guess again.

The numbers shown on the outgoing roads are the toll (in dollars) charged for using each road. Clearly, a lucky driver could get out of the system for $3; but most drivers would have to spend much more. How much would it cost the average driver?

Q255 Consider the equation

$$29x + 30y + 31z = 366,$$

where x, y, z are positive integers with $x < y < z$.

(a) Without any writing or computer assistance, find x, y, z which satisfy these conditions.

(b) Prove that your solution from (a) is the only possibility.

Q256

(a) Let n, a, b be positive integers such that

$$n^2 < a < b < (n+1)^2.$$

Prove that ab cannot be a perfect square.

(b) Find infinitely many examples of positive integers n, a, b, c such that

$$n^3 < a < b < c < (n+1)^3$$

and abc is a perfect cube.

Q257 A *polyomino* is a figure consisting of unit squares joined along their edges. Every join must involve the full edge of both squares. We can give a polyomino a "chessboard" colouring (alternately light and dark) and calculate the ratio of light to dark squares. The diagram shows some possibilities with ratios $\frac{4}{1} = 4$, $\frac{5}{4}$ and $\frac{3}{3} = 1$.

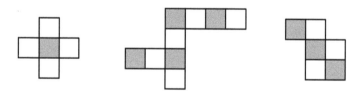

Prove that if $\frac{m}{n}$ is a rational number between $\frac{1}{3}$ and 3, then there is a chessboard–coloured polyomino such that the ratio of light to dark squares equals $\frac{m}{n}$.

Q258 We have a pattern of 34 dots arranged as shown.

It is permitted to remove any three dots, provided that one of them is exactly midway between the other two (the three dots may form a line in any direction – horizontal, vertical, diagonal or oblique); then to remove another three dots under the same condition; and so on. If we remove 33 dots, then which are the possibilities for the remaining dot?

Q259 Let m and n be positive integers with $m \neq n$. Prove that $m^4 + 3n^4$ can be written as the sum of the squares of three non–zero integers.

Q260 Eric is playing a game in which he rolls a (normal, six–sided) die three times, and wins if his three rolls are all different and in increasing order. For example, $1, 4, 5$ wins, but $4, 1, 5$ loses, and so does $4, 5, 5$. In the middle of the game, Eric calls you on the phone and tells you that his second roll was bigger than his first. If the game continues, what is Eric's chance of winning?

Q261 A shape consisting of a regular hexagon and two regular pentagons is cut out of cardboard; the pentagons are bent upwards along the lines AY and AZ until the two points marked B meet. What is then the angle $\angle XAB$?

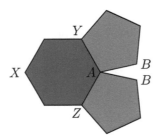

Q262 Consider all positive integers up to 2018 which are a power of 2 times a power of 5. These numbers can be arranged into sets of three numbers, each set consisting of a geometric progression, with one number left over. What are the possibilities for the leftover number?

Q263 Let a, b, c be positive numbers such that

$$\frac{a+b}{c} = 2018 \quad \text{and} \quad \frac{b+c}{a} = 2019.$$

Evaluate $\dfrac{a+c}{b}$.

Q264 Find the smallest positive integer n such that if n is divided by 61, then the 21st digit after the decimal point is 1, and the 41st digit is 9. You may find it useful to know that 10^{20} divided by 61 yields the remainder 13.

Q265 This puzzle is inspired by the "Breaklock" game, a combination of Mastermind and the Android pattern lock, www.mathsisfun.com/games/breaklock.html. We have a square pattern of nine dots, here supplemented by letters for easy reference.

$$
\begin{array}{ccc}
\bullet & \bullet & \bullet \\
\bullet & \bullet & \bullet \\
\bullet & \bullet & \bullet
\end{array}
\qquad
\begin{array}{ccc}
a & b & c \\
d & e & f \\
g & h & i
\end{array}
$$

A code consists of four different dots connected by straight lines, for example, *gdch*. Order is important, for example, *hcdg* is different from *gdch*. It is not possible to skip over a dot which has not – or not yet – been used: for example, *agei* and *agdc* are illegal. However, it is permitted to skip over a dot that has already been used: for example, *aecg* is allowed.

You guess the code *abcf* and are told that two of those letters are part of the code and are in the correct position in your guess, while another one is correct but not in the correct position. You then guess *dghi* and are told that one of these letters is correct and is in the correct position. You will be given similar information for future guesses. Can you find the code for certain in at most two more guesses?

Q266 The diagram below shows a rectangular grid with varying row heights and column widths. The areas of six sub–rectangles are shown: find the value of x.

x		$x+1$
$x+2$	$x+3$	
	$x+4$	$x+5$

Q267 Find the points of inflection on the graph of $y = f(x)$, and the tangents at these points, where

$$f(x) = x^4 - 2x^3 - 36x^2 + 28x + 99.$$

This is a routine problem if you have studied calculus, so do it **without calculus**.

Q268 Three questions are asked, and three answers given:

(A)	How many of these answers are correct?	**Answer:** a
(B)	Of the first and last answers, how many are correct?	**Answer:** b
(C)	Of the last two answers, how many are correct?	**Answer:** c

Here a is a number from $\{0, 1, 2, 3\}$, and b, c are numbers from $\{0, 1, 2\}$. If the values of a, b, c are known, then it is possible to tell for sure that one of the answers is correct and another is incorrect, but it is impossible to tell whether the remaining answer is correct or incorrect. What are the numbers a, b, c? Which answer was correct, and which was incorrect?

Q269 Let $f(x) = x^4 + ax^3 + bx^2 + cx + d$, and suppose that

$$d = \frac{c^2}{a^2}.$$

Find b in terms of a, c, such that $f(x)$ is the square of a quadratic. Hence, find all solutions of the equation

$$x^4 + 2x^3 - 20x^2 + 6x + 9 = 0.$$

Q270 A sequence of numbers $a_0, a_1, a_2, a_3, \ldots$ starts with $a_0 = 1$; each subsequent number in the sequence is given in terms of the previous one by the rule

$$a_n = a_{n-1}^2 - 2^{2^n+1}.$$

For example, you can check that

$$a_1 = -7, \quad a_2 = 17, \ldots, \quad a_6 = -32999768863368713983, \ldots.$$

Find a formula for a_n in terms of n.

Q271 For

$$f(x) = x^4 + 2x^3 - 7x^2 + 11,$$

find a line which is tangent to the graph $y = f(x)$ twice.

Q272 A 3×5 chessboard has three red counters in the leftmost column and three blue counters in the rightmost column. A counter can move to an adjacent square vertically or horizontally. Moves alternate between the two colours, and no two counters may occupy the same square simultaneously. Move the red counters into the right column and the blue counters into the left column in the minimum possible number of moves.

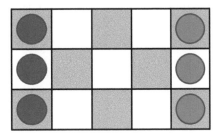

Q273 Briana arranges square unit tiles in a special way. She begins with a single tile, then puts one tile on its right to form a rectangle; and then a row of tiles from left to right along the top to form a square. She then puts two tiles on the right (from bottom to top) to form a rectangle, and another row on top to form a square; and so on. For example, the diagram indicates the order of placement of the first 11 tiles.

7	8	9	
3	4	6	11
1	2	5	10

What is the perimeter of Briana's figure after the first 2019 tiles have been placed?

Q274 Find the highest common factor of the integers

$$m = 2^{20} + 3^{19} \quad \text{and} \quad n = 2^{19} + 3^{20}.$$

Q275 In the following 4×4 array of numbers,

3	3	3	2
1	2	2	3
3	2	2	3
1	1	2	1

find a path which includes every square and does not visit any square more than once. You may start in any square you wish and finish in any square you wish; from a square containing the number n you must move to a square which is n squares away (either horizontally, vertically or diagonally).

Q276 If the expression

$$\left(1 + x^2 + \frac{1}{x}\right)^{10}$$

is expanded and like terms collected, find the coefficient of x^3.

Q277 Prove that any given string of decimal digits occurs (consecutively and in the given order) among the digits of n^2 for some integer n.

Q278

(a) Show that it is possible to choose a point O inside a square and to draw three rays from O, all separated by equal angles, in such a way that the square is divided into three regions of equal area.

(b) Show that in (a), the point O cannot be the centre of the square.

Q279 Take a point O inside a square; from this point, draw six rays, all spaced at equal angles. This will divide the square into six regions. Is it possible that all these regions have equal area?

Q280 Find the sum of the digits of

$$S = 1 + 11 + 111 + 1111 + \cdots + \overbrace{11\cdots 11}^{999 \text{ digits}},$$

where the last term on the right–hand side has 999 digits, all equal to 1.

Q281 Find the smallest multiple of 4321 which ends in the digits 1234.

Q282 I want to join at right angles two iron roofs pitched at an angle θ (see the diagram: the upper and lower edges of each roof are to be parallel). At what angles do I need to cut the iron pieces?

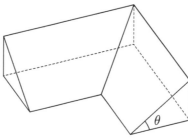

Q283 We have a bag and seven slips of paper on which are written

- at least one of the statements in the bag is true;
- at least two of the statements in the bag are true;
- at least three of the statements in the bag are true;
- at least one of the statements in the bag is false;
- at least two of the statements in the bag are false;
- at least three of the statements in the bag are false;
- at least four of the statements in the bag are false.

One of the slips is removed; the other six are placed in the bag and we determine whether they are true or false. How many are false?

Q284 Take a point O inside a square; from this point draw 13 rays, all spaced at equal angles. This will divide the square into 13 regions. Is it possible that all these regions have equal area?

Q285 A right circular cone (with a closed base) is partially filled with water. The base of the cone is placed on a table and the depth of water in the cone is found to be 10cm. The cone is then inverted so that its vertex is on the table and its base is parallel to the table, and the depth of water is found to be 11cm. What height of empty space is there now above the water's surface?

Q286 What is the largest integer that *cannot* be expressed in the form $99a + 100b + 101c$, where a, b, c are non–negative integers?

Q287 Let a, b, c, d be positive real numbers such that $a + b + c + d = 4$. Prove that

$$\left(\frac{16}{a^2} - 1\right)\left(\frac{16}{b^2} - 1\right)\left(\frac{16}{c^2} - 1\right)\left(\frac{16}{d^2} - 1\right) \geq 15^4.$$

Q288 A regular n–gon is rotated by some angle about its centre O and the result is superimposed upon the original; the diagram illustrates the situation for $n = 5$. Let A_0 be the area of the original polygon and P_0 its perimeter. Let A be the area common to both polygons (grey in the figure) and P the perimeter of the combined polygons (the whole red and grey region in the figure). Prove that

$$\frac{P}{P_0} + \frac{A}{A_0} = 2.$$

Q289 Of the students in a senior maths class, the proportion who read *Parabola* is 66%, *to the nearest percent*. What is the smallest possible number of students in the class?

Q290 Triangle ABC has a right angle at B, and D is a point on the hypotenuse AC. The perpendicular to AC at D intersects AB at E, and we draw the line EC.

Use this diagram to prove the "cosine of a sum" formula

$$\cos(x + y) = \cos x \cos y - \sin x \sin y.$$

Problems in Parabola *Volume 57 Issue 2, 2021, were dedicated to Thomas Britz, editor of* Parabola, *and his partner Ania, in celebration of the arrival of their twin sons Alex and Ben. The ten problems of that issue all had to do, one way or another, with twins. We reproduce five of them here.*

Q291 Alex and Ben are playing in their local park. This park consists of an n by n array of square gardens, separated by paths. Alex starts at the south–west corner of the park and walks along the paths at a speed which takes him along the side of any garden square in exactly one minute, and always heads north or east. Ben walks at the same speed, but starts at the north–east corner and always walks south or west. Find the probability that Alex and Ben meet after n minutes.

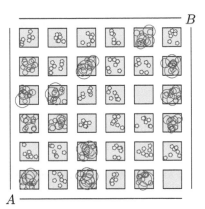

Q292 Alex and Ben are visiting the nation of Twinnia. In this country, there is a rule that on any given day, twins must behave alike in terms of telling the truth: that is, both must tell the truth or both must lie; it is forbidden for one to tell the truth and the other to lie. You overhear a conversation between four people. Two of them are Alex and Ben, but you cannot decide which is which, though one of them is wearing a yellow jumper and one is wearing a red jumper. The other two are Ellie and Fiona: they look very similar, and you are not sure whether or not they are twins. The following statements are made.

> Ellie: Fiona and I are twins.

> Fiona: The boy in the yellow jumper is Ben.

> Boy in yellow: The boy in the red jumper is Alex.

> Boy in red: Ellie and Fiona are not twins.

Can you determine which of the boys is which? Can you decide whether Ellie and Fiona are twins or not?

Q293 Thomas is designing a new nursery for his twins Alex and Ben. They will each have a cradle in the shape of an ellipse, placed side by side. In suitable coordinates, the ellipses have equations

$$(x+1)^2 + \frac{y^2}{4} = 1 \quad \text{and} \quad (x-1)^2 + \frac{y^2}{4} = 1.$$

The cradles will be surrounded by a wooden floor. As a mathematician, Thomas is very keen on symmetry, so the surround will also be an ellipse, in this case having the equation

$$\frac{x^2}{a^2} + \frac{y^2}{b^2} = 1;$$

but he is also conscious of using space efficiently, so he wants this ellipse to have the smallest possible area. What values of a and b should he choose?

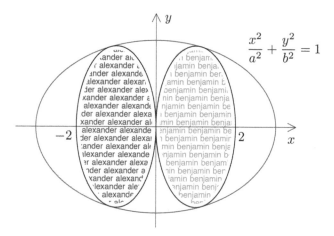

Q294 Alex and Ben want to access their favourite computer game. Each has to enter a password, which will be a string of letters a and b. If their words can be converted to each other by substituting aab for bba or vice versa, more than once if necessary, then the game app will agree that the passwords match and will let them access the game. For example, $aaaabab$ and $bbabbba$ match because of the chain of substitutions

$$aa\underline{aab}ab \sim aabb\underline{aab} \sim \underline{aab}bbba \sim bbabbba.$$

The twins enter their passwords and hit return... nothing happens! They have made a typing error. Even worse, the backspace and delete keys have frozen!! The only hope is to keep typing and see if the passwords match at some time in the future.

(a) Is this ever possible? That is, are there two non–matching passwords which can be extended to give matching passwords?

(b) Suppose that after realising their mistake, Alex and Ben are very careful to type exactly the same in the future. Now is it possible for them to gain access? In other words, are there two non–matching passwords which can be extended *in the same way* to give matching passwords?

Q295 Looking ahead a few years... On their first day at school, Alex and Ben are amazed to find that their class consists entirely of twins! – nine pairs of twins, to be exact. The teacher wants to split the class up for three different activities: 7 of the children will do music, 6 will do reading and 5 will do painting. Each pair of twins will do two different activities. In how many ways can the teacher allocate children to activities?

Q296 Solve Problem 293 without using calculus. Specifically,

(a) find the gradient of the ellipse $(x^2/a^2) + (y^2/b^2) = 1$ at the point (p, q);

(b) find the minimum value of $2a^4 - 2a^3\sqrt{a^2 - 4}$ for $a \geq 2$.

Q297 A variation of Problem 295. Now we have eight pairs of twins, and there are four activities, music, painting, reading and dancing, with four children to do each activity. Once again, each pair of twins is to do two separate activities. In how many ways can children be allocated to activities?

Q298 A closed path consists of lines from the centre of a square to the centre of an adjacent square on a $2n$ by $2n$ grid. The curve visits every square exactly once. An example is shown in the accompanying diagram.

There are a number of intersections of gridlines outside the path, shown as blue dots in the diagram. How many, on a $2n$ by $2n$ grid?

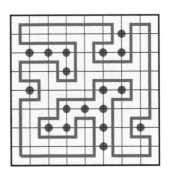

Q299 A sequence of numbers $x_1, x_2, x_3, \ldots$ is generated as follows. We begin with $x_1 = 1$; then we take the cosine and sine of x_1; then the cosine and sine of x_2; and so on. To clarify:

$$x_1 = 1$$
$$x_2 = \cos(x_1) = \cos(1)$$
$$x_3 = \sin(x_1) = \sin(1)$$
$$x_4 = \cos(x_2) = \cos(\cos(1))$$
$$x_5 = \sin(x_2) = \sin(\cos(1))$$
$$x_6 = \cos(x_3) = \cos(\sin(1))$$
$$x_7 = \sin(x_3) = \sin(\sin(1))$$
$$x_8 = \cos(x_4) = \cos(\cos(\cos(1)))$$

and so on. Find the smallest $n > 1$ such that $x_n > 0.99$.

Q300 For this question, a *knockout contest* among n entrants means the following. Let k be the integer for which $2^{k-1} < n \le 2^k$. Then $2^k - n$ players (chosen at random) are "given a bye" in the first round: that is, they progress to the next round without playing a match. The remaining $2n - 2^k$ players play in pairs (once again chosen at random), and the $n - 2^{k-1}$ winners also progress. This leaves 2^{k-1} entrants who will play in pairs, leaving 2^{k-2} winners in the next round; and so on; until the overall winner is decided by a match between the last 2 players. Note that there are no byes after the first round, and so the eventual winner will play either $k - 1$ or k matches, depending on whether they do or do not receive a bye in the first round.

(a) If n players enter the competition, how many matches will be played altogether?
Comment. This is a well-known problem and there is a very easy solution. Do not try to consider the number of byes, the number of rounds or other details!

(b) A football club wishes to rank the three strongest arm–wrestlers from a pool of 100 candidates by using a knockout contest to determine the best arm–wrestler; then another knockout contest to determine the second–best; then another to determine the third–best. Show that if the contests are carefully organised, then this can be accomplished in 113 matches altogether; but that fewer than 105 matches will never be enough.

Q301 Last Christmas, I pulled a Christmas cracker, and out popped the traditional paper crown.

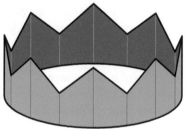

While inside the cracker, it had been flattened out between two opposite points A and E, and then folded right half over left three times, as in the diagrams. It's clear that if the

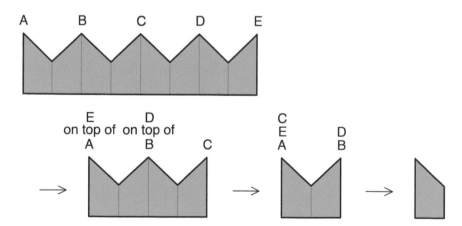

crown is unfolded, then some of the creases that have been made will point towards the outside of the crown, and some will point towards the inside. Is it possible to now refold the crown in the same way as before, but starting with a *different* pair of opposite points instead of A and E, and *without* reversing any of the folds already made?

Q302 The recently popular game "Wordle" challenges you to guess a secret five–letter word. You may enter any word from the official "Wordle" word list, and you will be given some information in return. You are allowed a maximum of six guesses.

In the not–at–all well–known game "Squardle", you have to guess a secret square number, and you may enter any five–digit square. Here is an example of the start of a game.

Two guesses have been entered so far: $124^2 = 15376$ and $175^2 = 30625$. When a digit is highlighted green in the diagram, it indicates that the digit occurs in the secret square in the same location as it is in the guess; a yellow highlight indicates a digit which occurs in the secret square, but not in the same location as in the guess; and a grey highlight indicates a digit which does not occur in the secret square at all. The secret square may contain the same digit more than once.

In Squardle, only three attempts are allowed. Can you win the game which was started above?

Q303

(a) Let $p(x) = 1 + 2x + 3x^2 + 4x^3$. Find a polynomial $q(x)$ with integer coefficients, not all zero, such that when $p(x)q(x)$ is expanded and terms collected, there will be no terms in x^k unless k is a square number. (Note that 0 is a square: so we want a product polynomial that looks like $a + bx + cx^4 + dx^9 + \cdots$.)

(b) Prove that if we replace $p(x)$ by any polynomial with integer coefficients, then a polynomial $q(x)$ with this property can always be found.

Q304

(a) Show how to arrange the numbers $1, 2, 3, 4, 5, 6, 7, 8, 9$ around a circle in such a way that the sum of two neighbouring numbers is never a multiple of 3 or 5 or 7. In how many ways can this be done?

(b) Given any 9 consecutive integers, is the same task always possible?

Q305 Simplify

$$\frac{\sqrt[3]{560 + 158\sqrt{2} + 324\sqrt{3} + 90\sqrt{6}}}{\sqrt[3]{560 - 158\sqrt{2} + 324\sqrt{3} - 90\sqrt{6}}}.$$

Q306

(a) A line with gradient m intersects the ellipse $x^2 + 2y^2 = 3$ at the point $(1, 1)$ and another point. Find the other point.

(b) Find all triples of positive integers a, b, c with no common factor so that $a^2 + 2b^2 = 3c^2$.

Q307 A one–person game is played as follows. Begin with a stack of n coins. Split them into two (non–empty) stacks with say a and b coins; this move gives a score of ab. Keep splitting the remaining stacks until all stacks consist of a single coin, and add all the scores. For example, starting with a stack of 30 coins, we might split it into stacks of 20 and 10, scoring 200; then into 20 and 7 and 3 scoring 21, total score so far 221; and so on until we have 30 stacks each containing one coin.

Prove that, no matter how the coins are split, the final total score is always the same.

Q308 An ant walking across the floor noticed a grain of ant poison and a grain of sugar. Hating poison and loving sugar, the ant decided to walk in such a way that its distance from the poison increases at the same rate at which its distance to the sugar decreases. The ant was surprised to discover that no matter how fast it walked, it could not reach the sugar this way.

(a) Explain why the ant could not reach the sugar as long as it moved in the way described.

(b) Are there any exceptional cases when the ant could reach the sugar?

(c) Describe the path of the ant in case (a).

Q309

(a) Five points are drawn on a page. Two points u and v are joined by both a red curve and a blue curve. All other pairs are joined by one line (or curve) which is shown in

the diagram as grey, and will be coloured either red or blue. Prove that, no matter how this colouring is done, the resulting diagram will contain three of the five original points mutually joined by red lines, or three points mutually joined by blue lines.

(b) Eight points a, b, c, d, e, f, g, h are drawn on a page. Four pairs are joined by red and blue curves, as shown in the diagram. All other pairs are joined by one line, which will be coloured either red or blue.

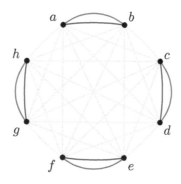

Prove that, no matter how this colouring is done, the resulting configuration will contain three points mutually joined by red lines, or four points mutually joined by blue lines.

Q310 In how many different ways can 10^{100} (a googol) be factorised as xyz, where x, y, z are positive integers and

(a) the order of the factors does matter; for example, $2^{50} \times 5^{50} \times 10^{50}$ is regarded as a different factorisation from $5^{50} \times 10^{50} \times 2^{50}$?

(b) the order of the factors does not matter; for example, $2^{50} \times 5^{50} \times 10^{50}$ is the same factorisation as $5^{50} \times 10^{50} \times 2^{50}$?

Q311 Prove that the sum of two different powers of 2 can never be a cube or higher power of an integer. That is, there are no solutions of

$$2^a + 2^b = m^p$$

in which a, b, m, p are non–negative integers, $a \neq b$ and $p \geq 3$.

Q312 In a semicircle on diameter AB, we have $|AX| = 3$ and $|XY| = 2$. As shown in the diagram, two isosceles triangles have AX and XY as their bases, and their third vertices are on the semicircle. If the triangles have equal area, find the diameter of the semicircle.

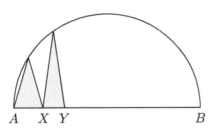

Q313 Three groups of 2, 3 and 4 passengers arrive independently and at random times at a railway station where a train departs every 12 minutes. What is the probability that the average waiting time per person is more than 8 minutes?

Q314 Let n be an integer, $n \geq 2$, and let $p(x)$ be a polynomial with degree at most n, having integer coefficients. Suppose that the values of $p(x)$, where x is an integer, include all of the numbers $0, 1, 2, \ldots, n$. Prove that $p(x) = x + c$ for some constant c.

> *A stray "twins" problem that did not make it into Issue 2 from 2021 (see page 43 above)!*

Q315 A school class consists entirely of twins: $2n^2 + 2n$ pairs of them, where $n \geq 2$. Together with the teacher, this means that there are $4n^2 + 4n + 1$ people in the class, so they can stand in a $2n + 1$ by $2n + 1$ square array. Prove that, however they arrange themselves in this array, it will be possible to find $2n + 1$ of the children (excluding the teacher) in such a way that no two of the chosen children are standing in the same row, no two are standing in the same column, and no two are twins.

Q316 Consider the sequence of numbers obtained by stringing together the digits of the positive integers, namely

$$1, \ 12, \ 123, \ 1234, \ 12345, \ 123456, \ldots$$
$$\ldots, \ 12345678910, \ 1234567891011, \ 123456789101112$$

and so on. Are any of these numbers multiples of 11? If so, find the smallest example.

Q317 In the game of chess, a knight can move from its current square to any square reached by going two squares horizontally or vertically, then one square in a perpendicular direction.

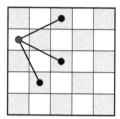

The diagram shows all the moves available on a 5×5 chessboard to a knight on the white square marked with a red dot. Since a single move takes a knight from a white square to a black square, or *vice versa*, no two of the 13 black squares are separated by a single knight's move. Is there any other choice of 13 squares on a 5×5 board for which this is true?

Q318 Arrange a hundred digits (digits are $0, 1, \ldots, 9$) on a circle in such a way that, reading clockwise, each of the pairs $00, 01, \ldots, 99$ occurs once.

Q319 Consider all numbers which can be formed by choosing eleven different positive integers whose sum is 82 and finding the product of the eleven integers. For example, one of the products to be considered is

$$1 \times 2 \times 3 \times 4 \times 5 \times 6 \times 7 \times 8 \times 9 \times 10 \times 27.$$

Find the greatest common divisor of all these products.

Q320 How many paths of length $m + n + 2$ are there from $(0,0)$ to (m,n) on an $m \times n$ grid, if the path may never visit the same grid point more than once?

Q321 Find the largest example of an 8–digit number which uses the digits $1, 2, 3, 4, 5, 6, 7, 8$ once each and is a multiple of 101.

Q322 Find the smallest perfect square whose decimal representation consists of the same block of digits twice over. (An example of such a number would be 123123 – but, of course, that's not a square.) As usual, the first digit of a number may not be zero.

Q323 In Problem 182, we proved the following fact about the set of integers $S = \{0, 1, 2\}$: if $f(x) = ax^2 + bx + c$ is any quadratic with real coefficients such that $f(x)$ is an integer for all values of x in the set S, then $f(x)$ is an integer for all integer values of x.

(a) Find all possible sets of three integers which have the same property.

(b) Can you find a set of four or more integers which does not include S or any of the other sets you found in (a), and which still has the same property?

Q324 In Problem 169, we described an apartment block consisting of 120 apartments. Every day, the inhabitants of an "intolerable" apartment – one having 15 or more residents – all go off to other apartments in the same block, no two to the same apartment. We showed that if there is a total of 119 residents, then sooner or later there will be no more intolerable apartments. Now suppose that one more person moves in, so that there are 120 residents in all. Devise a scenario in which there will always be an intolerable apartment.

Q325 Find an infinite set S of positive integers such that for any finite non–empty subset A of S, the sum of all elements of A is never a perfect power. By a *perfect power*, we mean a positive integer a^b, where a, b are positive integers and $b > 1$.

Q326 The numbers $1, 2, \ldots, 64$ are written onto the squares of an 8×8 chessboard, one to a square.

(a) Find such an arrangement in which the minimum difference between numbers in squares which are adjacent horizontally, vertically or diagonally is 15.

(b) Prove that there is no such arrangement in which the minimum difference is greater than 15.

Q327 In Problem 319, we considered all products of eleven different positive integers having sum 82, and found the greatest common divisor (highest common factor) of all these products. Now change the sum to s, where s is an integer not less than 66. (If $s < 66$, then there is no collection of eleven different positive integers with sum s, and so the problem does not make much sense.) Find the *smallest* value of s for which the greatest common divisor of all the corresponding products of eleven numbers is 1. If $s_{\min}$ is this smallest value and we consider a sum $s > s_{\min}$, then does it necessarily follow that the greatest common divisor of all products of eleven different positive integers with sum s is still 1?

Q328 In how many ways can one select 5 points from the 64 shown, such that at least three of the chosen points lie in a straight line? Here, a "straight line" means one of the horizontal or vertical lines shown in the diagram, and the points in a line *do not* need to be adjacent points on the grid.

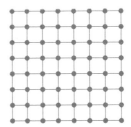

Q329 If x is a rational number, then we define $f(x)$ to be the denominator of x. That is, if $x = p/q$ in lowest terms, then $f(x) = q$. Part of the graph of $y = f(x)$ is shown below. Can you explain the "dotted curves" appearing in the image? Or any other notable features?

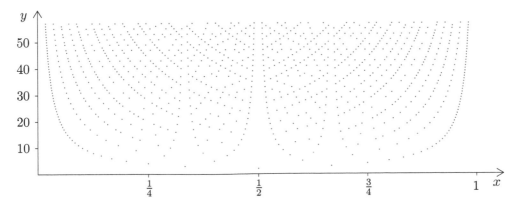

CHAPTER 2

Solutions

Q1 To solve (a), let $a = 1 - x$, $b = 1 + y$; then x and y are positive and we have

$$S - P = a + b - ab = (2 + y - x) - (1 + y - x - xy) = 1 + xy > 1.$$

For (b), if both numbers are 1, then $S = 2$. If one number is less than 1, then the other is greater than 1, and applying (a) with P equal to 1 gives $S > 2$.

Finally, the area of the right–angled isosceles triangle with base and height equal to k is $\frac{1}{2}k^2$. Let $\sqrt{u}\,k$ and $\sqrt{v}\,k$ be the base and height of a second right–angled triangle of the same area. Then $uv = 1$. Applying (b), the minimum value of the square on the hypotenuse (equal to $(u + v)k^2$) is $2k^2$, attained only when $u = v = 1$.

Q2 Of the allowable factorisations of 2450 into 3 numbers, the sums of the three factors are all different, except for $5, 10, 49$ and $7, 7, 50$, both of which add up to 64. Hence, one of these groups must give the age of the parishioners, the verger being 32, since if he had been any other allowable age (e.g., 41), he would have been able to find immediately the only possible factors summing to twice his age $(5, 7, 70)$. The bishop's age, being greater than any of the parishioners, certainly exceeds 49. If it also exceeded 50, then the verger would still have been unable to distinguish between the two groups already mentioned. Therefore, the bishop was exactly 50 years old.

Q3 First observe that
$$n^2 + n + 2 = (n + 4)^2 - 7(n + 2).$$

If $n^2 + n + 2$ is divisible by 49, then it is certainly divisible by 7, and therefore $(n + 4)^2$ is divisible by 7. But in this case, 7 is a factor of $n + 4$ and therefore not of $n + 2$. Therefore, $(n + 4)^2$ is divisible by 49 but $7(n + 2)$ is not, so their difference $n^2 + n + 2$ is also not.

Q4 Colour the map according to the following rule: if a region lies inside an even number of circles (including 0 circles), then use colour A; if it lies inside an odd number of circles, then use colour B. Every region is coloured with one of the colours. Adjacent regions are separated by a circular arc, one region lying inside the circle, and one outside it. Hence, the numbers of circles in which the adjacent regions lie differ by one, and they are given different colours by the rule.

Q5 The number of animals in the herd is $20k \pm x$ where k and x are integers and $0 \le x < 10$ (so that $x^2 < 100$). The sum of money obtained by the sale of the herd is $\pounds(20k \pm x)^2$, that is, $\pounds(400k^2 \pm 40kx + x^2)$. Since $400k^2 \pm 40kx$ is divisible by 40, a stage will be reached in the sharing out when $\pounds x^2$ remains, it being A's turn to take $\pounds 10$. It is clear from the data that x^2 lies between 30 and 40, or else between 70 and 80. Since x is an integer, the only possibility is $x^2 = 36$. Only $\pounds 6$ remains when it is D's final turn. The others must return $\pounds 1$ each to D, reducing their last share from $\pounds 10$ to $\pounds 9$, and increasing D's from $\pounds 6$ to $\pounds 9$.

DOI: 10.1201/9781003396413-2

Q6 For part (a), the "sums–to–products" formulae (Section 3.7) give

$$\sin \angle AOX + \sin \angle BOX = 2\sin \frac{\angle AOX + \angle BOX}{2} \cos \frac{\angle AOX - \angle BOX}{2}$$
$$= 2\sin \frac{\angle AOB}{2} \cos \frac{\angle AOX - \angle BOX}{2}.$$

Only the last factor is variable, and its maximum value of 1 occurs when $\angle AOX = \angle BOX$. The proof applies equally well if $90° \leq \angle AOB \leq 360°$, and the result can be generalised as follows to the case when a given angle is split into n angles.

Lemma. *If* $x_1 + x_2 + \cdots + x_n = \alpha$, *where* $0 < \alpha \leq 2\pi$ *and every* x_i *is positive, then* $\sin x_1 + \sin x_2 + \cdots + \sin x_n$ *is maximal when* $x_1 = x_2 = \cdots = x_n = \frac{\alpha}{n}$.

Proof. We show this by induction on n. Assume that the result is true for $n = k - 1$. After choosing x_1 in any fashion, we see by the induction hypothesis that $\sin x_2 + \cdots + \sin x_k$ is maximal when $x_2 = \cdots = x_n = (\alpha - x_1)/(k - 1)$. Hence,

$$\max\left(\sin x_1 + \sin x_2 + \cdots + \sin x_k\right) = \max\left(\sin x_1 + (k - 1)\sin \frac{\alpha - x_1}{k - 1}\right).$$

We can find the maximum value of the function on the right–hand side by the use of calculus. Set

$$y = \sin x_1 + (k - 1)\sin \frac{\alpha - x_1}{k - 1}.$$

Then $\dfrac{dy}{dx_1} = \cos x_1 - \cos \dfrac{\alpha - x_1}{k - 1}$, and this vanishes when

$$x_1 = \frac{\alpha - x_1}{k - 1}, \quad \text{that is, when} \quad x_1 = \frac{\alpha}{k}.$$

For this value of x_1,

$$\frac{d^2 y}{dx_1^2} = -\left(1 + \frac{1}{k - 1}\right)\sin \frac{\alpha}{k} < 0,$$

so that the stationary value is a maximum. Hence, the truth of the lemma for $n = k$ follows from the truth for $n = k - 1$, and since in (a) we have proved it when $n = 2$, the lemma must be true for all positive integers $n \geq 2$.

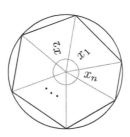

Since the area of an inscribed n–gon whose sides subtend angles $x_1, x_2, \ldots, x_n$ at the centre (so that $x_1 + x_2 + \cdots + x_n = 2\pi$) is equal to $\frac{1}{2}r^2(\sin x_1 + \sin x_2 + \cdots + \sin x_n)$, it follows that the n–gon of maximum area is that for which $x_1 = x_2 = \cdots = x_n = \frac{2\pi}{n}$, that is, the regular n–gon.

Q7 The sufficiency of the condition is obvious: if a divides x, say, $x = ka$, then the floor can be divided into k strips of size $a \times y$, each of which can be covered by a row of y tiles.

To prove that the condition is necessary, divide the floor into 1×1 squares, and colour the squares using a different colours, as shown in the diagram.

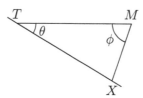

Thus, use the colour 1 for the lower left square, the colour 2 for the two squares next to it, and so on, forming diagonals, up to the colour a. Then use colour 1 and repeat the colours in the same order until every square has been coloured. Whichever of the two possible ways a tile is laid, it covers a squares, one of each colour. Hence, a necessary condition for the tiling to be possible if that there should be equal numbers of squares of each colour. There are certainly equal numbers of squares of each colour in the top a rows, since every column contains one square of each colour in these rows. We may omit these rows and, similarly, we may omit a columns along the right–hand side of the figure.

In fact, if $x = q_1 a + r_1$ and $y = q_2 a + r_2$ where $0 \leq r_1, r_2 < a$, by omitting q_1 blocks of a row and then q_2 blocks of a column, then we are left with a small rectangle in the left–hand bottom corner of the floor, with sides of length r_1 and r_2, respectively (unless one of these numbers is zero). But such a figure cannot have the same number of squares of each colour. In fact, the number of squares bearing colour a is $r_1 - r_2 - a$, whilst if $r_1 \geq r_2$, then the number bearing the colour corresponding to r_1 is r_2, and these numbers cannot be equal since $r_1 < a$. Hence, unless either $r_1 = 0$ or $r_2 = 0$ the floor cannot have equal numbers of squares of each colour and the tiling is impossible.

Q8 The house number must clearly be a perfect square; for if A had told B that it was not a square, then the further information of being odd or even cannot "determine it definitely". The perfect squares less than 30 are $1, 4, 9, 16, 25$.

Suppose that A had answered "less than 15" to B's first question; then again the number will not be definitely determined if A should answer "odd" to B's third question; thus, A in fact answered "greater than 15". But A lied; thus the number is an even perfect square less than or equal to 15. The answer is therefore 4.

Q9 If the man travels in a direction $\phi°$ south of west, then he will catch the train provided that $|MX|/|TX|$ is less than the ratio of his speed to the train's speed.

Hence, to give himself the greatest possible chance of survival, he should choose ϕ to make $|MX|/|TX|$ as small as possible. By the Sine Rule,

$$\frac{|MX|}{|TX|} = \frac{\sin \theta}{\sin \phi},$$

and, whatever value θ has, this will achieve its minimum value when the denominator achieves its greatest value, namely 1.

Hence, the man should travel due south, making ϕ equal to $90°$.

Q10 It follows from the Factor Theorem (Section 3.10) that, if $P(x)$ has the same value, say a, for k different values of x, say $x_1, \ldots, x_k$, then the degree of $P(x)$ is at least k. In fact, $x - x_1, \ldots, x - x_k$ are all factors of $P(x) - a$, so we see that

$$P(x) = a + Q(x)(x - x_1) \cdots (x - x_k)$$

for some polynomial $Q(x)$.

Now, suppose that the given polynomial factorises as $P(x) = F(x)G(x)$, where $F(x)$ and $G(x)$ are both polynomials with integral coefficients. There exist seven integers $x_1, \ldots, x_7$ for which the value $P(x_i) = \pm 1$. That is, $F(x_i)G(x_i) = \pm 1$ for each i. But $F(x_i)$ and $G(x_i)$ are themselves integers, so each must be ± 1 for each value of i. Then $F(x)$ has one of the values 1 or -1 at the least four times, so the degree of $F(x)$ is at least 4. Similarly, the degree of $G(x)$ is at least 4, so their product $P(x)$ must be a polynomial of degree at least 8. This contradiction shows that no such factorisation is possible.

Q11 Let $ABCDEF$ be a regular hexagon of side 1 unit. This hexagon has area $\frac{3}{2}\sqrt{3}$. We shall show how to transform it into a $\sqrt{3}$ by $\frac{3}{2}$ rectangle, and how to transform the latter into a square. The basic ideas are illustrated by the following diagrams.

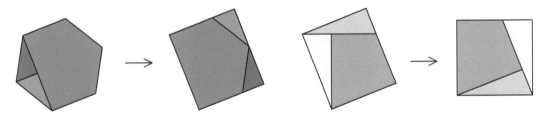

In more detail: join EA, bisect it at G and join FG. If the triangle FGA is moved to position DWC and triangle EFG to a position CBX, then the hexagon is transformed into the rectangle $AXWE$. This may be transformed into a square by constructing K on WC with $EK = \sqrt[4]{27}/\sqrt{2}$ units. Then construct AY parallel to EK, and AV and YZ perpendicular to EK produced. This gives a dissection of the hexagon into six pieces which reassemble to form the square. However, two of these, $\triangle CBX$ (which is congruent to $\triangle EFG$), and

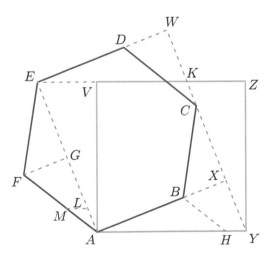

$BXYH$ (which is congruent to $FGLM$) can be joined up again in both figures (delete the cuts FG and BX) to obtain the desired result.

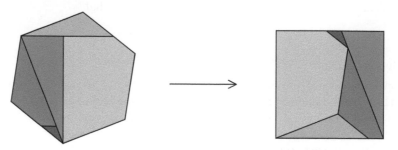

Q12 If the triangle is itself acute–angled, then joining the mid-points of the sides produces a dissection into 4 triangles each similar to the original. Repeating this operation on one of the smaller triangles yields a dissection into seven acute–angled triangles.

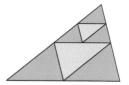

If the triangle is right–angled or obtuse–angled, then it is not possible to dissect it into fewer than seven triangles which are all acute–angled.

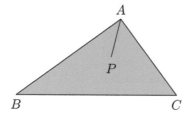

This is because the obtuse angle would need to be divided by at least one cut AP. If this were to reach the opposite side BC, then a further obtuse angled triangle would result (or two right–angled triangles) and we would be no nearer a dissection into acute–angled triangles than when we started. Hence, the cut AP must terminate at a point P in the interior of $\triangle ABC$. The 360° angle at P must be divided into at least 5 angles by further cuts. If two of these extra cuts emanating from P were to intersect AB, then a further obtuse–angled (or right–angled) triangle would result. Hence, only one of the cuts intersects AB, and similarly only one cut intersects AC, leaving two cuts to intersect BC, say at points R and S.

Completing cuts QR and ST now produces a dissection into seven triangles which appear to have a chance of being all acute–angled if P, Q, R, S and T are suitably chosen.

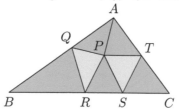

We proceed to show that such a choice is always possible. We use the fact that, if one vertex of a triangle is moved continuously along a path, then the angles of the triangle vary continuously.

Suppose that $\angle BAC \geq 90°$ and $\angle B \leq \angle C$. Construct the
point D such that $AD \perp BC$ and the point Q such that
$DQ \perp AB$ and, finally, the point F such that $QF \perp AD$.
Now, extend QF to intersect AC in a point T. Join TD.
It is easy to see that $\angle CTD \leq 90°$.

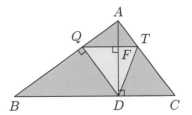

Using the above–mentioned fact, it is clear that points R on BD and S on DC can be chosen
so close to D that $\triangle BQR$, $\triangle TSC$, $\triangle QFR$, $\triangle FTS$ and $\triangle FRS$ are all acute–angled; see
the left-hand triangle below. If now F is moved a sufficiently small distance down AD to
a new position P, then the triangles $\triangle QPR$, $\triangle PTS$ and $\triangle PRS$ will still be acute–angled
and $\triangle QPA$ and $\triangle TPA$ will now also be acute–angled; see the right-hand triangle below.

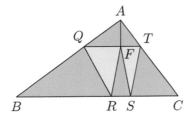 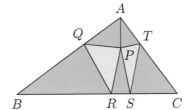

Q13 Suppose that the largest circuit starting (and ending) at T_1 visits k towns. By relabelling if necessary, we may assume that $T_1, T_2, \ldots, T_k, T_1$ is the order of towns visited along
this circuit. Now, assume that $k < n$; then there is some town, say A, not in the circuit.

Suppose that traffic flows from T_1 to A along the road $T_1 A$. Let $\ell \geq 1$ be the largest
number such that the direction of traffic flow along each of the roads $T_1 A, T_2 A, \ldots, T_\ell A$ is
towards A. If $\ell < k$, then the flow of traffic along the road $T_{\ell+1} A$ is directed from A to $T_{\ell+1}$,
so we have a circuit $T_1, T_2, \ldots, T_\ell, A, T_{\ell+1}, \ldots, T_k, T_1$. Since this circuit visits $k + 1$ towns,
we have a contradiction.

Therefore, $\ell = k$; in other words, all of the roads $T_1 A, T_2 A, \ldots, T_k A$ are directed towards A. It is nevertheless possible to travel from A to each of the towns in the circuit,
possibly by first visiting other towns not in the circuit. Let T_m be a town in the circuit
which can be reached from A by a sequence of roads that visits fewest possible towns along
the way. Then this sequence of roads cannot visit any other town in the circuit, and so
$T_1, T_2, \ldots, T_{m-1}, A, \ldots, T_m, \ldots, T_k, T_1$ is a circuit that visits more than k towns, a contradiction.

A similar contradiction arises if we suppose that the traffic flows from A to T_1 along the
road $T_1 A$.

Comment. Different questions regarding a similar setup may be found in Problems 238,
240 and 244.

Q14 We have been told the following facts.

(1) No club contains the entire community.

(2) Each pair of distinct clubs has at least two members in common.

(3) No two clubs have exactly the same members.

(4) For each any three people, there is exactly one club with all three as members.

(5) There are at least two clubs.

Conclusion (a) follows easily from (2) and (4).

By (5), there are at least two clubs, X and Y. By (a), there are two people, A and B, who belong to both X and Y. By (3), at least one of X and Y must contain a third member C; let X be that club which contains C. By (1), there is some member D of the community that is not in X. Hence, (b) is proved.

By (2) and (5), we see that every club has at least two members. Assume that club Y has only two members, A and B. By (b), there is a third person, C, in the community. By (4), A, B and C are contained in some club X. By (1), there is a person D not in X and, by (4), there exists a club Z having A, B and D as members. By (4), Z cannot also contain C. Now consider the club W containing A, C and D. By (4), it cannot also contain B. But then W and Y only have member A is common, contradicting (2). Therefore, Y must have at least three members, which proves (c).

More facts about the community and its clubs structure can also be proved, but we shall let you, dear reader, have the pleasure of discovering them for yourself.

Q15 Consider two consecutive circles C_{k-1} and C_k and let O_{k-1} and O_k be their centres. Also, let P_{k-1} and P_k be their points of contact with the line L. Let O be the centre of C and let P be its point of contact with L. Let M and N be the points of intersection between the line through O_k parallel to L and the line segments OP and $O_{k-1}P_{k-1}$, respectively. By Pythagoras' Theorem,

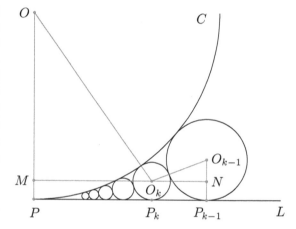

$$|OO_k|^2 = |OM|^2 + |MO_k|^2.$$

This can be written in terms of the diameters of the various circles as

$$\left(\frac{D}{2} + \frac{D_k}{2}\right)^2 = \left(\frac{D}{2} - \frac{D_k}{2}\right)^2 + |PP_k|^2,$$

which simplifies nicely to

$$DD_k = |PP_k|^2; \tag{2.1}$$

similarly, we have

$$DD_{k-1} = |PP_{k-1}|^2. \tag{2.2}$$

By applying Pythagoras' Theorem to the right–angled triangle $\triangle O_{k-1}NO_k$, we obtain the relation $|O_kN|^2 = |O_{k-1}O_k|^2 - |NO_{k-1}|^2$, that is,

$$\left(|PP_{k-1}| - |PP_k|\right)^2 = \left(\frac{D_{k-1}}{2} + \frac{D_k}{2}\right)^2 - \left(\frac{D_{k-1}}{2} - \frac{D_k}{2}\right)^2 = D_{k-1}D_k.$$

We therefore have

$$|PP_{k-1}| - |PP_k| = \sqrt{D_{k-1}D_k},$$

where we have taken the positive square root since $|PP_k| < |PP_{k-1}|$, as is clear from the diagram. Now using (2.1) and (2.2) and then dividing both sides by $\sqrt{D_{k-1}D_k}$ gives

$$\sqrt{\frac{D}{D_k}} = \sqrt{\frac{D}{D_{k-1}}} + 1$$

and therefore

$$\sqrt{\frac{D}{D_k}} = \sqrt{\frac{D}{D_0}} + k.$$

A little algebra suffices to rearrange this equation into the form

$$\sqrt{D_k} = \frac{\sqrt{DD_0}}{\sqrt{D} + k\sqrt{D_0}},$$

from which we may easily obtain a formula for D_k.

Comment. Instead of drawing circles in the diagram successively to the left of C_0, we could have drawn them successively to the right. Readers are invited to find a formula for D_k in this case (hint: it is very similar to the one we have just found), and use it to determine how many circles can be drawn in this way before the construction is no longer possible.

Q16 Let the speeds of A, B and C be a, b and c, respectively, measured in miles per minute. On the first occasion, let A be x miles behind at the instant that B overtakes C. Since A overtakes C at a rate of $a-c$ miles per minute, we have $x = 5(a-c)$ and similarly $x = 8(a-b)$ since A takes altogether 8 minutes to gain miles on B. Hence, $(a-b)/5 = (a-c)/8 = k$, say, $b - c = (a-c) - (a-b) = 8k - 5k = 3k$.

On the second occasion, let C be y miles ahead at the instant that A overtakes B. Then $y = 9(a-c) = 72k$. If B takes t minutes to overtake C, then we have also $y = t(b-c)$; that is, $72k = t(3k)$, so $t = 24$ minutes. In other words, B passes C 24 minutes after A passes B, or 15 minutes after A passes C.

Q17 Unroll the semi–cylindrical cover into a plane rectangle. Since the thread has minimum length, it must lie along the straight line joining two opposite corners, say A and C.

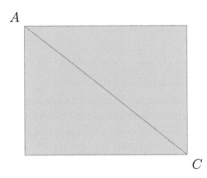

By scaling, we can assume that the semicircle has radius 1 and let ℓ be the length of the rectangle after this scaling. Now, draw a coordinate system on this rectangle, with horizontal

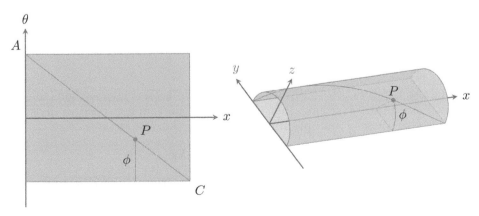

coordinates x and vertical coordinates θ, with the origin $(0,0)$ at the middle point of the left edge of the rectangle. Also, introduce an (x, y, z)–coordinate system to describe each spatial point P of the elastic thread, with x–axis through the middle of the rectangular base, y–axis along the left side of the base, and z–axis pointing vertically upwards.

In the first coordinate system, the corners A and C have coordinates $(0, \pi/2)$ and $(\ell, -\pi/2)$, respectively. Therefore, any point P on the thread has coordinates of the form $(x, \theta) = \left(x, \pi(\frac{1}{2} - x/\ell)\right)$. In the second coordinate system, the point P has coordinates $\left(x, -\cos(\phi), \sin(\phi)\right)$, where

$$\phi = \theta - \left(-\frac{\pi}{2}\right) = \left(\frac{\pi}{2} - \frac{\pi x}{\ell}\right) + \frac{\pi}{2} = \pi - \frac{\pi x}{\ell}$$

or in other words, $\left(x, \cos(\pi x/\ell), \sin(\pi x/\ell)\right)$. The shadow of P is the point P' projected from its (x, y, z) coordinates to its (x, y) coordinates. The shadow of the thread on the rectangular base is therefore the curve

$$y = \cos\left(\frac{\pi x}{\ell}\right)$$

for $0 \le x \le \ell$.

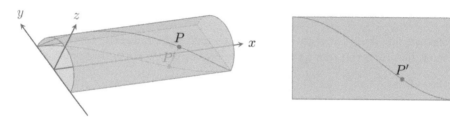

Q18 We shall prove the result by mathematical induction on n. We first observe that if $n = 3$, then the result is true, since, as three handshakes have been made, each person has shaken hands with both of the others. Now suppose that the statement is true when $n = k$, and consider a gathering of $n = k + 1$ people at which n handshakes have been made. Let P_1 be a person at the party who has shaken hands, let P_2 be any person who has shaken hands with P_1, and so on, until we have found a chain of distinct people, $P_1, P_2, \ldots, P_r$, which cannot be further extended, in which each person P_i has shaken hands with their predecessor P_{i-1} for $i = 2, \ldots, r$. This chain cannot be extended either because P_r has not shaken hands with anyone except P_{r-1} or because every other person with whom P_r has shaken hands is already in the chain. In the second case, we have the desired cycle of people shaking hands. In the first case, we simply omit from consideration P_r and the one handshake that P_r has made. The remaining k people have made k handshakes amongst themselves and, by our induction supposition, the desired result is true. By induction, the result holds for any $n \ge 3$.

Q19 Let the two numbers be $a = dh$ and $b = dk$ where d is their greatest common divisor. Then h and k are relatively prime; that is, they have no common positive divisor other than 1. The lowest common multiple of a and b is then dhk. We are given that

$$dhk = (a - b)^2 = d^2(h - k)^2,$$

and so $hk = d(h - k)^2$. Since $(h - k)^2$ is a factor of hk, either $(h - k)^2 = 1$ or any prime factor p of $h - k$ is also a divisor of at least one of h and k. However, the latter is impossible since, if p is a prime factor of $h - k$ and of, say, h, then it is also a factor of $k = h - (h - k)$, and this contradicts the fact that h and k are relatively prime. Hence, $(h - k)^2 = 1$, and so $h - k = \pm 1$. Therefore, h and k are consecutive integers and $d = hk$ is their product.

Q20 By the Sine and Cosine Rules,

$$\frac{2}{\tan B} = \frac{1}{\tan A} + \frac{1}{\tan C}$$

$$\iff \quad 2\frac{\cos B}{\sin B} = \frac{\cos A}{\sin A} + \frac{\cos C}{\sin C} = \frac{\sin(A+C)}{\sin A \sin C} = \frac{\sin B}{\sin A \sin C}$$

$$\iff \quad 2\cos B = \frac{\sin^2 B}{\sin A \sin C} = \frac{b^2}{ac}$$

$$\iff \quad 2ac\cos B = b^2$$

$$\iff \quad a^2 + c^2 - b^2 = b^2$$

$$\iff \quad a^2 + c^2 = 2b^2 \,.$$

Since all implications are true in both directions, the converse is also true.

Q21 In the figure, XY is the line through the centroid G which is parallel to the base BC, and M and N are the midpoints on the sides AB and BC, respectively. By symmetry, we may assume without loss of generality that the triangle is cut along a line PQ where P lies between X and M.

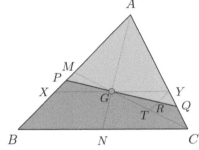

Suppose that $P = X$. Then since the triangle $\triangle APQ = \triangle AXY$ is similar to $\triangle ABC$, being scaled by $2/3$ since G trisects the median AG, area(APQ) is $(2/3)^2 = 4/9$ of area(ABC).
Hence, the ratio of the smaller area to the larger area is $\frac{4}{9}/\left(1 - \frac{4}{9}\right) = \frac{4}{5}$.

On the other hand, if $P = M$, then the two shapes are the triangles $\triangle APQ = \triangle AMC$ and $\triangle BMC$, and these have the same area since their base length along AB and their height are the same. Hence, the ratio of these areas is 1.

Now suppose that P lies strictly between X and M. Construct the line through Y parallel to AB and let R and T be the points of intersection between this line and the lines PQ and MC. Then the triangles $\triangle PGM$ and $\triangle RGT$ are congruent, as are the triangles $\triangle PGX$ and $\triangle RGY$. Hence,

$$\text{area}(PGM) < \text{area}(QGC) \qquad \text{and} \qquad \text{area}(PGX) < \text{area}(QGY).$$

Therefore,

$$\text{area}(APQ) = \text{area}(AXY) + \text{area}(QGY) - \text{area}(PGX)$$
$$> \text{area}(AXY) = \tfrac{4}{9}\,\text{area}(ABC);$$
$$\text{area}(APQ) = \text{area}(AMC) + \text{area}(PGM) - \text{area}(QGC)$$
$$< \text{area}(AMC) = \tfrac{1}{2}\,\text{area}(ABC).$$

Q22 By inspection, $p \neq 2$. Suppose that $2^{p-1} - 1 = py^2$ for some integer y. Then

$$\left(2^{(p-1)/2} - 1\right)\left(2^{(p-1)/2} + 1\right) = py^2 \,.$$

The factors $2^{(p-1)/2} - 1$ and $2^{(p-1)/2} + 1$ differ by 2 and, as they are both odd, are relatively prime. We must therefore have one of them be equal to y_1^2 and the other be equal to py_2^2 where $y_1 y_2 = y$.

Suppose firstly, then, that $2^{(p-1)/2} - 1$ is the square of an odd number. If $p > 3$, then $2^{(p-1)/2} - 1$ is one less than a multiple of 4, whereas every odd square is one more than a multiple of 4:

$$(2k+1)^2 = 4k(k+1) + 1 .$$

So $p = 3$ is the only possibility in this case. Similarly, if $2^{(p-1)/2} + 1$ is an odd square then we have $4k(k+1) + 1 = 2^{(p-1)/2} + 1$, which implies that

$$k(k+1) = 2^{(p-5)/2} .$$

Here, $k(k+1)$ is the product of two consecutive integers but is also a power of 2; this is only true when $k = 1$, in which case, $p = 7$. Therefore, the only prime numbers p for which $(2^{p-1} - 1)/p$ is a perfect square are 3 and 7.

Q23 Let $\angle CFE = \theta$, and let $x = |AE|$. After verifying the angles marked on the figure, we see that triangle ACE is right–angled, with $|AC| = 1$, so $|CE| = \sqrt{x^2 - 1}$. Now apply the Sine Rule in triangles FEC and FAC. This gives

$$\frac{\sqrt{x^2 - 1}}{\sin \theta} = \frac{1}{\sin 30°} = 2$$

and

$$\frac{1}{\sin \theta} = \frac{1 + x}{\sin 120°} = \frac{2}{\sqrt{3}}(1 + x);$$

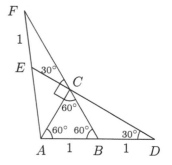

eliminating $\sin \theta$ from these equations leads after a little easy algebra to $(1 + x)\sqrt{x^2 - 1} = \sqrt{3}$ and finally to

$$x^4 + 2x^3 - 2x - 4 = 0 .$$

The left–hand side factorises into $(x^3 - 2)(x + 2)$, whence the solutions of the equation are $x = \sqrt[3]{2}$ and $x = -2$. Since the negative root is inadmissible, we have $x = \sqrt[3]{2}$ as required.

Q24 Suppose that Joe first lost $n - 1$ games and then won the nth game. Joe paid

$$B = 1 + 3 + \cdots + 3^{n-1} = \frac{3^n - 1}{3 - 1} = \frac{1}{2}(3^n - 1)$$

dollars in bets for these n games and won twice the last betted amount, $W = 2 \times 3^{n-1}$, which is then put in the bag. This round of losing after a number of bets and then winning occurs 100 times. For $i = 1, \ldots, 100$, let n_i be the number of bets made during the ith betting round; similarly, let $B_i = \frac{1}{2}(3^{n_i} - 1)$ and $W_i = 2 \times 3^{n_i - 1}$ be the amount spent by Joe and the amount won by Joe in the ith betting round, respectively. Note that $2B_i = 3^{n_i} - 1 = \frac{3}{2}W_i - 1$, so $W_i = \frac{2}{3}(2B_i + 1)$. Since $B_1 + \cdots + B_{100} = \$2500$, it follows that the amount of money in the bag is

$$W_1 + \cdots + W_{100} = \frac{2}{3}(2(B_1 + \cdots + B_{100}) + 100) = \frac{2}{3}(2 \times 2500 + 100) = \$3400 .$$

Q25 Since $-1 \le \cos x \le 1$ for all x, all solutions to the equation $\cos x = x/50$ lie in the interval $[-50, 50]$.

A quick graph sketch of the two functions $\cos(x)$ and $x/50$ in this interval shows that $\cos x = x/50$ for $2 \times 8 = 16$ negative values of x. When $x \geq 0$, the sketch shows that there are either 15 or 16 values of x for which $\cos x$ and $x/50$ are equal, depending on whether these two functions are equal for some value of $x \approx 50$.

Let us look closer at the function values of these functions around $x = 50$. Since $\cos 50 = 0.964966\ldots < 1 = 50/50$ and since these functions are both strictly increasing at $x = 50$, we see that there is no value x close to, but smaller than, 50 for which $\cos x = x/50$. Therefore, there are 15 positive solutions to the equation and thus $16 + 15 = 31$ solutions in total.

Q26 Dudeney's "buttons and string" method of solution enables the problem to be easily solved. The clock face is imagined replaced by a loop of string with the numbers in the order 1, 6, 11, 4, 9, 2, 7, 12, 5, 10, 3, 8; these being the destinations in order of a button starting at 1 and moving clockwise in accordance with the requirements of the problem.

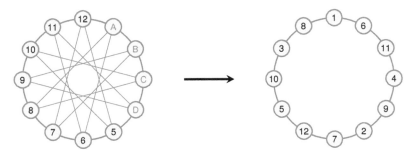

We start with the four buttons **A**, **B**, **C** and **D** on numbers 1, 2, 3 and 4, respectively. These buttons cannot pass one another since no two are allowed to occupy the same position simultaneously. Thus, their order round the loop of string never changes. Any one of them can be moved round to the destination 1; but, once in position, there is no further choice if the four destinations 1, 2, 3 and 4 are all filled. Hence, there are only four possible arrangements.

Q27 By the Binomial Theorem,

$$\left(2+\sqrt{2}\right)^n + \left(2-\sqrt{2}\right)^n = \sum_{i=0}^{n} \binom{n}{i} 2^{n-i} \left(\sqrt{2}\right)^i + \sum_{i=0}^{n} \binom{n}{i} 2^{n-i} \left(-\sqrt{2}\right)^i$$

$$= \sum_{i=0}^{n} \binom{n}{i} 2^{n-i} \left(\left(\sqrt{2}\right)^i + \left(-\sqrt{2}\right)^i\right).$$

Now the expression $\left(\sqrt{2}\right)^i + \left(-\sqrt{2}\right)^i$ is zero when i is odd, and is the sum of two integers when i is even, so it is always an integer; therefore,

$$\left(2+\sqrt{2}\right)^n + \left(2-\sqrt{2}\right)^n$$

is an integer for all n. Since $2 - \sqrt{2} < 1$, we see that $(2-\sqrt{2})^n < 0.000001$ for all sufficiently large values of n. It follows that, if n is a sufficiently large integer, then $(2+\sqrt{2})^n$ falls short of an integer by less than 0.000001, and hence has a fractional part greater than 0.999999.

Q28 The probability P_r that A first picks up an ace on the rth draw is the probability that no aces were picked up on the first $r - 1$ draws, namely

$$\frac{48}{52} \times \frac{47}{51} \times \cdots \times \frac{48 - (r-1) + 1}{52 - (r-1) + 1} = \frac{48!(53-r)!}{52!(49-r)!},$$

multiplied by the probability

$$\frac{4}{52 - (r - 1)} = \frac{4}{53 - r}$$

of drawing an ace on the rth draw. In each game, A receives 10 cents for each of the $r - 1$ non–ace cards picked, or in other words $\$0.1(r - 1)$. Therefore, A pays $\$1$ and expects to receive

$$\sum_{r=1}^{49} 0.1(r - 1)P_r = \sum_{r=1}^{49} 0.1(r - 1)\frac{48!(53 - r)!}{52!(49 - r)!}\frac{4}{53 - r}$$

$$= \sum_{r=1}^{49} 0.4(r - 1)\frac{48!(52 - r)!}{52!(49 - r)!}$$

$$= \frac{0.4}{52 \times 51 \times 50 \times 49} \sum_{r=1}^{49}(r - 1)(52 - r)(51 - r)(50 - r).$$

We can calculate this by use of a computer or by many – but not impossibly many – hand and calculator calculations, to get very nearly $\$0.96$, which is less than $\$1$ that A pays. The game is therefore not quite fair; it favours player B slightly but noticeably.

Q29 First of all note that, as the expression starts as $1 \div 2$, the 2 must end up in the denominator of the answer. To get the value $7/10$, we must put the 7 in the numerator and the 5 in the denominator.

Now, if the 8 is placed in the numerator, it must be cancelled by the even digits 4 and 6 in the denominator. The factor 3 in the 6 will then need to be cancelled by putting the 9 in the numerator and the 3 in the denominator; that is,

$$\frac{1 \times 7 \times 8 \times 9}{2 \times 3 \times 4 \times 5 \times 6} = (((1 \div 2) \div 3) \div 4) \div 5) \div (((6 \div 7) \div 8) \div 9).$$

If the 8 is instead placed in the denominator, then placements of the digits 3, 4, 6 and 9 described above will be reversed; that is,

$$\frac{1 \times 3 \times 4 \times 6 \times 7}{2 \times 5 \times 8 \times 9} = 1 \div (((2 \div 3) \div (4 \div (5 \div 6))) \div ((7 \div 8) \div 9)).$$

The largest value is obtained by placing as many digits as possible in the numerator (remember that 2 must go in the denominator!); that is,

$$\frac{1 \times 3 \times 4 \times 5 \times 6 \times 7 \times 8 \times 9}{2} = 1 \div (((((((2 \div 3) \div 4) \div 5) \div 6) \div 7) \div 8) \div 9).$$

Similarly, the smallest value is

$$\frac{1}{2 \times 3 \times 4 \times 5 \times 6 \times 7 \times 8 \times 9} = (((((((1 \div 2) \div 3) \div 4) \div 5) \div 6) \div 7) \div 8) \div 9.$$

Q30 If n is a prime number, then it is clear that n and thus n^2 cannot be a factor of $(n - 1)!$. Similarly, if $n = 2p$ for any prime number p, then p^2 and thus $n^2 = 4p^2$ cannot be a factor of $(n - 1)! = 1 \times 2 \times \cdots \times p \times \cdots \times (2p - 1)$.

Now suppose that $n = ab$ where a and b are relatively prime and both greater than 2. Since the integers a, $2a$, b and $2b$ all occur as factors of $(n - 1)!$, it follows that n^2 divides $(n - 1)!$.

This leaves for consideration only integers n of the form 2^k, p^k or $2p^k$ where p is an odd prime and k is an integer greater than 1.

The largest power of 2 that divides $(2^k)!$ is 2^{2^k-1}; for instance, $8! = (2^3)!$ is divisible by $2^{2^3-1} = 2^7$. To prove this, observe that every second factor of $1 \times 2 \times 3 \times 4 \times \cdots \times 2^k$ is even, and removing one factor of 2 from each of these 2^{k-1} terms contributes 2^{k-1} factors of 2. Every 4th factor is divisible by 4, so a further 2^{k-2} factors of 2 are obtained by taking one more factor from each of these. Similarly, we can find an extra 2^{k-3} factors of 2 from the multiples of 8, and so on, yielding eventually a total of $2^{k-1} + 2^{k-2} + \cdots + 2 + 1 = 2^k - 1$ factors of 2, as claimed above. Omitting the last factor, 2^k, we see that if $n = 2^k$, then $(n-1)!$ is divisible by 2^{2^k-1-k}. For such n, n^2 divides $(n-1)!$ provided that $2k \leq 2^k - 1 - k$, that is, $3k + 1 \leq 2^k$ which is satisfied if only if $k \geq 4$. We can quickly verify that the numbers $n = 2^1 = 2$, $n = 2^2 = 4$ and $n = 2^3 = 8$ also satisfy that $(n-1)!$ is not divisible by n^2.

A similar (but simpler) analysis shows that the largest power of p (an odd prime) that divides $(p^2 - 1)!$ is p^{p-1}. Hence, if $n = p^2$, then $n^2 = p^4$ divides $(n-1)!$ provided that $p - 1 \geq 4$. We also quickly verify that $n = 3^2$ satisfies that $(n-1)!$ is not divisible by n^2. If $n = 2p^2$, then p^{2p} divides $(n-1)!$, and, as $4 < 2p$ for any odd prime p, n^2 always divides $(n-1)!$. The same result is easily found for any n of the form p^k or $2p^k$ where $k > 2$.

We have now found the positive integers n for which $(n-1)!$ is not divisible by n^2: they are the numbers 8, 9, any prime number $n = p$ and any prime number times two, $n = 2p$.

Q31 Without loss of generality, suppose that $m < n$. None of the given numbers is an integer since m and n are relatively prime. Clearly, the smallest number in either list is greater than 1, and the largest number in either list is

$$\frac{(n-1)(m+n)}{n} = m + n - 1 - \frac{m}{n}.$$

Each of the given fractions is therefore a number between 1 and $m + n - 1$ and thus lies in one of the given intervals.

The distance between any two adjacent numbers in the first list is $(m+n)/m > 1$, so no two of these can both lie in the same interval $(k, k+1)$. A similar observation applies to the second list. Now consider a number $r(m+n)/m$ from the first list and a number $s(m+n)/n$ from the second list. Since $(r+s) - r(m+n)/m = (sm - rn)/m$ and $s(m+n)/n - (r+s) = (sm - rn)/n$ have the same sign, it follows that the integer $r + s$ lies between the numbers $r(m+n)/m$ and $s(m+n)/n$. Therefore, they do not lie in the same unit interval.

Since there are the same number of intervals as numbers, and since no two of the numbers lie in the same interval, there is exactly one number in each interval.

Q32 First note that the diameter of a circle subtends an obtuse angle at any point within it, and an acute angle at any point outside it. Let P be any point inside the quadrilateral. The sum of the four angles at P is $360°$, so the largest such angle ($\angle APB$, say), is at least equal to $90°$. Hence, P cannot lie outside the circle with side AB as diameter. Note, however, that if the two diagonals meet at right angles, then the point of intersection lies on all of the circles.

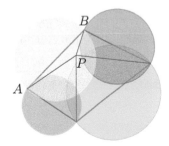

Q33 An obvious solution is $x = 2$ and $y = 4$. This is the only solution.

To see this, let us look first for solutions with $y > 4$, say $y = 4 + h$ and $x = 2 - h$ where $h > 0$. Then $y = 4 + h < 4 + 4h + h^2 = (2 + h)^2$ and so $x + \sqrt{y} < 2 - h + 2 + h = 4$. Hence, $x + \sqrt{x + \sqrt{y}} < 2 - h + 2 < 4$. Continuing in this way, we can show that the right–hand side of the first equation is seen to be less than 4, whilst the left–hand side is greater than 4, a contraction. A similar argument shows that there are no solutions with $3 < y < 4$. It is even more obvious that no solutions exist with $y < 3$, since the first equation implies that $x < y$, whereas, if $y < 3$, then the second equation gives $x = 6 - y > 3 > y$.

Q34 There are 28 squares that lie along the edge of the chessboard, and the king's circuit must visit them sequentially (in either a clockwise or anticlockwise direction) around the board. To see why this is true, consider the left–hand diagram below. Suppose that the king visits a square **A** on the edge, that the next edge square visited is **B**, and that **B** is *not* adjacent to **A**. Then there are squares **C** and **D** which are separated from each other by the path from **A** to **B**. The segment of the king's circuit from **C** to **D** cannot cross this path, so it must incorporate it by taking either the route **C–A–B–D** or **C–B–A–D**; but then it is impossible to get from **D** back to **C** again.

Since the 28 squares on the boundary are visited sequentially, they alternate in colour between light and dark. It is impossible for the king's path to change colour by making only diagonal moves; therefore there must be at least one horizontal or vertical move between any two successive boundary squares, and so at least 28 such moves in all.

By trial and error, one can find various circuits which use no more than 28 horizontal and vertical moves: an example is given in the right–hand diagram.

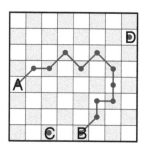
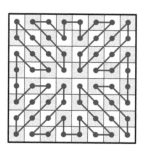

Q35 Let the nth interval be (a_n, b_n) where the intervals have been ordered so that $a_m \leq a_n$ whenever $m < n$. Suppose the largest number of intervals with a common point is r. Imagine r parallel copies $L_1, L_2, \ldots, L_r$ of the real number line drawn above each other. Now plot the 50 intervals on these lines as follows. First plot (a_1, b_1) on L_1. If (a_2, b_2) overlaps (a_1, a_1), then plot it on L_2; otherwise, plot it on L_1. More generally, plot each interval (a_n, b_n) on the lowest line L_k that does not already contain an interval that overlaps (a_n, b_n). Continue in this way until all 50 intervals have been plotted.

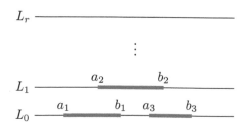

Now suppose that statements (a) and (b) are both false. Then $r \leq 7$ and at most 7 intervals can be found that are pairwise disjoint. This means that our diagram contains at most 7 lines, and no line has more than 7 intervals plotted on it. This would be possible only if the total number of intervals was at most 49. Since there are in fact 50 intervals, at least one of the statements (a) and (b) is true.

Q36 Write $T_i \to T_j$ to denote that T_j can be reached by a direct route from T_i. Also, write $T_i + T_j$ to denote the town T_k with the smallest value of $k \neq i, j$ for which $T_i \to T_k$ and $T_j \to T_k$.

(a) By renaming the towns, we may suppose that $T_1 \to T_2$ and $T_3 = T_1 + T_2$. Since there is only one road between T_2 and T_3, it follows that $T_3 \not\to T_2$, so $T_1 + T_3 \neq T_2$. Thus, there are at least three roads directed away from T_1, namely $T_1 \to T_2$, $T_1 \to T_3$ and $T_1 \to T_1 + T_3$. Similarly, there are at least three roads directed away from each town T_i. We have now accounted for all $7 \times 3 = 21$ roads in the network, and no fourth road can be directed away from any town.

(b) Suppose without loss of generality that $T_1 + T_3 = T_4$. Since there are only 3 towns that can be reached directly from T_1, it follows that $T_1 + T_4$ is either T_2 or T_3. Since $T_3 \to T_4$, it follows that $T_1 + T_4 \neq T_3$, and so $T_1 + T_4 = T_2$. Therefore, $T_2 \to T_3 \to T_4 \to T_2$, so the roads between these three towns form a circuit of directed roads.

(c) One possible orientation of traffic on the 21 roads is as follows:

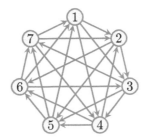

Q37

Q38 The largest possible value of N is $5n - 1$. The set $S = \{n, n+1, \ldots, 5n - 1\}$ can be partitioned into the sets

$$A = \{n, n+1, \ldots, 2n - 1, 4n, 4n + 1, \ldots, 5n - 1\}$$
$$B = \{2n, 2n + 1, \ldots, 4n - 1\}.$$

These sets satisfy the requirements of the problem; in particular, if $x, y \in A$, then $x - y \notin A$, and if $x, y \in B$, then $x - y \notin B$. Hence, $N \geq 5n - 1$.

Also, it is impossible to appropriately partition the set $S = \{n, n+1, \ldots, 5n\}$, since the five numbers $n, 2n, 3n, 4n, 5n$ cannot be apportioned to sets A and B. Indeed, if $n \in A$, then $2n \in B$, so $4n \in A$, and then $3n \in B$. But then $5n$ cannot belong to either set.

Q39 Since $b_1, b_2, \ldots, b_7$ are the same integers as $a_1, a_2, \ldots, a_7$, just rearranged,

$$(a_1 - b_1) + (a_2 - b_2) + \cdots + (a_7 - b_7) = (a_1 + a_2 + \cdots + a_7) - (b_1 + b_2 + \cdots + b_7) = 0.$$

If all seven terms $(a_i - b_i)$ were odd, then their sum would also be odd. However, their sum is 0 and thus even, so at least one of these terms is even. Therefore, the product of these terms is even.

Q40 It is possible that some of the outlaws form pairs each of whom points at the other and such that neither is pointed at by anyone else. However, as the total number of outlaws is odd, they cannot all be paired off in this fashion. Hence, we must be able to find a "chain" $A, B, C, \ldots, J, K, L$ with the most number of outlaws in it, each outlaw pointing to the next one in the chain. Note that $|AB| > |BC| > |CD| > \cdots > |KL|$. Since the chain has the most number of men in it, L cannot point to an outlaw not already in the chain. Also, L cannot point to the outlaw A since, if $|LA| < |LK| < |AB|$, then A could not have pointed to B. By similar arguments, L cannot point to the outlaws $B, C, \ldots, J$. Therefore, L must point to K.

There are now two easy ways to finish the solution to this problem. Firstly, the outlaw A cannot be pointed at by anyone since this would increase the number of people in this already–largest chain. Secondly, at least two people accuse the second–last member of any chain of length greater than 2, and since the total number of accusations is equal to the number of outlaws, at least one outlaw goes unaccused.

Q41 First, cut the 1×1 square into 2 squares of side length $\frac{1}{2}$, 4 squares of side length $\frac{1}{4}$, 8 squares of side length $\frac{1}{8}$, 16 squares of side length $\frac{1}{16}$, and so on, as shown below.

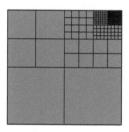

Next, place

the squares of side lengths $\frac{1}{2}$ and $\frac{1}{3}$ into the 2 squares of side lengths $\frac{1}{2}$,

the squares of side lengths $\frac{1}{4}, \frac{1}{5}, \frac{1}{6}, \frac{1}{7}$ into the 4 squares of side lengths $\frac{1}{4}$,

the squares of side lengths $\frac{1}{8}, \frac{1}{9}, \ldots, \frac{1}{15}$ into the 8 squares of side lengths $\frac{1}{8}$,

and, more generally for each $n \geq 1$,

the squares of side lengths $\frac{1}{2^n}, \ldots, \frac{1}{2^{n+1}-1}$ into the 2^n squares of side lengths $\frac{1}{2^n}$:

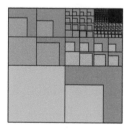

Comment. This problem shows that the infinite series

$$\sum_{n=1}^{\infty} \frac{1}{n^2} = 1 + \frac{1}{4} + \frac{1}{9} + \cdots$$

is less than 2. Are you possibly able to find the actual value? This is not easy!

Q42 Suppose on the contrary that there is a point P which lies inside each of the six circles and let the centres of the circles be $O_1, \ldots, O_6$. Without loss of generality, we may assume that these centres lie in clockwise order around P.

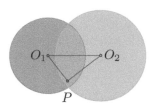

Since P is in the first two circles, $|O_1P| < r_1$ and $|O_2P| < r_2$ where r_1 and r_2 are the radii of the two circles. Since O_2 is not inside the first circle, we see that $|O_2O_1| > r_1 > |O_1P|$; similarly, $|O_2O_1| > r_2 > |O_2P|$. Thus, O_2O_1 is the longest side of the triangle O_1O_2P, which is therefore not equilateral, and it follows that $\angle O_1PO_2$ exceeds $60°$. Similarly, $\angle O_2PO_3$, $\angle O_3PO_4$, $\angle O_4PO_5$, $\angle O_5PO_6$ and $\angle O_6PO_1$ all exceed $60°$, and their sum is therefore greater than $6 \times 60° = 360°$, which is impossible.

Q43 The radius r_n of the nth circle is $\dfrac{1}{2n(n+1)}$: we shall prove this by induction.

First, consider $n = 1$. Let O be the centre of the circle, let C_1 be one of the centres C_1 of the biggest circles and let A be one of the petal points lying on the circle, as shown in the picture to the right. The triangle $\triangle AOC_1$ is right–angled, so $|AC_1|^2 = |AO|^2 + |OC_1|^2$ or in other words, $(1 + r_1)^2 = 1^2 + (1 - r_1)^2$. Therefore,

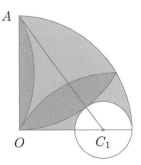

$$r_1 = \frac{1}{4} = \frac{1}{2n(n+1)},$$

as claimed.

Now let $n \geq 2$, and assume that $r_k = \dfrac{1}{2k(k+1)}$ for all $k = 1, 2, \ldots, n-1$. Then

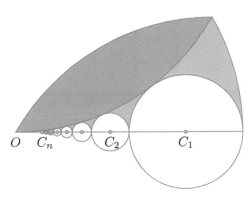

$$|OC_n| = 1 - 2r_1 - 2r_2 - \cdots - 2r_{n-1} - r_n$$
$$= 1 - \frac{1}{1 \times 2} - \frac{1}{2 \times 3} - \cdots$$
$$- \frac{1}{(n-1)n} - r_n$$
$$= 1 - \left(1 - \frac{1}{2}\right) - \left(\frac{1}{2} - \frac{1}{3}\right) - \cdots$$
$$- \left(\frac{1}{n-1} - \frac{1}{n}\right) - r_n$$
$$= \frac{1}{n} - r_n.$$

The triangle $\triangle AOC_n$ is right–angled, so $|AC_n|^2 = |AO|^2 + |OC_n|^2$, and therefore

$$(1 + r_n)^2 = 1^2 + \left(\frac{1}{n} - r_n\right)^2 \quad \Rightarrow \quad 2\left(1 + \frac{1}{n}\right)r_n = \frac{1}{n^2} \quad \Rightarrow \quad r_n = \frac{1}{2n(n+1)}.$$

Q44 The interior angles of the polygon are made up from angles of size 60° or 90°. Thus, their sizes are 60°, 90°, 120° or 150° and so the exterior angles, such as α in the figure are 30°, 60°, 90° or 120°. Since the sum of the exterior angles is 360°, it is clear that the largest number of sides possible is 12. A little experimentation enables one to construct polygons with 5, 6, 7, 8, 9, 10, 11 or 12 sides. Examples are given below.

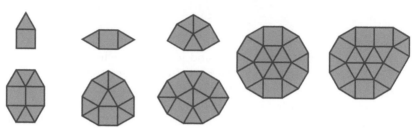

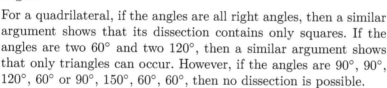

Neither any triangle, nor any quadrilateral, can be dissected into a mixture of squares and equilateral triangles.

The triangle must have all its angles 60°. The corner piece at vertex A must be a triangle but then the angles $\angle CBX$ and $\angle BCY$ are both 120°, which forces another layer of triangles $\triangle DEB$, $\triangle BEC$ and $\triangle CEF$. This argument may be repeated for the angles $\angle FDX$ and $\angle DFY$, and so on, until the whole figure is covered with triangles.

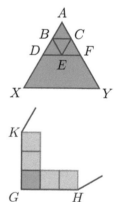

For a quadrilateral, if the angles are all right angles, then a similar argument shows that its dissection contains only squares. If the angles are two 60° and two 120°, then a similar argument shows that only triangles can occur. However, if the angles are 90°, 90°, 120°, 60° or 90°, 150°, 60°, 60°, then no dissection is possible.

For let G be a 90° angle in the quadrilateral in the above-right. The corner piece at G must be a square. A row of identical squares must take us to each of the neighbouring vertices H and K of the quadrilateral. We observe that $\angle H$ and $\angle K$ must both be either 90° or 150°.

Q45 Let $A = x^2 + y^2$ and $B = xy$. Then

$$1 = 1^2 = (x + y)^2 = (x^2 + y^2) + 2xy = A + 2B\,;$$

therefore, $A = 1 - 2B$. Also,

$$1 = (x + y)^4 = (x^4 + y^4) + 4xy(x^2 + y^2) + 6(xy)^2 = 7 + 4AB + 6B^2$$
$$= 7 + 4(1 - 2B)B + 6B^2$$
$$= 7 + 4B - 2B^2\,,$$

so $B^2 - 2B - 3 = 0$. Solving this equation gives the solutions $B = -1, 3$. The solution $B = 3$ implies that $A = 1 - 2B = -5$, which is impossible since $A = x^2 + y^2 \geq 0$. Therefore, $B = -1$ and so $A = 3$. Then

$$1 = (x + y)^3 = (x^3 + y^3) + 3xy(x^2 + y^2) = (x^3 + y^3) + 3AB = (x^3 + y^3) - 3\,.$$

Thus, $x^2 + y^2 = 3$ and $x^3 + y^3 = 4$.

Q46 Let A be the set of non–negative integers whose expressions in base 3 have no more than n digits, with the digit 2 not occurring. All of these numbers are less than 3^n and therefore fall in the specified range; moreover, there are two ways (0 or 1) of choosing each of the n digits, so there are exactly 2^n integers in A.

To confirm that A has the desired property, suppose that three different numbers a, b, c are in A and that a is the arithmetic mean of b and c. Then $b+c = 2a$. But this is impossible. For since b and c have digits 0 and 1 only in base 3, each digit of $b+c$ is obtained by adding the corresponding digits in b and c: there are no "carries" since the maximum sum is less than 3. Since b and c are different, one of the digits in the sum on the left–hand side must be $0 + 1 = 1$. However, since a has digits 0 and 1 only, $2a$ has digits 0 and 2 only, and there cannot be a digit 1 on the right–hand side. Thus, no three of the numbers in A are in arithmetic progression.

Q47 In the figure below, O is the centre of the square; A, B, C and D are the points of the square; K, L, M and N are the four inner triangle points; and P and Q are the midpoints of CM and KN, respectively.

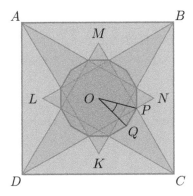

We will show that $|OP| = |OQ|$ and that $\angle POQ = 30°$. By the symmetries of the figure, it follows that the twelve points have the same distance from O and, indeed, that the dodecagon is regular.

As $|BK| = |BC|$ and $\angle KBC = 30°$, we have $\angle BCK = \angle BKC = 75°$, so $\angle KCD = 15°$. By symmetry about AC, we have $\angle BCN = 15°$, leaving $\angle KCN = 90° - 15° - 15° = 60°$. It follows that $\triangle CKN$ is equilateral, since $|CK| = |CN|$.

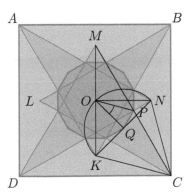

Now both O and P are the midpoint of two sides of $\triangle MKC$, so the line segment OP is parallel to CK and has length $\frac{1}{2}|CK|$. Therefore, $\angle NOP = \angle KCD = 15°$. Now, Q is the centre of the semicircle on KN which, because of the right angle $\angle KON$, passes through O. Therefore, $|OQ| = \frac{1}{2}|KN| = \frac{1}{2}|CK| = |OP|$ and $\angle NOQ = 45°$. It follows that $\angle POQ = \angle NOQ - \angle NOP = 45° - 15° = 30°$, which is what we wanted to prove.

Q48 A convenient notation for describing the situation is to use "disjoint cycles" as illustrated by, say $(ADE)(CF)(B)$. This notation means that A took D's coat, D took E's and E took A's (completing one "cycle"); that C and F took each other's coats; and that B took their own coat. Each letter occurs exactly once; also, the sum of the lengths of the cycles is 6; for instance for $(ADE)(CF)(B)$, the sum of lengths is $3 + 2 + 1 = 6$.

Now to the problem. By (1), there is no cycle of length 1. By (2), the cycle containing A also contains B with just one letter in between. By (3), the cycle containing C also contains D with two letters in between. If these two cycles are different, then we have used 7 letters, and this is not allowed. Therefore, A, B, C and D occur in the same cycle. There is no way of arranging these letters to form a cycle of length 4 that satisfies all requirements, so the cycle containing these letters must be of length at least 5. But it cannot have length 5 since that would leave one letter to form a cycle of length 1, contradicting (1).

Hence, the letters are contained in a single cycle of length 6, either of the form $(ACBxDx)$ or the form $(xCxADB)$ where the letters E and F have yet to be inserted into the positions given by x. By (4), F cannot appear two places to the left of E, so the possibilities are $(ACBEDF)$ or $(ECFADB)$.

In either case, it was F who took A's coat, and A took the coat belonging to either C or to D.

Q49 We shall solve this problem by using ideas of modular arithmetic – see Section 3.3. Let n be a non–negative integer, and write $k = \lfloor \frac{n}{p} \rfloor$; then $n = kp + r$ where $0 \le r < p$. We can write the numbers $n, n - 1, \ldots, n - p + 1$ as

$$kp + r,\ kp + (r - 1),\ \ldots,\ kp + 1,\ kp,\ (k - 1)p + (p - 1),\ \ldots,\ (k - 1)p + (r + 1);$$

exactly one of these numbers, namely kp, is congruent to 0 modulo p, and the remaining $p - 1$ are congruent modulo p in some order to $1, 2, \ldots, p - 1$. Therefore,

$$\frac{n(n - 1)\cdots(n - p + 1)}{p} \equiv n(n - 1)\cdots(kp + 1)(kp - 1)\cdots(n - p + 1)k$$

$$\equiv 1 \times 2 \times \cdots \times (p - 1)k$$

$$\equiv 1 \times 2 \times \cdots \times (p - 1)\left\lfloor \frac{n}{p} \right\rfloor \quad (\bmod\ p).$$

It follows that

$$\frac{n(n - 1)\cdots(n - p + 1)}{p} - 1 \times 2 \times \cdots \times (p - 1)\left\lfloor \frac{n}{p} \right\rfloor \equiv 0 \quad (\bmod\ p)$$

or, rearranged,

$$1 \times 2 \times \cdots \times (p - 1)\left(\binom{n}{p} - \left\lfloor \frac{n}{p} \right\rfloor\right) \equiv 0 \quad (\bmod\ p).$$

In other words, the expression on the left–hand side is a multiple of p; since p is relatively prime to each of the factors $1, 2, \ldots, p - 1$, it follows that $\binom{n}{p} - \lfloor \frac{n}{p} \rfloor$ is a multiple of p.

Q50 Set $a = |BC|$, $b = |AC|$ and $c = |AB|$. Applying the Sine Rule to $\triangle ARB$ gives

$$\frac{|AR|}{\sin 15°} = \frac{|AB|}{\sin 150°} = 2c.$$

By the double–angle formula for sine, $2 \sin 15° \cos 15° = \sin 30° = \frac{1}{2}$, so

$$|AR| = \frac{c}{2 \cos 15°}.$$

Now apply the Sine Rule to the triangles $\triangle AQC$ and $\triangle BPC$:

$$\frac{|AQ|}{\sin 30°} = \frac{|AC|}{\sin 105°} = \frac{|CQ|}{\sin 45°} \quad \text{and} \quad \frac{|BP|}{\sin 30°} = \frac{|BC|}{\sin 105°} = \frac{|CP|}{\sin 45°}.$$

Then

$$|AQ| = \frac{b}{2 \cos 15°}, \quad |CQ| = \frac{b}{\sqrt{2} \cos 15°}, \quad |BP| = \frac{a}{2 \cos 15°}, \quad |CP| = \frac{a}{\sqrt{2} \cos 15°}.$$

Set $\alpha = \angle CAB$, $\beta = \angle ABC$ and $\gamma = \angle BCA$. The area of $\triangle ABC$ is given by three expressions which are therefore all equal:

$$bc \sin \alpha = ab \sin \gamma = ca \sin \beta.$$

Dropping perpendiculars from A, B, C to the opposite sides of $\triangle ABC$ gives

$$a = b \cos \gamma + c \cos \beta, \quad b = c \cos \alpha + a \cos \gamma, \quad c = a \cos \beta + b \cos \alpha.$$

Now the Cosine Rule applied to $\triangle ARQ$ gives

$$|QR|^2 = |AR|^2 + |AQ|^2 - 2|AR||AQ| \cos(60° + \alpha),$$

and using the above relations, we obtain

$$\begin{aligned}
4 \cos^2 15° |QR|^2 &= c^2 + b^2 - 2bc\big(\cos(60°) \cos \alpha - \sin(60°) \sin \alpha \big) \\
&= c^2 + b^2 - bc \cos \alpha + \sqrt{3} bc \sin \alpha \\
&= c^2 + b^2 - b(b - a \cos \gamma) + \sqrt{3} ab \sin \gamma \\
&= c^2 + ab \cos \gamma + \sqrt{3} ab \sin \gamma.
\end{aligned}$$

Applying the Cosine Rule similarly to the triangles $\triangle BRP$ and $\triangle CPQ$ yields

$$\begin{aligned}
4 \cos^2 15° |PR|^2 &= c^2 + ab \cos \gamma + \sqrt{3} ab \sin \gamma = 4 \cos^2 15° |QR|^2 \\
2 \cos^2 15° |PQ|^2 &= a^2 + b^2 - ab \cos \gamma + \sqrt{3} ab \sin \gamma,
\end{aligned}$$

so $|PR| = |QR|$ and

$$\begin{aligned}
2 \cos^2 15° \big(|PR|^2 + |QR|^2\big) &= c^2 + ab \cos \gamma + \sqrt{3} ab \sin \gamma \\
&= (a^2 + b^2 - 2ab \cos \gamma) + ab \cos \gamma + \sqrt{3} ab \sin \gamma \\
&= a^2 + b^2 - ab \cos \gamma + \sqrt{3} ab \sin \gamma \\
&= 2 \cos^2 15° |PQ|^2.
\end{aligned}$$

Therefore, $|PR|^2 + |QR|^2 = |PQ|^2$. By Pythagoras' Theorem, $\triangle PQR$ is a right–angled triangle with equal side lengths $|PR| = |QR|$.

Q51 Let $k \geq 1$ be the smallest number of acquaintances of any student with the students in one of the other classes. Label the classes A, B and C, so that there is a student a in class A who has exactly k acquaintances in class B and let b be one of these acquaintances. Now,

a has $(n+1) - k$ acquaintances in class C, so there are $k - 1$ students in C not acquainted with a. By our definition of k, student b knows at least k students in C and at least one of these, c say, must be known to a. Thus, a, b and c are mutual acquaintances from the three classes.

Q52 (a) Below are squares of area 8 and 10, respectively.

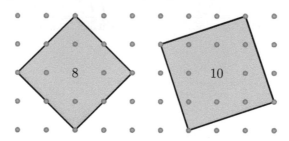

For part (b): by Pythagoras' Theorem, the area of the square in the figure below is $a^2 + b^2$.

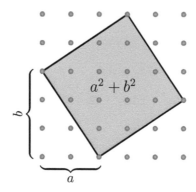

If a and b are both even, then a^2 and b^2 are both divisible by 4, so the area $a^2 + b^2$ is also divisible by 4. If a and b are both odd, then a^2 and b^2 both have remainder 1 when divided by 4, so area $a^2 + b^2$ has remainder 2 when divided by 4. Finally, if one of a and b is even and the other is odd, then the area $a^2 + b^2$ has remainder 1 when divided by 4.

In other words, when dividing the area $a^2 + b^2$ by 4, the remainder is either 0, 1 or 2; it is never 3.

Q53 Let O be the centre of the circle and let L, M and N be the points of contact between the circle and the triangle, as shown in the figure on the right. Since the angles at L, M and C are right angles and since $|OL| = |OM|$, we see that $CMOL$ is a square. Hence, $|CL| = |CM| = \frac{1}{2}d$, and it follows that $|BL| = a - \frac{1}{2}d$ and $|AM| = b - \frac{1}{2}d$.

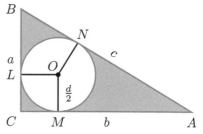

The right–angled triangles $\triangle OAM$ and $\triangle OAN$ have a common hypotenuse OA and equal side lengths $|ON| = |OM|$, so they are congruent; therefore, $|AN| = |AM| = b - \frac{1}{2}d$. Similarly, $|BN| = |BL| = a - \frac{1}{2}d$. It follows that

$$c = |AB| = |AN| + |BN| = \left(a - \frac{1}{2}d\right) + \left(b - \frac{1}{2}d\right) = a + b - d.$$

Q54 Let S be a set of 10 distinct positive integers less than 100. There are $2^{10} = 1024$ ways of choosing a subset T of S (or 1023 if we discount the empty subset, having no elements). To see this, note that to choose T, we can take each of the 10 elements of S in turn and do one of two things with it, namely decide to put it in T, or to leave it out of T. Now, each such subset T contains at most 10 integers all less than 100, so the sum of its elements, sum(T) say, is always an integer less than 1000. Since there are $1023 > 1000$ possible subsets of S, we can find two subsets, A and B say, with sum$(A) =$ sum(B). Let A' and B' be obtained from A and B by omitting any common elements. Then both sums are decreased by the same amount, namely by sum$(A \cap B)$, so we have sum$(A') =$ sum(B'), as required.

Q55 The diagram shows the surface of the cone flattened out into a plane. The summit is S, and the points A, B, P are as specified in the question. Note that when we draw the cone in this way, the shortest path AB becomes a straight line; and the point P at which the path is horizontal will be the closest point on the path to the summit, so that PS and AB are perpendicular.

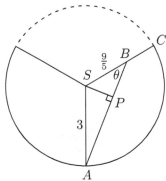

The point B is $\frac{3}{5}$ of the way down from the summit, so the distance SB is $\frac{9}{5}$. The curved distance from A to C is half the base circumference of the mountain; that is, 2π; the circumference of the circle in the diagram is 6π; so $\angle ASB$ is one third of a full circle. We can now apply the Cosine Rule from Section 3.7 to $\triangle ASB$ to give

$$|AB|^2 = |AS|^2 + |SB|^2 - 2|AS||SB|\cos 120° = \frac{441}{25},$$

and so the length of the path AB is $\frac{21}{5}$. A different instance of the Cosine Rule also gives

$$|AS|^2 = |SB|^2 + |AB|^2 - 2|AS||SB|\cos\theta,$$

so $\cos\theta = \dfrac{11}{14}$ and hence

$$|PB| = \frac{9}{5}\cos\theta = \frac{99}{70}.$$

Q56 Let X, Y be points on BC such that $|BX| = |BA|$ and $|BY| = |BD|$.

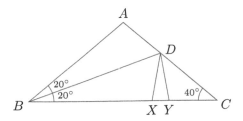

Then

- $\triangle BAD$ and $\triangle BXD$ are congruent, so $|DX| = |DA|$;
- $\triangle BDY$ is isosceles, so $\angle BYD = 80°$, so $\angle YDC = 40°$ and $\triangle YDC$ is isosceles;
- $\angle BXD = \angle BAD = 100°$, so $\angle YXD = 80°$, so $\triangle DXY$ is also isosceles.

Therefore, $|CY| = |DY| = |DX| = |DA|$; and $|BY| = |BD|$ by construction; so

$$|BC| = |BY| + |CY| = |BD| + |DA|$$

as claimed.

Q57 Suppose that we can find k of the thirty–two counters lying in k different rows and k different columns. (It is clear that we can do this when $k = 1$.) If $k = 8$, then we are finished. If $k < 8$, then we are going to show that we can find $k+1$ counters occupying $k+1$ different rows and columns.

So, suppose that we have k counters lying in k different rows and k different columns, and colour these k counters blue. Choose any row R containing no blue counters and any column C containing no blue counters. Firstly, suppose that there is a counter in R which is not in the same column as any blue counter; then this counter can also be coloured blue, and we have $k + 1$ blue counters occupying $k + 1$ different rows and columns. Secondly, suppose likewise that there is a counter in C which is not in the same row as any blue counter: then it can be our $(k + 1)$th blue counter.

Finally, suppose that neither of the previous cases holds. Then the four counters X_1, X_2, X_3, X_4 in R are in the same columns as blue counters A_1, A_2, A_3, A_4 respectively, and the four counters Y_1, Y_2, Y_3, Y_4 in C are in the same rows as blue counters B_1, B_2, B_3, B_4 respectively. But since there are fewer than 8 blue counters, the As and Bs cannot all be different: say A_i is the same as B_j. Then by removing this blue counter and colouring two new counters X_i and Y_j blue, we have $k + 1$ blue counters in $k + 1$ different rows and columns.

Q58 Draw the perpendicular PM from P to line a; draw the angle bisector at P as shown in the diagram, meeting a at A; draw the perpendicular to a at A, meeting b at O.

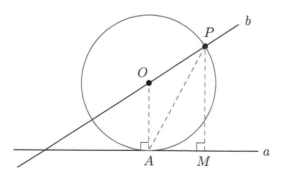

Then the circle with centre O, passing through A, clearly has centre on b and is tangent to a. To confirm that this circle passes through P, we simply note that $\angle OPA = \angle MPA$ by construction; and $\angle OAP = \angle MPA$ since $OA \parallel MP$; so $\triangle OAP$ is isosceles, and hence $OA = OP$.

In fact, we can bisect the "other" angle at P to obtain a second circle satisfying the conditions of the problem. Both circles are shown in the following diagram; the proof for the second case is virtually identical with the first.

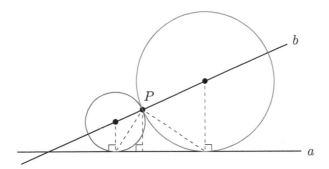

Comment. It was given in the question that lines a and b intersect; but in fact, if they are parallel, exactly the same constructions produce two circles with the desired properties.

Q59 Rewriting the given expression, we have

$$\frac{p}{q} = 1 + \frac{1-2}{2} + \frac{1}{3} + \frac{1-2}{4} + \cdots + \frac{1-2}{1318} + \frac{1}{1319}$$

$$= \left(1 + \frac{1}{2} + \frac{1}{3} + \frac{1}{4} + \cdots + \frac{1}{1318} + \frac{1}{1319}\right) - \left(\frac{2}{2} + \frac{2}{4} + \cdots + \frac{2}{1318}\right)$$

$$= \frac{1}{660} + \frac{1}{661} + \cdots + \frac{1}{1318} + \frac{1}{1319}$$

$$= \left(\frac{1}{660} + \frac{1}{1319}\right) + \left(\frac{1}{661} + \frac{1}{1318}\right) + \cdots + \left(\frac{1}{989} + \frac{1}{990}\right)$$

$$= \frac{1979}{660 \times 1319} + \frac{1979}{661 \times 1318} + \cdots + \frac{1979}{989 \times 990}.$$

If we now multiply both sides by q and by $660 \times 661 \times \cdots \times 1318 \times 1319$, then each side will be an integer; 1979 will clearly be a factor of the right–hand side, and therefore also of the left–hand side. That is,

$$1979 \quad \text{is a factor of} \quad (660 \times 661 \times \cdots \times 1318 \times 1319)\,p.$$

However, 1979 is a prime number, and is certainly not a factor of any of the numbers $660, 661, \ldots, 1318, 1319$; so 1979 must be a factor of p.

Q60 We seek to list all $2n$ people at the party in a sequence $P_1, P_2, \ldots, P_{2n}$ in such a way that P_1 is acquainted with P_2, and P_2 is acquainted with P_3, and so on, and P_{2n} is acquainted with P_1. We shall call such an arrangement a *closed acquaintance cycle*.

Suppose that the original party O has no closed acquaintance cycle. First, we show that by letting pairs of people introduce themselves to each other, thereby creating new acquaintances as well as those already existing, we can obtain a new party N (with the same $2n$ people) in which there is still no closed acquaintance cycle, but there is an *open acquaintance path* – that is, a sequence of people $P_1, P_2, \ldots, P_{2n}$ in which each is acquainted with the next, but the last is *not* acquainted with the first. To see that this is true, observe that O, by assumption, has no closed acquaintance cycle, but if we were to continue the introductions until all pairs were acquainted, there would certainly be a closed acquaintance cycle. Therefore, starting with O and making introductions one at a time, there must be a point at which we obtain a closed acquaintance cycle *for the first time*. Now back up one step by "unfriending" the pair of acquaintances just made: this will reduce the closed acquaintance cycle $P_1, P_2, \ldots, P_{2n}, P_1$ to the open acquaintance path $P_1, P_2, \ldots, P_{2n}$, where P_1 is not acquainted with P_{2n}, and so it is a party N of the type just described.

To solve the problem, we shall now prove that N, which we constructed so as to have no closed acquaintance cycle, does have a closed acquaintance cycle after all! As the whole argument is ultimately based on the assumption that the original party O had no closed acquaintance cycle, this assumption becomes untenable, and we deduce that O did have a closed acquaintance cycle. We note that N includes all the acquaintanceships in O, and possibly more besides, so in N, each person is still acquainted with at least n others.

So, let $P_1, P_2, \ldots, P_{2n}$ be the open acquaintance path in N described above. Let P be the set of people acquainted with P_1, excluding P_2, and let Q be the set of people immediately preceding these in the open acquaintance path. (See below for an illustrative example.) Since P_1 is not acquainted with P_{2n}, and since we excluded P_2 from P, the set Q will be a subset of $\{ P_2, P_3, \ldots, P_{2n-2} \}$. Also, P_1 has at least n acquaintances including P_2, and so Q contains at least $n - 1$ people. Next, let R be the set of people acquainted with P_{2n}, excluding P_{2n-1}. Then R contains at least $n - 1$ of the people $\{ P_2, P_3, \ldots, P_{2n-2} \}$.

The key point is that the sets Q and R contain a minimum of $n - 1$ people each, $2n - 2$ altogether; but they are subsets of $\{ P_2, P_3, \ldots, P_{2n-2} \}$, which contains only $2n - 3$ people. Therefore, there must be a person P_k who is in both Q and R. This means that P_k is

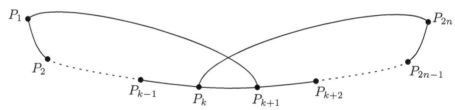

acquainted with P_{2n}, and P_{k+1} is acquainted with P_1, and so N has a closed acquaintance cycle $P_1, P_{k+1}, P_{k+2}, \ldots, P_{2n}, P_k, P_{k-1}, \ldots, P_2, P_1$. But by construction, N has no closed acquaintance cycle, so this is impossible; therefore, our assumption that O had no closed acquaintance cycle was false, and the proof is complete.

Comments.

- We illustrate the construction of the sets P, Q, R with an example. Consider the case $n = 5$, and let N have an open acquaintance path $P_1, P_2, \ldots, P_{10}$. Suppose that P_1 is acquainted with P_2, P_3, P_6, P_7, P_9 and that P_{10} is acquainted with $P_3, P_4, P_6, P_7, P_8, P_9$. Then

$$P = \{ P_3, P_6, P_7, P_9 \},$$
$$Q = \{ P_2, P_5, P_6, P_8 \},$$
$$R = \{ P_3, P_4, P_6, P_7, P_8 \}$$

and we see that Q and R do have people (P_6 and P_8) in common, as we know they must.

- What we have proved is in fact an important result of graph theory known as *Dirac's Theorem*: in any graph with $2n$ vertices, $n \geq 2$, if every vertex has degree at least n, then there is a Hamilton cycle. For more information on these topics, see Section 3.4.

Q61 Supposing the construction to have been completed, draw an equilateral triangle APQ in such a way that Q and P are on opposite sides of the line AC.

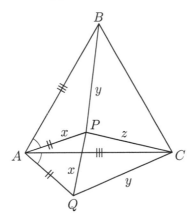

Then since $\angle BAC = 60° = \angle PAQ$, we have

$$\angle PAB = 60° - \angle PAC = \angle QAC \quad \text{and} \quad |AP| = |AQ| \quad \text{and} \quad |AB| = |AC|,$$

so triangles PAB and QAC are congruent. Therefore, $\triangle PQC$ has sides of length x, y, z, and we must have the largest of x, y, z less than the sum of the other two: this is the required condition.

Supposing that P, x, y, z are given and the condition holds, we may construct the triangle with sides $|PQ| = x$, $|QC| = y$, $|CP| = z$; then draw an equilateral triangle on PQ to find A; then complete the equilateral triangle $\triangle ABC$.

Comment. Alternatively, we may draw $\triangle APQ$ so that Q and P are on the same side of AC. In this case, it is triangles QAB and PAC which are congruent, and a similar argument works.

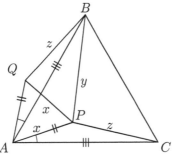

Q62 Let $\alpha_1, \alpha_2, \alpha_3, \alpha_4, \alpha_5$ be five members of S such that $P(x) = x$. By the Factor Theorem (noting that $P(x) - x$ has degree 5 and leading coefficient 1), we have

$$P(x) - x = (x - \alpha_1)(x - \alpha_2)(x - \alpha_3)(x - \alpha_4)(x - \alpha_5).$$

Let β be the element of S for which $P(\beta) = 0$. Then

$$-\beta = (\beta - \alpha_1)(\beta - \alpha_2)(\beta - \alpha_3)(\beta - \alpha_4)(\beta - \alpha_5).$$

Since S consists of seven consecutive integers and β is neither the least nor the greatest, every factor on the right–hand side is a non–zero integer from -5 to $+5$, and the difference between the greatest factor and the least is exactly 6. Moreover, the product of the five factors is negative, so an odd number of them must be negative. There are many factorisations which meet these conditions. For one, take

$$-\beta = (-4)(-2)(-1)(1)(2).$$

This gives

$$\beta = 16, \quad \alpha_1 = 20, \quad \alpha_2 = 18, \quad \alpha_3 = 17, \quad \alpha_4 = 15, \quad \alpha_5 = 14$$

so that we have $S = \{\, 14, 15, 16, 17, 18, 19, 20 \,\}$ and

$$P(x) = x + (x - 14)(x - 15)(x - 17)(x - 18)(x - 20).$$

Q63 We shall draw a graph to represent the party – see Section 3.4 for basic terminology and results in this topic. There will be a vertex for each person at the party, and a coloured edge joining each pair to indicate their relationship: red if the two people know each other, blue if they do not. The given information means that any triangle in the graph contains at least one blue edge: that is, there is no red triangle. We have to prove that there is a blue quadrilateral: that is, a set of four vertices in which all six edges joining these vertices are blue.

For a first case, suppose that there is a vertex u which is an endpoint of *more than three* red edges, in other words, a person who knows at least four others at the party. Consider the four red–neighbours of u; that is, the vertices at the other ends of these four red edges. No two of them can be joined by a red edge, as then these two, together with u, would form a red triangle, contrary to the data. Therefore, all four are joined by blue edges, and we have our blue quadrilateral.

Secondly, suppose that there is a vertex v which is an endpoint of *fewer than three* red edges. Since every vertex has eight neighbours altogether, v has six or more blue–neighbours. Among these six vertices, there must be either a red triangle or a blue triangle. (Readers may care to prove this for themselves; if required, there is a proof in Section 3.4.) A red triangle, once again, is ruled out by the data; so there is a blue triangle, which together with v makes a blue quadrilateral.

To finish the proof, we show that there are no further cases to consider: any configuration of red and blue edges must fall into one of the categories already examined. If neither case were to hold, then every vertex would have *exactly* three red–neighbours. Now delete all the blue edges. What remains is a graph consisting of 9 vertices, each having degree 3. The total degree is therefore 27, an odd number, and this contradicts the Handshaking Lemma, Theorem 3.4. Therefore, this situation can never occur; the two cases we considered above are the only possibilities; and, in each case, the graph contains a blue quadrilateral.

Q64 The diagram shows the positions P, Q of the ships at some stage of the chase. The starting positions were the points P_0 and Q_0, and M is the point due east of P on the line Q_0Q. We shall show that the instantaneous rate at which the distance $|PQ|$ is decreasing is exactly the same as the rate at which $|MQ|$ is increasing. It then follows that the sum of the distances $|PQ|$ and $|MQ|$ remains, throughout the chase, unchanged at the starting value of $20 + 0$ miles. Eventually, when P and M are virtually the same point, we must have $|PQ| = |MQ| = 10$, and so the pirates are (just about) 10 miles behind their quarry.

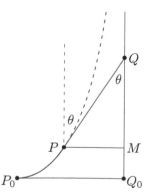

Let the speed of both ships be v, and denote the angle PQM by θ. The northward component of P's speed is then $v \cos \theta$, and this is also M's speed northwards. Hence, the distance $|MQ|$ is increasing at a speed of $v - v \cos \theta$. In the PQ direction, Q's component of speed is $v \cos \theta$, so that PQ is *decreasing* in length at a rate of

$$(P\text{'s speed}) - (Q\text{'s speed in this direction}) = v - v \cos \theta.$$

Hence, these two rates are equal, as promised, and this completes our argument.

Comment. The curve followed by the pirate ship is, unsurprisingly, known as a *pursuit curve*. It is an excellent (but difficult!) problem in calculus to find the equation of this curve.

Q65 Let
$$F_1 = 1, \quad F_2 = 1, \quad \text{and} \quad F_n = F_{n-1} + F_{n-2} \text{ for } n \geq 3.$$

This generates the Fibonacci sequence

$$1, \, 1, \, 2, \, 3, \, 5, \, 8, \, 13, \, 21, \, 34, \, 55, \, 89, \, 144, \, 233, \, 377, \, 610, \, 987, \, 1597, \, 2584, \ldots.$$

We claim that all pairs (m, n) of positive integers satisfying $n^2 - mn - m^2 = \pm 1$ are given by $(m, n) = (F_k, F_{k+1})$ for some positive integer k.

To validate our claim, consider the equation

$$y^2 - xy - x^2 = \pm 1. \tag{2.3}$$

Firstly, if $(x, y) = (m, n)$ is a solution of (2.3), then

$$(m + n)^2 - n(m + n) - n^2 = -(n^2 - mn - m^2) = \mp 1,$$

so $(x, y) = (n, m+n)$ is another solution. Clearly, $(x, y) = (F_1, F_2) = (1, 1)$ is a solution, and so by mathematical induction, $(x, y) = (F_k, F_{k+1})$ is a solution for every positive integer k.

Now we need to prove that there are no solutions of (2.3) in positive integers, other than those just found. Suppose that there do exist solutions (x, y) which are not of the above form; among all these solutions, let (m, n) be the one with the smallest possible value of y. Note that this means that any solution (x, y) in positive integers with $y < n$ must be of the form found above, $(x, y) = (F_k, F_{k+1})$ for some k. Then we have

$$n(n - m) = m^2 \pm 1$$

which is non–negative; and we know n is positive; so $n - m$ is a non–negative integer. Since

$$m^2 - (n - m)m - (n - m)^2 = \mp 1$$

(proof: as in the previous paragraph), the pair $(x, y) = (n - m, m)$ is also a solution of (2.3). There are two possibilities.

If $n - m = 0$, then we easily find $m = 1$, $n = 1$; so our solution (m, n) was, after all, one of those we knew already.

If $n - m > 0$, then $(x, y) = (n - m, m)$ is a solution in positive integers with $y < n$; so we have $(n - m, m) = (F_k, F_{k+1})$ for some k and hence

$$(m, n) = (F_{k+1}, F_k + F_{k+1}) = (F_{k+1}, F_{k+2}).$$

Once again, it turns out that (m, n) is a solution of the above type, and, therefore, it is impossible for there to be any solutions not of this type.

We have shown that any pair of positive integers (m, n) satisfying the required condition $(n^2 - mn - m^2)^2 = 1$ must consist of consecutive Fibonacci numbers. Since in this question it was also specified that m and n are at most 1981, we see by consulting the above list of Fibonacci numbers that the maximum value of $m^3 + n^3$ is

$$m^3 + n^3 = 987^3 + 1597^3 = 5034507976.$$

Q66 Pam, given the product P, could not initially work out its factors x and y; but after being told that the sum $S = x + y$ was less than 23, she was able to do so. The only way to make sense of this is that P must have exactly one factorisation

$$P = xy, \quad x \neq y, \quad x \geq 2, \quad y \geq 2$$

with $S < 23$, and at least one with $S > 22$. We need to find products P satisfying these conditions. Now, numbers less than 2×21 have no factorisations with sum exceeding 22, and numbers greater than 11×11 have no factorisations with sum less than 23. The possible values of P can be found with the expenditure of a little effort, and prove to be the following:

$$120, \ 117, \ 105;$$
$$110, \ 108, \ 104, \ 98, \ 68;$$
$$99, \ 75, \ 64;$$
$$88, \ 78;$$
$$66, \ 52;$$
$$63;$$
$$50, \ 44.$$

They have been placed in groups as follows: in each group, if the only allowable factorisation with $S < 23$ of each of the numbers is found, the sum of the factors is always the same. For example, in the first group, the only permissible factorisations are 10×12, 9×17, 7×15, and in each case the sum of factors is 22. The sums of factors in the other groups are $21, 20, 19, 17, 16, 15$, respectively.

Which of these sums was given to Sam?

Could it have been 22? If so, then after Pam's declaration that she knew the numbers, Sam would have been able to conclude that the product given to Pam was one of $120, 117, 105$; but he would not have been able to decide which. Well, it is now clear that the only possible value of the sum given to Sam is 16, since this corresponds to the only group of products containing just one entry, namely, 63.

Thus, the two numbers were 7 and 9.

Q67 Of the four corners of the square, at least two must have the same colour, say red. Since diagonally opposite corners are greater than $\sqrt{65}/8$ apart, we may assume that exactly two adjacent corners, A and B in the diagram, are coloured red, and another corner, D, has a second colour, blue. There are two cases to consider, depending on whether C is blue or green.

Suppose C is coloured blue. Then N, the midpoint of BC, whose distances from A and from D are each greater than $\sqrt{65}/8$, can be assumed to be coloured green. But then the point X, one eighth of a unit from A on the side AD, has distances of $\sqrt{65}/8$ or more from each of B (red), C (blue), and N (green). Whichever colour is used on X, we have an appropriate pair of points.

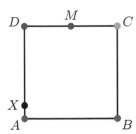

Suppose on the other hand that C is coloured green instead of blue. The midpoint M of DC, whose distance from A exceeds $\sqrt{65}/8$, can be assumed to be coloured blue. But then X has distances of $\sqrt{65}/8$ from B (red) and M (blue), and a greater distance from C (green). Hence, again, we have an appropriate pair of points, whichever colour is given to X.

Comment. The distance $\sqrt{65}/8$ is the largest that can be used in this question. Suppose that the square is coloured as in the third diagram ("boundary" points in black can take the colour on either side of them). Then there is no pair of similarly coloured points whose distance exceeds $\sqrt{65}/8$. Indeed, M and X are the only points at this distance which are similarly coloured.

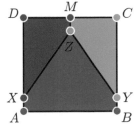

Q68 We shall prove that, in general, if there are n points lying on more than one circle, then the minimum number of regions is $n+2$. It is sufficient to prove this on the assumption that the circles are not in two or more disconnected "clumps" as in the figure, for if the two parts of the diagram are moved together until two of the circles touch (point P now coinciding with point Q), the number of "two–circle points" has increased but the number of regions has not changed.

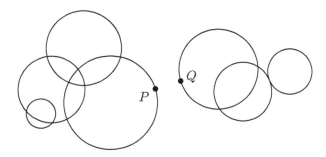

We shall solve the problem by using techniques from graph theory – see Section 3.4. We interpret the diagram as a graph, in which the vertices are the points of intersection of different circles and the edges are the circular arcs joining these points. Since *every* intersection of two circles is a vertex, the edges in the diagram never intersect each other, we have a planar map, and Euler's Formula

$$r - e + v = 2$$

applies, where r is the number of regions in the diagram, e is the number of edges and v is the number of vertices. Now, each vertex lies on at least two circles and therefore has degree at least 4; so by the Handshaking Lemma (Theorem 3.4), the number of edges satisfies $2e \geq 4n$. Hence,

$$r = e - v + 2 \geq 2n - n + 2 = n + 2.$$

(In fact, our argument shows that, for every such connected map, if no point lies on more than two circles, then the number of regions is exactly $n + 2$.) Taking $n = 12$, the smallest number of regions possible is 14.

Q69 In the diagram, P and Q are the points of intersection with ℓ of a circle through A and B, and M is the midpoint of PQ.

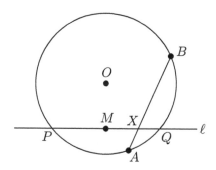

It is a well–known theorem of geometry that, if the chords PQ and AB intersect at X, then $|PX||XQ| = |AX||XB|$. This can be written as

$$(|PM| + |MX|)(|PM| - |MX|) = |AX||XB|,$$

and so

$$|PM|^2 = |AX||XB| + |MX|^2.$$

Since A, B and X are fixed by the conditions of the problem, $|AX||XB|$ is invariable, and so the smallest value of $|PM|$ is obtained when $|MX| = 0$; that is, when the chord PQ has its midpoint at X. The centre O of the circle can be readily constructed as the point of intersection of the perpendicular bisectors of the chords AB and PQ: that is, the intersection of the perpendicular bisector of AB with the line perpendicular to ℓ at the point X.

Q70 We establish two properties of the function f.

(a) For any k, we have $f(2k) = f(k) + f(k) + \langle 0 \text{ or } 1\rangle = 2f(k) + \langle 0 \text{ or } 1\rangle$ and hence

$$f(3k) = f(2k) + f(k) + \langle 0 \text{ or } 1\rangle = 3f(k) + \langle 0 \text{ or } 1 \text{ or } 2\rangle.$$

Dividing both sides by 3, rounding down and remembering that $f(k)$ is an integer gives

$$f(k) = \left\lfloor \frac{f(3k)}{3} \right\rfloor,$$

where the notation $\lfloor x \rfloor$ denotes x rounded to the nearest integer downwards, for example, $\lfloor \pi \rfloor = 3$.

(b) We show that $f(3k) \geq k$ for all $k \geq 1$. This was given for $k = 1$, and we proceed by mathematical induction. Suppose that $f(3k) \geq k$ is true for some specific k; then we have

$$f(3(k+1)) = f(3k+3) = f(3k) + f(3) + \langle 0 \text{ or } 1\rangle \geq k+1.$$

Hence, the inequality is true for all $k \geq 1$. A virtually identical argument shows that if $f(3k) > k$ for some specific k, then $f(3(k+1)) > k+1$. Combining these two results proves that,

$$\text{if} \quad f(3(k+1)) = k+1, \quad \text{then} \quad f(3k) = k.$$

Now, it was given that $f(3 \times 3333) = 3333$, so applying the result of (b) with $k = 3332$ shows that $f(3 \times 3332) = 3332$. Applying the same result with $k = 3331$ shows that $f(3 \times 3331) = 3331$, and so on; eventually, we reach $f(3 \times 1982) = 1982$, and then (a) above gives the final result

$$f(1982) = \left\lfloor \frac{1982}{3} \right\rfloor = 660.$$

Q71 Note that

$$x^3 - 3xy^2 + y^3 = (y-x)^3 - 3x^2y + 2x^3$$
$$= (y-x)^3 - 3(y-x)x^2 - x^3 = X^3 - 3XY^2 + Y^3,$$

where

$$X = y - x, \quad Y = -x. \tag{2.4}$$

Hence, if $(x,y) = (x_1, y_1)$ is one solution in integers of the equation, another solution is

$$(x,y) = (y_1 - x_1, -x_1) = (x_2, y_2).$$

The same transformation performed on (x_2, y_2) gives a third solution

$$(x,y) = (y_2 - x_2, -x_2) = (-y_1, x_1 - y_1) = (x_3, y_3).$$

One can check immediately that performing the transformation (2.4) on (x_3, y_3) returns us to the original solution (x_1, y_1).

It can be verified easily that when the transformation (2.4) is performed, it is never true that $(X, Y) = (x, y)$, except when $(x, y) = (0, 0)$. It follows that when n is a positive integer, the three solutions are all different.

Now take $n = 2891 = 7^2 \times 59$, and suppose that our equation

$$x^3 - 3xy^2 + y^3 = 2891$$

has an integer solution (x, y). If 7 is a factor of x, then 7 is also a factor of x^3, of $3xy^2$ and of 2891; therefore, 7 is a factor of y^3 and hence of y; so 7^3 is a factor of 2891. But this is not the case; so 7 cannot be a factor of x, and an identical argument shows that 7 cannot be a factor of y. We now consider the solutions modulo 7 (see Section 3.3 on modular arithmetic), noting that the non–zero cubes modulo 7 are 1 and -1 only. As seen in the first part of this solution, the equation has three solutions

$$(x, y) \quad \text{and} \quad (y - x, -x) \quad \text{and} \quad (-y, x - y).$$

In at least one of these pairs, the cubes of the two elements must be different modulo 7; for if not, then we should have

$$(x - y)^3 \equiv (-y)^3 \equiv (-x)^3 \equiv (y - x)^3 \pmod 7;$$

that is, $1 \equiv -1 \pmod 7$, an impossibility. By symmetry, we may assume that x^3 and y^3 are different non–zero cubes modulo 7, so that $x^3 \equiv -y^3 \pmod 7$; but then we have

$$3xy^2 = x^3 + y^3 - 2891 \equiv 0 \pmod 7,$$

which is impossible, since we have already shown that neither x nor y can be a multiple of 7. Because of this contradiction, we must deduce that no solution exists.

Q72 For each corner X of the square, let $N(X)$ be the set of all points on L which have minimum possible distance from X. Note that $N(X)$ may consist of a single point, as shown at X in the example; or of many points, as shown at Y; and that it may include points other than the "corner points" of L, as at Z.

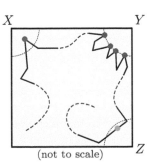

(not to scale)

Now suppose we traverse L from A_0 to A_n. Let Q_1 be the first point we reach which is in one of the sets $N(B)$ for some corner B; let Q_2 be the first point which is in $N(C)$ for some corner C different from B; and let Q_3 be the first point in some $N(D)$, where D is a corner adjacent to B and different from C. Denote by L_1 the portion from A_0 to Q_2 of L, and by L_3 the portion from Q_2 to A_n of L, as shown in the following (schematic) diagram; note that L_1 contains Q_1 and L_3 contains Q_3.

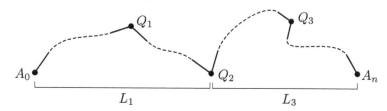

Now consider a point P which moves from B along the edge BD to D. Initially, P is closer to L_1 than it is to L_3, because the closest point to B is Q_1, which is on L_1. When it reaches D, the point P is closer to L_3 than it is to L_1, because the closest point to D is Q_3. Since P moves continuously and its (minimum) distances from L_1 and L_3 change continuously, there must be a position of P at which these distances are equal. That is, there is a point P on BD, a point X on L_1 and a point Y on L_3 such that

- the distances $|PX|$ and $|PY|$ are equal;
- X is the closest point on L_1 to P, and Y is the closest point on L_3 to P.

We shall show that the pair of points X, Y satisfies the conditions of the problem.

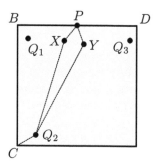

First, either X or Y must be the closest point (or one of the equal closest points) on L to P; by assumption, one of the distances $|PX|$ and $|PY|$ is not more than $\frac{1}{2}$; and since the distances are equal, this is actually true for both of them. As any side of a triangle is no longer than the sum of the other two sides, the straight line distance $|XY|$ satisfies

$$|XY| \leq |XP| + |PY| \leq \tfrac{1}{2} + \tfrac{1}{2} = 1.$$

Finally, the point Q_2 lies on the path L between X (on L_1) and Y (on L_3), and so the distance along L between X and Y is at least the sum of the straight line distances $|XQ_2|$ and $|Q_2Y|$. We have

$$|PX| + |XQ_2| + |Q_2C| \geq |PC| \geq 100 \quad \text{and} \quad |CQ_2| + |Q_2Y| + |YP| \geq |CP| \geq 100;$$

therefore,

$$|XQ_2| + |Q_2Y| \geq 200 - |PX| - 2|Q_2C| - |YP| \geq 200 - 4(\tfrac{1}{2}) = 198,$$

and the proof is complete.

Q73 The triangles may be coloured using just two colours, say, red and blue, in such a way that triangles with a common edge have different colours. Of course, this isn't possible for every triangle dissected into triangles – see the figure on the right, for example – but

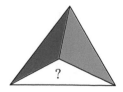

when the dissection is performed with non–intersecting diagonals, we can prove by induction on the number of triangles that such a colouring is possible. Choose any one of the triangles and colour it red. Removing this triangle from the diagram leaves one, two or three separate configurations, and by the induction hypothesis, each of these can be two–coloured. By reversing the colours if necessary, we can choose blue for the triangle in each of these configurations which borders our chosen red triangle; and then reassembling the pieces gives a valid colouring of our original figure.

In our 2–colouring of the dissected polygon, all triangles which include an edge (or edges) of the polygon among their sides bear the same colour. To see this, consider, for example, the triangle ABC and suppose that it is red.

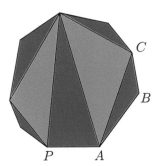

Since there are an odd number of triangles with vertex A, and the colours alternate from each to the next, we are back at red when we reach the triangle with the next edge AP. Hence, the sides of the blue triangles, taken together, consist of all the diagonals of the polygon, once each. The sides of the red triangles consist of all the diagonals, and also all the sides of the polygon. Therefore, the number of sides is the difference between the total number of sides of the red triangles and the total number of sides of the blue triangles. Since triangles have three sides, each of these numbers is a multiple of 3, and, hence, so is their difference.

Q74 From the definition, if $a_1, a_2, a_3, \ldots, a_n$ is a sequence in which every term is the harmonic mean of its neighbours, then

$$\frac{1}{a_1}, \frac{1}{a_2}, \frac{1}{a_3}, \ldots, \frac{1}{a_n} \tag{2.5}$$

is a sequence in which every term is the arithmetic mean of its neighbours: that is, this is an arithmetic sequence, in which the difference of successive terms is constant. An example of such a sequence is

$$\frac{1}{d}, \frac{2}{d}, \frac{3}{d}, \ldots, \frac{n}{d}.$$

If we choose (2.5) to equal this sequence, then we have

$$a_1 = d, \quad a_2 = \frac{d}{2}, \quad a_3 = \frac{d}{3}, \quad \ldots, \quad a_n = \frac{d}{n}.$$

To answer the question, we need all the a_k to be integers, and we can achieve this by choosing

$$d = \mathrm{lcm}(1, 2, 3, \ldots, n) = L_n,$$

the least common multiple of $1, 2, 3, \ldots, n$. That is,

$$L_n, \frac{L_n}{2}, \frac{L_n}{3}, \ldots, \frac{L_n}{n}$$

is a sequence of the required type having n terms. Since n could be any positive integer whatsoever, such sequences can have as many terms as we like.

Now suppose that we have an infinite sequence

$$a_1,\ a_2,\ a_3,\ \ldots \tag{2.6}$$

of the required type. By the same reasoning as above,

$$\frac{1}{a_1},\ \frac{1}{a_2},\ \frac{1}{a_3},\ \ldots \tag{2.7}$$

is an infinite arithmetic sequence. Denote its common difference by D. Then

$$\frac{1}{a_{n+1}} = \frac{1}{a_1} + nD,$$

which can be rewritten as

$$D = \frac{1}{n}\left(\frac{1}{a_{n+1}} - \frac{1}{a_1}\right).$$

Now, all a_k are positive integers. Therefore, $1/a_{n+1} \le 1$ and $1/a_1 > 0$; this shows that $D < 1/n$. Together with a similar argument, this yields

$$-\frac{1}{n} < D < \frac{1}{n}.$$

But since we are investigating an *infinite* sequence, this must be true for all values of n, no matter how large; and since D is a fixed number, the same for all n, the only possibility is that $D = 0$. Therefore, all the terms in (2.7) are the same, and so all the terms in (2.6) are the same; but this is contrary to what was required.

Q75 We have $a_n = k$ if and only if $k - \frac{1}{2} < n < k + \frac{1}{2}$; that is, if and only if

$$k^2 - k + \tfrac{1}{4} < n < k^2 + k + \tfrac{1}{4}.$$

There are $2k$ integer values of n in this range, namely, $n = k^2 - k + m$ for $m = 1, 2, 3, \ldots, 2k$: for example, $a_n = 44$ for the 88 values of n from 1983 to 1980 inclusive. The given sum can therefore be written

$$\overbrace{\frac{1}{1} + \frac{1}{1}}^{2\,\text{terms}} + \overbrace{\frac{1}{2} + \frac{1}{2} + \frac{1}{2} + \frac{1}{2}}^{4\,\text{terms}} + \overbrace{\frac{1}{3} + \cdots + \frac{1}{3}}^{6\,\text{terms}} + \cdots + \underbrace{\frac{1}{44} + \cdots + \frac{1}{44}}_{88\,\text{terms}} + \frac{1}{45} + \frac{1}{45} + \frac{1}{45} + \frac{1}{45}.$$

Since the sum of the terms in each bracketed group is 2, the final total of all terms is $44 \times 2 + \frac{4}{45} = 88\frac{4}{45}$.

Similarly, $b_n = k$ if and only if $k - \frac{1}{2} < \sqrt[3]{n} < k + \frac{1}{2}$; that is, if and only if

$$k^3 - \frac{3k^2}{2} + \frac{3k}{4} - \frac{1}{8} < n < k^3 + \frac{3k^2}{2} + \frac{3k}{4} + \frac{1}{8}.$$

This is an interval of length $3k^2 + \frac{1}{4}$, and the number of integer values of n within it is $3k^2$, except when k is a multiple of 4. In that case, the left–hand end point is $\frac{1}{8}$ less than an integer, and the right–hand end point is $\frac{1}{8}$ more than an integer, and there are $3k^2 + 1$ integer values in the interval, namely,

$$n = k^3 - \frac{3k^2}{2} + \frac{3k}{4} + m$$

for $m = 0, 1, 2, \ldots, 3k^2$. The given sum therefore includes 3×1^2 terms $1/1^2$; and 3×2^2 terms $1/2^2$; and so on; and 3×12^2 terms $1/12^2$, with an extra term in groups $4, 8, 12$. This accounts for all values of n less than $(12\frac{1}{2})^3 = 1953\frac{1}{8}$, and so there remain $1984 - 1953 = 31$ terms $1/13^2$. Therefore, the total sum is

$$(12 \times 3) + \frac{1}{4^2} + \frac{1}{8^2} + \frac{1}{12^2} + \frac{31}{13^2} = 36\frac{26137}{97344}.$$

Q76 We call a positive number t a *fixed point* of f if $f(t) = t$. Setting $y = x$ in the first condition gives $f(xf(x)) = xf(x)$, so f certainly has at least one fixed point. Moreover, if t is a fixed point, then

$$x = t,\ y = t \quad \Rightarrow \quad f(t^2) = f(tf(t)) = tf(t) = t^2$$
$$x = t^2,\ y = t \quad \Rightarrow \quad f(t^3) = f(t^2 f(t)) = tf(t^2) = t^3$$

and so on, so that t^n is also a fixed point of f for all positive n. Again, setting $x = t^{-n}$ and $y = t^{n+1}$, we have

$$t^{n+1} f(t^{-n}) = f(t^{-n} f(t^{n+1})) = f(t) = t;$$

so $f(t^{-n}) = t^{-n}$, and t^{-n} is also a fixed point.

We can now prove that the *only* fixed point of f is $t = 1$. For if there is a fixed point $t > 1$, then $t, t^2, t^3, \ldots$ is an increasing sequence of x values with $x \to \infty$ and $f(x) = x \to \infty$; this contradicts the requirement that $f(x) \to 0$ as $x \to \infty$. If $t < 1$, then the sequence $t^{-1}, t^{-2}, t^{-3}, \ldots$ gives the same contradiction.

Hence, for every positive x, the fixed point $xf(x)$ must be equal to 1; so the only function satisfying the stated conditions is the one given by

$$f(x) = \frac{1}{x}.$$

Q77 The required number is $99999 \times 11112 = 1111188888$. To prove it, suppose that $99999n$ has no digit equal to 9. If the last digit of n were a zero, then $99999n/10$ would be a smaller multiple with the same property. Now note that

$$99999n = 100000n - n. \tag{2.8}$$

If the last digit of n were 1, then the last digit of the product would be 9. So the last digit must be at least 2. If any other of the last five digits of n were a zero, then a 9 would occur in the corresponding place in the answer when the subtraction on the right–hand side of (2.8) is performed. Hence, n must be, at least, a five–digit number in which all digits are at least 1 and the last is at least 2. Thus, the multiple stated above is the smallest possible.

Q78 Note that $\left(\sqrt{a} + \sqrt{b}\right)^2 = a + b + 2\sqrt{ab}$ and hence

$$\sqrt{\frac{a+b}{2} + \sqrt{ab}} = \frac{\sqrt{a} + \sqrt{b}}{\sqrt{2}}.$$

Using this,

$$\sqrt{5 + \sqrt{21}} = \sqrt{\frac{3+7}{2} + \sqrt{3 \times 7}} = \frac{\sqrt{3} + \sqrt{7}}{\sqrt{2}};$$

similarly,

$$\sqrt{8 + \sqrt{55}} = \frac{\sqrt{5} + \sqrt{11}}{\sqrt{2}}, \qquad \sqrt{7 + \sqrt{33}} = \frac{\sqrt{3} + \sqrt{11}}{\sqrt{2}}$$

and

$$\sqrt{6 + \sqrt{35}} = \frac{\sqrt{5} + \sqrt{7}}{\sqrt{2}}.$$

Therefore, each of the given expressions is equal to $\dfrac{\sqrt{3} + \sqrt{5} + \sqrt{7} + \sqrt{11}}{\sqrt{2}}$.

Q79 Let ℓ represent the length of AB, and b the length of BC. Then

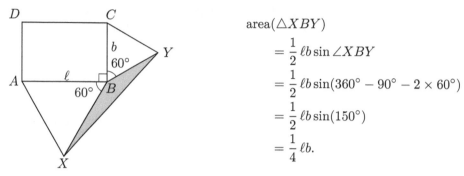

$$\text{area}(\triangle XBY)$$

$$= \frac{1}{2}\,\ell b \sin \angle XBY$$

$$= \frac{1}{2}\,\ell b \sin(360° - 90° - 2 \times 60°)$$

$$= \frac{1}{2}\,\ell b \sin(150°)$$

$$= \frac{1}{4}\,\ell b.$$

Since all four triangles are obviously congruent, their areas total ℓb, the area of the rectangle.

Alternative solution, without trigonometry.

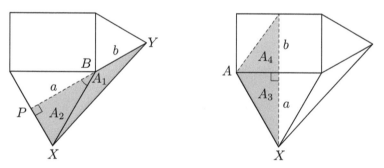

If two triangles have the same altitude, then their areas are proportional to their bases. Since $\triangle ABX$ is equilateral, all its altitudes a are equal. Therefore, (noting that $\angle PBX = 30°$), the first diagram shows that $A_1/A_2 = b/a$; the second shows that $A_3/A_4 = a/b$; and since $A_2 = A_3$, we have

$$\text{area}(\triangle XBY) = A_1 = A_4,$$

which is one quarter of the rectangular area.

Q80 Let $A, B, C, D, E, F, G, H, I, J$ have $9, 9, 9, 8, 8, 8, 7, 6, 4, 4$ acquaintances. Then A is acquainted with all nine other people at the party. Thus, both I and J are acquainted with A, and similarly with B and with C. Next, D, having eight acquaintances, must know at least one of I and J; similarly for E and F. Hence, together, I's and J's acquaintances total at least $2 + 2 + 2 + 1 + 1 + 1 = 9$. Since the given numbers 4 and 4 total less than 9, the list of figures given is impossible.

Alternatively, we may observe that $A, B, C, D, E, F, G, H, I, J$ have respectively $0, 0, 0, 1, 1, 1, 2, 3, 5, 5$ "non–acquaintances". So, each of I and J must be unacquainted with at least two of D, E, F; so, one of the latter must be unacquainted with both I and J. This is impossible.

Q81 Note that
$$a(a + b + c + d) = a^2 + ab + ac + ad$$
$$= a^2 + cd + ac + ad$$
$$= (a + c)(a + d).$$

If $a + b + c + d$ were a prime number, then it would necessarily be a factor of either $a + c$ or $a + d$, neither of which is possible since $a + b + c + d$ is larger than either of them. The result follows.

Q82 Since the volume of a pyramid is

$$\frac{1}{3} \text{ (area of base)(perpendicular height)},$$

and since for all of our six pyramids, the base area is the same, we require a point P whose perpendicular distances from the six faces of the cube are in the given ratios. Such is a point inside the cube whose distances from the three faces meeting at one vertex are $\ell/7$, $2\ell/7$ and $3\ell/7$, where ℓ denotes the length of a side of the cube, since its distances from the opposite faces are respectively $6\ell/7$, $5\ell/7$ and $4\ell/7$.

Q83 We begin by recalling the Pascal's triangle property (page 259) of the binomial coefficients, $C(m + 1, r) = C(m, r) + C(m, r - 1)$. In particular for all n,

$$C(n + 2, 2) = C(n + 3, 3) - C(n + 2, 3).$$

Now subtract from the numbers in the $(n + 1)$th diagonal those in the nth diagonal:

$$S_{n+1} - S_n = 1 + 2 + \cdots + (n + 1) = \frac{(n + 1)(n + 2)}{2} = C(n + 2, 2)$$
$$= C(n + 3, 3) - C(n + 2, 3).$$

Hence, if $S_n = C(n + 2, 3)$, then it follows immediately that $S_{n+1} = C(n + 3, 3)$. But $S_1 = 1 = C(3, 3)$, so the former is true when $n = 1$, and hence, by mathematical induction, it is true for all $n \geq 1$. That is,

$$S_n = C(n + 2, 3) = \frac{n(n + 1)(n + 2)}{6}.$$

Q84 The three players' total score after the first playoff round is 3 points. Since Andy never loses, he is certain to score a minimum of 1 point (from two draws). If he scores more than 1 point, then either he will be the outright winner already; or he will be the equal winner, will progress to the next round with one other player, will eventually beat this player and become champion.

Therefore, Betty will be the overall winner if and only if the results of the first round are Andy v Betty draw, Andy v Colin draw, Betty beats Colin; and the probability of this happening is $(0.9)(0.8)(0.6) = 0.432$. Similarly, Colin's chances are $(0.9)(0.8)(0.4) = 0.288$; and Andy's winning probability will be 1 minus both of these. **Answer:** the winning probabilities are Andy 28%, Betty 43.2%, Colin 28.8%.

Q85 Since there are only 90000 different five–digit numbers, and infinitely many powers of 3, there must be at least one five–digit number giving the first five digits of infinitely many different powers of 3. This answers (a).

Now let 3^s and 3^t, with $s < t$, be powers of 3 having the same first five digits, and let a be the number between 10000 and 99999 composed of those digits. If 3^s has m further digits after these five and 3^t has n further digits, then

$$a10^m < 3^s < (a+1)10^m \quad \text{and} \quad a10^n < 3^t < (a+1)10^n.$$

Dividing these inequalities yields

$$\frac{a10^n}{(a+1)10^m} < \frac{3^t}{3^s} < \frac{(a+1)10^n}{a10^m}$$

or

$$\left(1 - \frac{1}{a+1}\right)10^{n-m} < 3^{t-s} < \left(1 + \frac{1}{a}\right)10^{n-m};$$

and, since $a \geq 10000$, we have

$$0.9999 \times 10^{n-m} < 3^k < 1.0001 \times 10^{n-m},$$

where $k = t - s$. Dividing by 9 gives

$$0.1111 \times 10^{n-m} < 3^{k-2} < 0.1112 \times 10^{n-m}.$$

Finally, writing out the left– and right–hand sides in full, we have

$$11110\cdots 0 < 3^{k-2} < 11120\cdots 0$$

with the same number of zeros on each side; hence, the first four digits of 3^{k-2} are 1111.

Q86 Suppose that $p_n = 5$. Then

$$x = p_1 p_2 \cdots p_{n-1} + 1$$

has no prime factor greater than 5. Since neither $p_1 = 2$ nor $p_2 = 3$ is a factor of x, it follows that x has no prime factor smaller than 5 either: that is, $x = 5^m$ for some m. Note that after p_1, every p_k is odd since it is a factor of the odd number $p_1 p_2 \cdots p_{k-1} + 1$. Therefore, x is p_1 times an odd number, plus 1: that is, $x = 2(2t+1) + 1$ for some integer t. But this is

$$5^m = 4t + 3,$$

which is impossible since $5^m \equiv 1 \pmod{4}$ whereas $4t + 3 \equiv 3 \pmod{4}$. (See the section on *modular arithmetic* in Chapter 3.) It follows that 5 can never occur in the list.

Q87 Divide the board into quarters in both directions, giving 16 "cells" each consisting of 2×2 chessboard squares. By the pigeonhole principle (Section 3.5.5), at least two of the 17 numbers $1, 2, \ldots, 17$ must lie in the same cell. These two numbers are in adjacent squares on the chessboard, and their difference is at most 16.

Comment. This does not mean that it is possible to place the numbers on the board in such a way that the minimum difference is *equal* to 16. In Problem 326, readers are invited to find an arrangement of numbers which is "best possible" in this respect.

Q88 Denote the side length of the square with a corner at A by s, and the side length of the other square by t. We observe that every triangle in the figure is similar to the original triangle $\triangle ABC$ and therefore has sides in the ratio $3:4:5$. Looking at the interval AC and using this similarity, we have

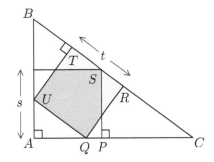

$$4 = |AP| + |PC| = s + \frac{4}{3}s \quad \Rightarrow \quad s = \frac{12}{7}$$

$$4 = |AQ| + |QC| = \frac{4}{5}t + \frac{5}{3}t \quad \Rightarrow \quad t = \frac{60}{37}.$$

The "bases" (along the side BC) of the two grey triangles are given by

$$|RS| = |CS| - |CR| = \frac{5}{3}s - \frac{4}{3}t = \frac{180}{7 \times 37} \quad \text{and} \quad |ST| = t - |RS| = \frac{240}{7 \times 37}.$$

The area of the overlap region, shaded in orange, is the area of the square $QRTU$ of side t, minus the areas of the two grey triangles. Since we know the base and the proportions of each of the triangles, we can write the required area as

$$\left(\frac{60}{37}\right)^2 - \frac{1}{2}\frac{4}{3}\left(\frac{180}{7 \times 37}\right)^2 - \frac{1}{2}\frac{3}{4}\left(\frac{240}{7 \times 37}\right)^2 = \left(\frac{60}{7 \times 37}\right)^2(49 - 6 - 6) = \frac{60^2}{7^2 \times 37}.$$

Q89 Put $y = x$, $z = 0$ in the given identity, obtaining

$$x \,\square\, (x + 0) = (x \,\square\, x) + (0 \,\square\, x),$$

that is,

$$x \,\square\, x = (x \,\square\, x) + (0 \,\square\, x).$$

Therefore,

$$0 \,\square\, x = 0 \tag{2.9}$$

for any x. Now putting $x = u$, $y = v$, $z = 0$ in the given identity and using (2.9), we obtain

$$u \,\square\, (v + 0) = (v \,\square\, u) + (0 \,\square\, u) = v \,\square\, u,$$

that is, $u \,\square\, v = v \,\square\, u$, as required.

Q90 This solution makes use of counting techniques which are explained in Section 3.5.1. First, we calculate the sum of $2^{f(k)}$ for all t–digit numbers having r zeros. To choose such a number,

- select any r of the last $t - 1$ digits to be zero (the first digit cannot be zero): this can be done in $C(t - 1, r)$ ways;
- make a choice from $\{1, \ldots, 9\}$ for each of the $t - r$ non–zero digits: this can be done in 9^{t-r} ways.

Therefore, there are $C(t - 1, r)9^{t-r}$ such numbers. The value of $2^{f(k)}$ is the same for all of them, namely, 2^r, and so the sum of the required terms corresponding to these numbers only is $C(t - 1, r)9^{t-r}2^r$. Now, r can have any value from 0 to $t - 1$; therefore, the total for all t–digit numbers is

$$\sum_{r=0}^{t-1} C(t - 1, r)9^{t-r}2^r = 9^t \sum_{r=0}^{t-1} C(t - 1, r)\left(\frac{2}{9}\right)^r = 9^t\left(1 + \frac{2}{9}\right)^{t-1} = 9 \times 11^{t-1},$$

where the second equality uses the Binomial Theorem. Finally, the question asks for the sum of $2^{f(k)}$ for all k having from 1 digit up to 10 digits, so our answer is found by adding up a geometric series:

$$S_n = \sum_{t=1}^{10} 9 \times 11^{t-1} = 9 \frac{11^{10} - 1}{11 - 1} = 23343682140.$$

Q91 From the figure, we deduce that

$$\gamma_2 = \angle APD$$
$$= 180° - \theta - \delta_1$$
$$= \gamma_1,$$
$$\delta_2 = \angle CPB$$
$$= 180° - \theta - \gamma_2$$
$$= 180° - \theta - \gamma_1$$
$$= \delta_1.$$

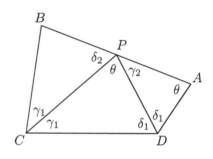

Therefore, triangles ADP and PDC and BPC are similar, so we have

$$\frac{|AP|}{|PD|} = \frac{|PC|}{|CD|} \quad \text{and} \quad \frac{|BP|}{|PC|} = \frac{|PD|}{|DC|},$$

and it follows that $|AP| = |BP|$.

Q92 Let $x + y = 10^{10}$. If the final digits of x and y are equal, then they are 0 or 5, and we are finished. Otherwise, let them be a and $10 - a$ respectively. Suppose that we find the sum $x + y$ "by hand", calculating one digit at a time and working from right to left. At each stage the total 10 is obtained, a 1 being carried; that is, $x_i + y_i = 9$ for every pair of digits in corresponding places in x and y, except for the units places. If there are m digits a in y, then they are "opposite" m digits b in x (where $b = 9 - a$), and y must also contain m digits b in some positions – but not in the units position, since $b \neq 10 - a$. The corresponding m digits in x are also as, so that when the units digit is included, the number of as in x is $m + 1$. But from the given information, x cannot contain more as than y. Hence, it is impossible that the final digits are not equal.

Q93 Suppose that the letter X occurs k times and O occurs $20 - k$ times in the string: then $7 \leq k \leq 13$. Write the Xs in a row; select the 7 of them which are to form part of an "OX" – there are $C(k, 7)$ ways to make this choice – and make a gap in the row to the left of each chosen X, to make room for the insertion of one or more Os.

Similarly, write the Os in a row, choose 7 of them – there are $C(20 - k, 7)$ different ways possible – to form the OXs, and make a gap in the row to the right of each chosen O. Finally, interleave the two rows to form the 7 pairs OX. An example of the procedure for $k = 9$ is shown, with the chosen Xs and Os in red.

X X X X XX X X X   and   OOO O OO O O O OO

$\rightarrow$ XOOOXOXOOXOXXOXOXOOX

This process can be performed in $C(k, 7)C(20 - k, 7)$ ways, and every possible sequence with k letters X and $20 - k$ letters O which includes 7 pairs OX will be obtained exactly once. Adding for all possible values of k from 7 to 13 gives our first result.

To obtain the second (and much simpler!) expression, we begin with the "template"

$$\text{SOXOXOXOXOXOXOXOXF}.\tag{2.10}$$

Here, we have our desired seven occurrences of OX. Any of the seven Os could have further Os added without disturbing the pattern or changing the number of OXs; likewise, any of the Xs could have further Xs added. In the above, S represents the start of the sequence, before the first O: this will be replaced by a number of Xs, possibly zero. Likewise, the F which finishes the sequence will be replaced by zero or more Os. So, to finalise our sequence of 20 letters, we have to choose how many extra letters are involved for each of the 16 letters in (2.10). Let the numbers of letters be $x_1, x_2, \ldots, x_{16}$, respectively; then each x_k is a non–negative integer, and since we need 6 further letters altogether, we have

$$x_1 + x_2 + \cdots + x_{16} = 6.$$

But this is a "dots and lines" problem of the kind explained on page 253; there are 6 dots and 15 lines, and the number of solutions is $C(21, 15)$.

Q94 Let us first answer a slightly easier question: Assuming that adjacent dots in the array are 1 unit of distance apart, how many squares of side length r have their sides horizontal and vertical? For instance, consider the possible such squares where $r = 2$ in the 4×6 array of dots below.

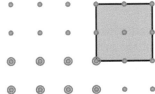

There are 8 such squares since their left–hand bottom vertex must be one of the $2 \times 4 = 8$ circled vertices. In general, there are $(m - r)(n - r)$ squares with sides of length r that are parallel to the rows and columns of the $m \times n$ grid.

Now, we turn our attention to the more difficult problem of counting squares whose sides are not necessarily parallel to the rows and columns of dots, such as the square below.

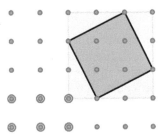

It is contained by a larger square that has edges parallel to the rows and columns. Here, that larger square has side lengths $r = 3$ and, as before, there are $2 \times 3 = 6$ positions in which to place this square. This larger square contains 3 largest–possible squares, as follows:

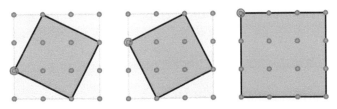

Each of these squares is determined by the position of their left–most vertex. Since there are r such positions, there are r squares within the larger square.

Hence, the total number of squares whose vertices are vertices in the $m \times n$ grid is

$$\sum_{r=1}^{m-1} r(m-r)(n-r) = mn \sum_{r=1}^{m-1} r - (m+n) \sum_{r=1}^{m-1} r^2 + \sum_{r=1}^{m-1} r^3 \,.$$

By using the formulas

$$\sum_{r=1}^{k} r = \frac{1}{2} k(k+1), \quad \sum_{r=1}^{k} r^2 = \frac{1}{6} k(k+1)(2k+1), \quad \sum_{r=1}^{k} r^3 = \frac{1}{4} k^2 (k+1)^2 \,,$$

which can be proved by induction (or by more elegant means), we find the total number of squares to be

$$mn \frac{1}{2}(m-1)m - (m+n)\frac{1}{6}(m-1)m(2m-1) + \frac{1}{4}(m-1)^2 m^2$$

$$= \frac{1}{12} m(m-1)\big(6mn - 2(m+n)(2m-1) + 3m(m-1)\big)$$

$$= \frac{1}{12} m(m-1)(m+1)(2n-m) \,.$$

Q95 If the numbers in the array all equal 0, then the problem is immediately solved, so assume that at least one number in the array is non–zero.

Let a_{ij} be the number in row i and column j of the array, and define

$$M = \sum_{i=1}^{m} \sum_{j=1}^{n} |a_{ij}| \,.$$

Also, consider all numbers obtained by choosing any row or column and adding or subtracting each entry in that row or column. There are finitely many such numbers, namely $m \, 2^n + n \, 2^m$, and they cannot all equal 0 since we assume that there is at least one non–zero number in the array. We can therefore find a smallest positive number among these numbers, say m.

Now, let S be the sum of all mn entries of the array, and note that $S \le M$. If some row or column has a negative sum, say $-\delta$, then change the sign of each entry in that row or column. This increases the total sum S of the entries by $2\delta \ge 2m$ but leaves it less than or equal to M. We can only perform such sign–changing operations at most $(M-S)/2m$ times until there are no more negative entries. At that point, $S = M$, and all row– and column–sums are non–negative.

Q96 If $p = 3$, then 111 is a multiple of p. Consider any prime $p > 5$. On dividing the powers $10, 10^2, \ldots$ by p, there are only finitely many possible remainders, namely $1, 2, \ldots, p-1$. Hence, some remainder must eventually occur twice. Suppose therefore that n and k are positive integers such that 10^n and 10^{n+k} leave the same remainder when divided by p. Then p is a factor of the difference $10^n(10^k - 1)$. Since p is not a factor of 10^n, p must be a factor of

$$10^k - 1 = \overbrace{99\cdots 99}^{k} \,.$$

Since 9 and p are coprime, p is a factor of $\underbrace{11\cdots 11}_{k}$.

Q97 Suppose, on the contrary, that the list $y_1, y_2, \ldots$ contains only finitely many composite numbers. Then from some point onwards, all numbers in the list are prime: say, y_n is prime whenever $n \geq N$. All digits x_n for $n > N$ must be 1, 3 or 7, because a number of more than one digit ending with $0, 2, 4, 5, 6$ or 8 cannot be prime.

Now for $n \geq N$, consider the remainders of the primes y_n modulo 3. We have

$$y_{m+1} \equiv y_m \quad (\text{mod } 3) \qquad \text{if } x_{m+1} \text{ is } 3$$
$$y_{m+1} \equiv y_m + 1 \quad (\text{mod } 3) \qquad \text{if } x_{m+1} \text{ is } 1 \text{ or } 7;$$

that is, at each step, the remainder either stays the same or increases by 1. But since y_n is prime, the remainder can never be zero, and therefore can increase by 1 at most twice. Hence, from some point onwards, all digits x_n must be 3: suppose that this is so for $n \geq M$.

By assumption, y_M is prime, so by Problem 96, it is a factor of the k–digit integer $11 \cdots 11$ for some k; but then y_M is also a factor of

$$10^k y_M + \overbrace{33 \cdots 33}^{k} = y_{M+k},$$

and this is a contradiction. Therefore, the list $y_1, y_2, \ldots$ contains infinitely many composite numbers.

Q98 Since
$$P(2) = a_0 + 2a_1 + 4a_2 + \cdots + 2^n a_n = 3,$$

the coefficient a_0 must be odd; and hence, if $x = 2m$ is any even integer, then

$$P(x) = a_0 + 2ma_1 + 4m^2 a_2 + \cdots + 2^n m^n a_n$$

which is also odd. Moreover, since

$$P(5) = a_0 + 5a_1 + 25a_2 + \cdots + 5^n a_n = 7,$$

an odd number of terms in the sum must be odd; and this will continue to be the case if 5 is replaced by any other odd integer.

We have therefore shown that $P(x)$ is odd when x is even; and is also odd when x is odd; therefore, $P(x)$ can never be zero for any integer x.

Q99 For a regular n–gon in the unit circle, the set L contains $h = \lfloor \frac{n}{2} \rfloor$ line segment lengths, each determined by a multiple of the angle $\theta = 2\pi/n$ at the circle centre, namely, $\theta, 2\theta, \ldots, h\theta$:

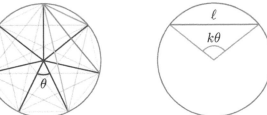

The squared length ℓ^2 of the line segment determined by the angle $k\theta$ for some integer k is given by the Cosine Rule from Section 3.7:

$$\ell^2 = 2 - 2\cos k\theta.$$

The sum of the squares of the elements of L is therefore

$$S = (2 - 2\cos\theta) + (2 - 2\cos 2\theta) + \cdots + (2 - 2\cos h\theta) = 2h - 2T,$$

where $T = \cos\theta + \cos 2\theta + \cdots + \cos h\theta$. By the trigonometric identity

$$2\cos\alpha\sin\beta = \sin(\alpha + \beta) - \sin(\alpha - \beta),$$

which can be found on page 261, we see that

$$2T\sin\frac{\theta}{2} = \left(\sin\frac{3\theta}{2} - \sin\frac{\theta}{2}\right) + \left(\sin\frac{5\theta}{2} - \sin\frac{3\theta}{2}\right) + \cdots + \left(\sin\frac{(2h+1)\theta}{2} - \sin\frac{(2h-1)\theta}{2}\right)$$

$$= \sin\frac{(2h+1)\theta}{2} - \sin\frac{\theta}{2};$$

substituting back $\theta = 2\pi/n$, we have

$$2T = \left(\sin\frac{(2h+1)\pi}{n} \Big/ \sin\frac{\pi}{n}\right) - 1.$$

If n is even, then $2h = n$ and so

$$2T = \left(\sin\left(\pi + \frac{\pi}{n}\right) \Big/ \sin\frac{\pi}{n}\right) - 1 = -2,$$

while if n is odd, then $2h + 1 = n$, and so $2T = -1$. Therefore, the sum of the squares of the elements of L is

$$S = 2h - 2T = \begin{cases} n + 2 & \text{if } n \text{ is even;} \\ n & \text{if } n \text{ is odd.} \end{cases}$$

Q100 We claim that $n = 2^m$ operations suffice.

Let $\mathbf{x} = (x_1, x_2, \ldots, x_n)$ and $\mathbf{y} = (y_1, y_2, \ldots, y_n)$ be two sequences of length n, and define the *cyclic shift* function $S(\mathbf{x}) = (x_2, \ldots, x_n, x_1)$ and the *entrywise product* $\mathbf{x} * \mathbf{y} = (x_1 y_1, x_2 y_2, \ldots, x_n y_n)$. The operation described in the problem can then be expressed as taking the entrywise product of the sequence $\mathbf{a} = (a_1, a_2, \ldots, a_n)$ and its cyclic shift $S(\mathbf{a}) = (a_2, \ldots, a_n, a_1)$ to get the new sequence $\mathbf{b} = \mathbf{a} * S(\mathbf{a}) = (a_1 a_2, a_2 a_3, \ldots, a_n a_1)$. We can visualise these operation by the following example in which $+1$ and -1 are represented simply as $+$ and $-$.

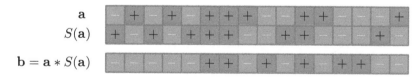

Before proceeding, we note three important properties of $*$ and S.

(1) For any $\mathbf{x}$ and $\mathbf{y}$, we have $S(\mathbf{x} * \mathbf{y}) = S(\mathbf{x}) * S(\mathbf{y})$, because multiplying two sequences and then shifting will give the same result as shifting and then multiplying.

(2) For any $\mathbf{x}$, we have $\mathbf{x} * \mathbf{x} = (1, 1, \ldots, 1)$, because every entry on the left–hand side is either $(1)(1)$ or $(-1)(-1)$.

(3) For any $\mathbf{x}$, we have $\mathbf{x} * (1, 1, \ldots) = \mathbf{x}$, because the kth entry on the left–hand side is $(x_k)(1) = x_k$.

Now define $f(\mathbf{x}) = \mathbf{x} * S(\mathbf{x})$ for all sequences $\mathbf{x}$ of length n. Also, for all $i \geq 0$, define $S^{i+1}(\mathbf{x}) = S(S^i(\mathbf{x}))$ and $f^{i+1}(\mathbf{x}) = f(f^i(\mathbf{x}))$ where $S^0(\mathbf{x}) = f^0(\mathbf{x}) = \mathbf{x}$. We claim that, for all integers $k \geq 0$ and all sequences $\mathbf{x}$ of n numbers each of which is -1 or 1,

$$f^{2^k}(\mathbf{x}) = \mathbf{x} * S^{2^k}(\mathbf{x}).$$

Let us prove this claim by induction. Certainly, this is true for $k = 0$ since $f^{2^0}(\mathbf{x}) = f^1(\mathbf{x}) = \mathbf{x} * S(\mathbf{x}) = \mathbf{x} * S^{2^0}(\mathbf{x})$. Now assume that the claim is true for some specific $k \geq 0$. Then

$$
\begin{aligned}
f^{2^{k+1}}(\mathbf{x}) &= f^{2^k}\left(f^{2^k}(\mathbf{x})\right) \\
&= f^{2^k}\left(\mathbf{x} * S^{2^k}(\mathbf{x})\right) \\
&= \mathbf{x} * S^{2^k}(\mathbf{x}) * S^{2^k}\left(\mathbf{x} * S^{2^k}(\mathbf{x})\right) \\
&= \mathbf{x} * \left[S^{2^k}(\mathbf{x}) * S^{2^k}(\mathbf{x})\right] * S^{2^k}\left(S^{2^k}(\mathbf{x})\right)) \qquad \text{by property (1)} \\
&= \mathbf{x} * (1, 1, \ldots, 1) * S^{2^{k+1}}(\mathbf{x}) \qquad \text{by property (2)} \\
&= \mathbf{x} * S^{2^{k+1}}(\mathbf{x}), \qquad \text{by property (3)}
\end{aligned}
$$

so the claim follows by induction. Therefore, for $n = 2^m$,

$$f^n(\mathbf{a}) = \mathbf{a} * S^n(\mathbf{a}) = \mathbf{a} * \mathbf{a} = (1, 1, \ldots, 1).$$

This is what we were asked to prove.

Q101 Since the sum of digits is a multiple of 9 both before and after deletion, the digit erased must be either 0 or 9. Also, it is not the final digit, whose deletion always causes a reduction by 10. Call the deleted digit d, and suppose it is in the nth place in x, so that

$$x = 10^{n+1}a + 10^n d + b,$$

where a is a positive integer, and where b is a positive integer less than 10^n that is not a multiple of 10. Then $y = 10^n a + b$ and $x = 9y = 9 \times 10^n a + 9b$. Equating the two expressions for x yields

$$8b - 10^n d = a(10 - 9)10^n = 10^n a > 0,$$

so $d \neq 9$. Therefore, $d = 0$, so $8b = 10^n a$ and b is divisible by 5^n; also, $a < 8$ since $b < 10^n$. Note also that $y = 10^n a + b = 9b$. Therefore, deleting the first digit a from y yields the integer $b = y/9$.

Now, suppose that $n > 3$; then b is divisible by 2^{n-3}, so b is a multiple of 10, a contradiction. Therefore, n is either 1, 2 or 3.

- If $n = 1$, then $8b = 10a$ and $b < 10$, so $b = 5$, $a = 4$, $x = 405$ and $y = 45$.
- If $n = 2$, then $8b = 100a$, so $b = 25$, $a = 2$, $x = 2025$ and $y = 225$.
- If $n = 3$, then $8b = 1000a$, so $b = 125a$ where $a \in \{1, 3, 5, 7\}$. This gives the four (x, y) pairs

$$(10125, 1125), \quad (30375, 3375), \quad (50625, 5625), \quad (70875, 7875).$$

Q102 Consider first the case in which $y = 2$ and assume that $u = \sqrt{x + \sqrt{x}}$ is a whole number. Then $x = a^2$ where $a = u^2 - x > 0$, so $u^2 = x + \sqrt{x} = a^2 + a = a(a + 1)$, a contradiction since a and $a + 1$ cannot both be square numbers. Therefore, $\sqrt{x + \sqrt{x}}$ is not a whole number.

Assume that the statement is true for some $y \geq 2$. Then

$$u = \underbrace{\sqrt{x + \sqrt{x + \sqrt{x + \cdots + \sqrt{x}}}}}_{y+1 \text{ square roots}}$$

cannot be a whole number since this would imply that

$$u^2 - x = \underbrace{\sqrt{x + \sqrt{x + \sqrt{x + \cdots + \sqrt{x}}}}}_{y \text{ square roots}}$$

would be a whole number, contradicting our assumption.

The proof now follows by induction.

Q103 There are only a finite number of ways, say N, of drawing 50 line segments between the 50 points labelled A and the points labelled B. For each of these N possible constructions, calculate L, the sum of the lengths of all 50 line segments. Of the N values of L, find the smallest value, call it $L_{\min}$, and consider any of the N constructions with $L = L_{\min}$. By definition, this construction satisfies condition (i). We will now prove that it also satisfies condition (ii).

Assume that the construction does not satisfy condition (ii). There are then two line segments, say A_1B_1 and A_2B_2, that intersect in some point X.

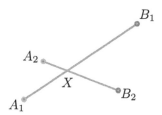

Then $|A_1X| + |XB_2| > |A_1B_2|$ and $|A_2X| + |XB_1| > |A_2B_1|$, so

$$|A_1B_1| + |A_2B_2| = |A_1X| + |XB_1| + |A_2X| + |XB_2| > |A_1B_2| + |A_2B_1|.$$

Thus, by replacing the line segments A_1B_1 and A_2B_2 by the line segments A_1B_2 and A_2B_1 would give a construction satisfying (i) with $L < L_{\min}$, a contradiction. Thus, no two line segments cross each other, and (ii) is satisfied.

Q104 Since 0 can be expressed as $0 = x_1 + x_2 + x_3 + x_4$ where $x_i \in X_i$ and $x_i \geq 0$ for each i, it follows that $x_1 = x_2 = x_3 = x_4 = 0$ is contained in all four sets $X_1, \ldots, X_4$. Now consider any number $n \in X_1$. Since the expression $n = x_1 + x_2 + x_3 + x_4 = n + 0 + 0 + 0$ is unique, n cannot be written as $n = x_1 + x_2 + x_3 + x_4 = 0 + n + 0 + 0$; therefore, $n \notin X_2$. By the same argument, no number in the four sets can appear in two or more sets. This answers part (i).

For each $i = 1, 2, 3, 4$, let n_i be the number of elements in X_i that are less than N. By (i), the four sets each contain 0 but otherwise share no elements with each other, so $n_1 + n_2 + n_3 + n_4 = n(N) + 3$. Also, each of the N numbers $0, 1, \ldots, N-1$ is expressible as

$x_1 + x_2 + x_3 + x_4$ where $x_i \in X_i$ and $x_i < N$ for each i. There are $n_1 n_2 n_3 n_4$ different such expressions, so $n_1 n_2 n_3 n_4 \geq N$. Then, by the Arithmetic–Geometric Mean Inequality,

$$\frac{n(N) + 3}{4} = \frac{n_1 + n_2 + n_3 + n_4}{4} \geq \sqrt[4]{n_1 n_2 n_3 n_4} \geq \sqrt[4]{N},$$

so $n(N) \geq 4\sqrt[4]{N} - 3$.

Finally, we solve part (iii). For each $i = 1, 2, 3, 4$, define X_i to be the set of non–negative integers whose digits are all 0 except in the decimal positions $i, i+4, i+8, \ldots$:

$$X_1 = \{0, \quad 1, \ldots, \quad 9, \quad 10000, \quad 10001, \ldots, \quad 90009, \ldots\}$$
$$X_2 = \{0, \quad 10, \ldots, \quad 90, \quad 100000, \quad 100010, \ldots, \quad 900090, \ldots\}$$
$$X_3 = \{0, \quad 100, \ldots, \quad 900, \quad 1000000, \quad 1000100, \ldots, 9000900, \ldots\}$$
$$X_4 = \{0, 1000, \ldots, 9000, 10000000, 10001000, \ldots, 90009000, \ldots\}.$$

Then, for example, the number $n = 5478267316425$ can be written as $x_1 + x_2 + x_3 + x_4$ thus:

$$n = 5478267316425$$
$$= 5000200010005 +$$
$$8000300020 +$$
$$70007000400 +$$
$$400060006000.$$

For $N = 10000$, we have $n_1 = n_2 = n_3 = n_4 = 10$ and so

$$n(N) = 37 = 4\sqrt[4]{10000} - 3.$$

Q105 Yes, it is possible. Consider the first row. Keep taking one bean from each cup in this row until at least one of the cups contains only one bean. If some cup in this row contains two or more beans, then find the cups with just one bean and double the number of beans in every cup in the columns containing those cups with just one bean. Then those cups containing one bean now contain two beans, while the other cups in the first row have the same number of beans as before. Continue in this fashion, by removing one bean from each cup in the first row and doubling the number of beans in the cups of any column containing a cup in row 1 with just one bean, until every cup in the first row contains a single bean. Now remove that bean from each cup, leaving no beans in any cup in the first row. Apply this procedure to each row of the grid to leave all cups in the grid empty.

Q106 This proof uses *Ptolemy's Theorem* which states that, for any four vertices A, B, C and D of a cyclic quadrilateral,

$$|AB||CD| + |AD||BC| = |AC||BD|.$$

Let r be the radius of the circle, let s be the length of each side of the polygon, and let t be the length of each chord subtending two sides of the polygon. Applying Ptolemy's Theorem to the quadrilateral $PA_{k-1}A_k A_{k-1}$, where $k = 2, 3, \ldots, n-1$, gives

$$sx_{k-1} + sx_{k+1} = tx_k.$$

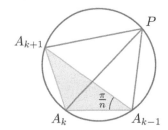

By basic trigonometry (see Section 3.7), we see that

$$s = 2\sin\frac{\pi}{n} \quad \text{and} \quad t = 2\sin\frac{2\pi}{n} = 4\sin\frac{\pi}{n}\cos\frac{\pi}{n} = 2s\cos\frac{\pi}{n},$$

so

$$sx_{k-1} + sx_{k+1} = \left(2s\cos\frac{\pi}{n}\right)x_k.$$

Dividing by s and adding the resulting equations for $k = 2, 3, \ldots, n-1$ gives

$$x_1 + x_2 + 2(x_3 + \cdots + x_{n-2}) + x_{n-1} + x_n = \left(2\cos\frac{\pi}{n}\right)(x_2 + \cdots + x_{n-1}). \qquad (2.11)$$

Now apply Ptolemy's Theorem to the two quadrilaterals $PA_1A_2A_n$ and $PA_1A_{n-1}A_n$

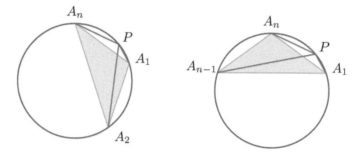

to get the equations

$$x_1t + x_ns = x_2s \quad \text{and} \quad x_1s + x_nt = x_{n-1}s.$$

As above, write t in terms of s and divide by s; then add the resulting equations and rearrange a little to get

$$-x_1 + x_2 + x_{n-1} - x_n = \left(2\cos\frac{\pi}{n}\right)(x_1 + x_n);$$

finally, adding this to equation (2.11) and cancelling 2 gives

$$\frac{x_2 + \cdots + x_{n-1}}{x_1 + x_2 + \cdots + x_{n-1} + x_n} = \cos\left(\frac{\pi}{n}\right).$$

as required.

Q107 Let s_1 be a line segment in L of minimum length, a. We shall prove that there cannot be any exceptional pair (s_2, s_3) where neither s_2 nor s_3 is s_1. Suppose on the contrary that b and c are the lengths of s_2 and s_3, respectively, and that $c > 2b$. Since $a + b \leq 2b < c$, there can be no triangle with sides s_1, s_2 and s_3, a contradiction. It follows that the only possible exceptional pairs in S are $(s_1, s_2), (s_1, s_3), \ldots, (s_1, s_n)$, a maximum of $n-1$ exceptional pairs. This maximum can certainly be achieved; for example, let s_1 have length 1 and $s_2, \ldots, s_n$ all have equal length $c > 2$.

Q108 To prove (a), we show first that each product $x = abc$ of any three numbers a, b, c in S is an integer. In particular, $x^2 = (ab)(bc)(ca)$ is an integer since ab, bc and ca are integers by the definition of S. Write $n = x^2$ and $x = m/k$ where m and $k > 0$ are coprime integers. Assume that $k > 1$; then k has a prime factor p, so p divides $nk^2 = m^2$ and therefore m, a contradiction. It follows that $k = 1$, and so x is an integer.

We have proved that the product of any three numbers in S is an integer. Also, the product of any two numbers in S is also an integer, by the definition of S. Therefore, any product of $k \geq 2$ numbers in S is an integer, since the k numbers can be partitioned into pairs and triples, each of which has an integer product.

To answer (b), consider the example

$$S = \{\, n\sqrt{2} \;:\; n = 1, 2, \ldots \,\}.$$

Products of any even number of elements of S are integral but a product containing an odd number of elements is an integer multiple of $\sqrt{2}$ which is not rational, let alone an integer.

Q109 Let $S = x_1 + \cdots + x_n$ be a sum of n numbers. Then taking $y_1 = y_2 = \cdots = y_n = 1$ and applying the Cauchy–Schwarz Inequality, Theorem 3.16, we have

$$\begin{aligned}
S^2 &= (x_1 y_1 + x_2 y_2 + \cdots + x_n y_n)^2 \\
&\leq (x_1^2 + x_2^2 + \cdots + x_n^2)(y_1^2 + y_2^2 + \cdots + y_n^2) \\
&= n(x_1^2 + x_2^2 + \cdots + x_n^2).
\end{aligned}$$

For $(x_1, \ldots, x_n) = (v, w, x, y)$, we have $n = 4$ and $S = v + w + x + y = 11 - z$, so

$$25 - z^2 = v^2 + w^2 + x^2 + y^2 \geq \frac{S^2}{n} = \frac{(11 - z)^2}{4}.$$

This simplifies to $5z^2 - 22z + 21 \leq 0$, and hence to $(5z - 7)(z - 3) \leq 0$, so that $\frac{7}{5} \leq z \leq 3$. Therefore, the smallest possible value of z is $\frac{7}{5}$, and the largest is 3.

Q110 Construct the line through P that is parallel with one half–line and find its point of intersection M with the other half–line. Construct point A to be on this half–line so that $|OM| = |MA|$. Draw a line through A and P and find its point of intersection B with the other half–line.

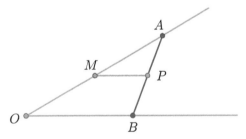

We claim that $\triangle AOB$ has minimum area. To prove this claim, consider any other possible lines through P such as CD in the figures below.

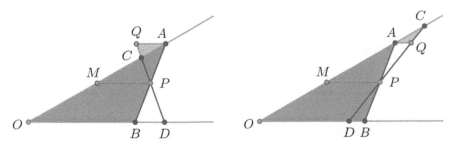

Let Q be the point of intersection between the line through C and D and the line through A that is parallel to OB. In both figures, the triangles $\triangle PAQ$ and $\triangle PBD$ are congruent and have the same area. In the first figure, area$(PAQ) >$ area(PAC), so

$$\text{area}(COD) = \text{area}(AOB) + \text{area}(PBD) - \text{area}(PAC) > \text{area}(AOB)\,,$$

whereas in the second figure, area$(PAC) >$ area(PAQ), so

$$\text{area}(COD) = \text{area}(AOB) + \text{area}(PAC) - \text{area}(PBD) > \text{area}(AOB)\,.$$

In both cases, the triangle $\triangle COD$ has greater area than $\triangle AOB$.

Q111 For each integer k, the graph of the function $g_k(x) = \frac{k}{x-k}$ is asymptotic to the line $x = k$. It is negative when $x < k$ and positive when $x > k$; it is decreasing in both regions since $g'(x) = -k/(x-k)^2$; and it approaches zero as $x \to \pm\infty$. The function $f(x) = g_{114}(x) + \cdots + g_{184}(x)$ must then be negative for all $x < 114$ and positive for all $x > 184$.

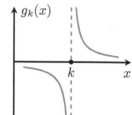

For values of c just exceeding some integer K in the range, the large positive value of will dominate the other terms in the sum; that is, $f(c) \to \infty$ when $x \to K^+$ for $K = 114, \dots, 184$. Similarly, $f(c) \to -\infty$ when $x \to K^-$ for $K = 114, \dots, 184$. The sum of decreasing functions is decreasing, and the sum of a finite set of functions tending to 0 at $\pm\infty$ also tends to 0 at $\pm\infty$. Hence, the graph of $f(x)$ is as follows:

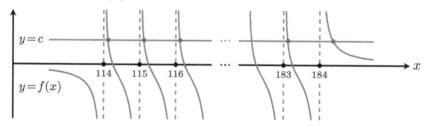

Also on the figure is the line $y = c$, with dots indicating the intersection points (x_k, c) between this line and the function $f(x)$. Here, $k < x_k < k+1$ for each $k = 114, \dots, 183$ and $x_{184} > 184$. We see that $f(x) > c$ in each of the 71 intervals (k, x_k) and nowhere else. So,

$$L(c) = \sum_{k=114}^{184} (x_k - k) = \sum_{k=114}^{184} x_k - \sum_{k=114}^{184} k\,. \tag{*}$$

Now, $x_{114}, \dots, x_{184}$ are the solutions to $f(x) = c$ or, in other words, to

$$\sum_{k=114}^{184} \frac{k}{x-k} = c\,.$$

By multiplying both sides of this equation by $(x - 114) \cdots (x - 184)$ and rearranging, we get the polynomial equation

$$c \prod_{k=114}^{184} (x - k) - \sum_{k=114}^{184} k \prod_{\substack{\ell=114 \\ \ell \neq k}}^{184} (x - \ell) = 0\,.$$

By dividing by c and expanding the terms, we see that

$$x^{71} - \frac{c+1}{c} \left(\sum_{k=114}^{184} k \right) x^{70} + h(x) = 0$$

where $h(x)$ is a polynomial of degree at most 69. The sum of roots of the polynomial on the left–hand side of the equation can therefore be written in two ways, giving us the identity

$$\sum_{k=114}^{184} x_k = \frac{c+1}{c} \sum_{k=114}^{184} k.$$

By (*),

$$L(c) = \frac{1}{c} \sum_{k=114}^{184} k = \frac{(184 + 1 - 114) \times (114 + 184)}{2c} = \frac{71 \times 149}{c}.$$

Therefore,

$$L\left(\frac{1988}{336}\right) = \frac{336}{1988} \times 71 \times 149 = 1788 \quad \text{and} \quad L\left(\frac{1788}{336}\right) = \frac{336}{1788} \times 71 \times 149 = 1988.$$

Q112 Now,

$$\ln 11 = A\left(\frac{1}{23} + \frac{1}{3 \times 23^3} + \frac{1}{5 \times 23^5} + \cdots\right) + B\left(\frac{1}{65} + \frac{1}{3 \times 65^3} + \frac{1}{5 \times 68^5} + \cdots\right)$$
$$- C\left(\frac{1}{485} + \frac{1}{3 \times 485^3} + \frac{1}{5 \times 485^5} + \cdots\right)$$
$$= \frac{A}{2} \ln\left(\frac{23+1}{23-1}\right) + \frac{B}{2} \ln\left(\frac{65+1}{65-1}\right) + \frac{C}{2} \ln\left(\frac{485+1}{485-1}\right)$$
$$= \frac{A}{2}(2\ln 2 + \ln 3 - \ln 11) + \frac{B}{2}(-5\ln 2 + \ln 3 + \ln 11) - \frac{C}{2}(-\ln 2 + 5\ln 3 - 2\ln 11)$$
$$= \frac{1}{2}\left(2A - 5B + C\right)\ln 2 + \frac{1}{2}\left(A + B - 5C\right)\ln 3 + \frac{1}{2}\left(-A + B + 2C\right)\ln 11,$$

so we want

$$2A - 5B + C = 0, \qquad A + B - 5C = 0, \qquad -A + B + 2C = 2.$$

We can solve this system of equations to get the solution $A = 48$, $B = 22$ and $C = 14$; performing the calculations gives

$$\ln 11 \approx 2.39789527.$$

Q113 First, note that all of the sec terms in the sum are defined, since an integer can never be an odd multiple of $\frac{\pi}{2}$. For any k, we have

$$\sec(k-1)\sec k = \frac{1}{\cos(k-1)\cos k}$$
$$= \frac{\sin(k - (k-1))}{\sin 1 \cos(k-1) \cos k}$$
$$= \frac{1}{\sin 1} \frac{\sin k \cos(k-1) - \cos k \sin(k-1)}{\cos(k-1) \cos k}$$
$$= \frac{1}{\sin 1}(\tan k - \tan(k-1)).$$

Therefore, the given sum is

$$\frac{1}{\sin 1}\left((\tan 1 - \tan 0) + (\tan 2 - \tan 1) + (\tan 3 - \tan 2) + \cdots + (\tan n - \tan(n-1))\right);$$

almost everything cancels, leaving the result

$$\sec 0 \sec 1 + \sec 1 \sec 2 + \sec 2 \sec 3 + \cdots + \sec(n-1)\sec n = \frac{\tan n}{\sin 1}.$$

Q114 Let n_s be the number of whole numbers $0, 1, \ldots, 9999$ with digit sum s. For any such number x, the number $y = 9999 - x$ has digit sum $36 - s$, and conversely. Therefore, $n_s = n_{36-s}$, so

$$10000 = n_0 + n_1 + \cdots + n_{36} = 2n_0 + 2n_1 + \cdots + 2n_{17} + n_{18}.$$

It follows that $n_0 + n_1 + \cdots + n_{17} = 5000 - \frac{1}{2}n_{18}$. The number N of whole numbers $0, 1, \ldots, 9999$ with digit sum less than 20 is therefore

$$N = n_0 + n_1 + \cdots + n_{19} = 5000 + \frac{1}{2}n_{18} + n_{19}.$$

Now, we know that n_{18} is the number of integers of the form $abcd$ where $0 \leq a, b, c, d \leq 9$ and $a + b + c + d = 18$. By the "dots and lines" method described in Section 3.5.2, the number of all non–negative solutions to $a + b + c + d = 18$ is

$$C(18 + 4 - 1, 4 - 1) = C(21, 3) = 1330.$$

However, some of these solutions have $a \geq 10$. How many? Write $a = 10 + A$ where $A \geq 0$; then $A + b + c + d = 8$. There are $C(11, 3) = 165$ solutions to this equation. Similarly, 165 of the solutions to $a + b + c + d = 18$ have $b \geq 10$; another 165 have $c \geq 10$; and yet another 165 have $d \geq 10$. Therefore,

$$n_{18} = 1330 - 4 \times 165 = 670.$$

An identical argument shows that

$$n_{19} = C(19 + 3, 3) - 4C(9 + 3, 3) = 1540 - 4 \times 220 = 660,$$

so the required answer is

$$N = 5000 + \frac{1}{2}n_{18} + n_{19} = 5995.$$

Q115 The set $\{49, 50, \ldots, 148\}$ consists of 100 consecutive positive whole numbers, no three of which add up to any fourth number in the set. For this set, $x = 148$.

We claim that 148 is the smallest possible value of $x = \max S$ for any set S of 100 different positive whole numbers, no three of which add up to any fourth number in the set. Let $m = \min S$ be the smallest number in S and note that $x \geq m + 99$ with equality exactly when the numbers in S are consecutive.

The number of pairs $\{a, b\}$ of positive integers with sum $a + b = x - m$ is $\frac{x-m-1}{2}$ if $x - m$ is odd and $\frac{x-m}{2} - 1$ if $x - m$ is even. Since S cannot contain both of the numbers a and b if $a + b = x - m$ unless $\{a, b\} = \{m, x - 2m\}$, the number of positive integers less than x that are not in S is at least $\frac{x-m-1}{2} - 1$ if $x - m$ is odd and at least $\frac{x-m}{2} - 2$ if $x - m$ is even. Therefore, $x \geq 100 + \frac{x-m-1}{2} - 1$ and so $x \geq 197 - m$ when $x - m$ is odd and $x \geq 100 + \frac{x-m}{2} - 2$ and so $x \geq 196 - m$ when $x - m$ is even. By adding the inequality $x \geq m + 99$ to these inequalities, we find that $2x \geq 296$, so $x \geq 148$ when $x - m$ is odd and $2x \geq 295$, so $x \geq \lceil 147.5 \rceil = 148$ when $x - m$ is even.

Q116 Suppose that N is an integer that cannot be expressed as $a_k + k$ for any k. If $N < a_1$, then $b_0 = b_1 = \cdots = b_N = 0$, so $N = b_N + N$. Otherwise, since $a_1 < a_2 < \cdots$, we can find k such that

$$a_{k-1} + (k-1) < N < a_k + k.$$

Define $t = N - k + 1$ and note that $a_1 < a_2 < \cdots < a_{k-1} < t < a_k + 1$. Therefore, $b_t = k - 1$, so

$$N = k - 1 + t = b_t + t.$$

Now suppose that $N = a_k + k$ for some k. Then $b_{a_n} + a_n = n - 1 + a_n = N - 1$, whereas $b_{a_n+1} + (a_n + 1) = n + (a_n + 1) = N + 1$. Therefore, N does not equal $b_l + t$ for any t.

Q117 Let d be the distance around the track, and suppose that the bodies' first meeting occurred at time t_1, at point B. Then d is the sum of the distances that each body travelled, namely

$$d = vt_1 + \frac{1}{2} at_1^2.$$

Their second meeting, at time t_2, occurred at A, so $d = vt_2$ and

$$d = \frac{1}{2} at_2^2 = \frac{1}{2} a\left(\frac{d}{v}\right).$$

Therefore, $d = 2v^2/a$, so

$$\frac{1}{2} at_1^2 + vt_1 - \frac{2v^2}{a} = 0.$$

The positive solution to this equation is the time at which the first meeting occurred, namely

$$t_1 = \frac{v}{a}\left(\sqrt{5} - 1\right).$$

Q118 We take the unit of length equal to the radius of the circle. Let O be the centre of the circle, let N be the foot of the perpendicular from C to AB, and let M be the intersection of CN and SR produced. Let $|OP| = |OQ| = x$; then we have $|PS| = 2x$ and $|OP|^2 + |PS|^2 = |OS|^2 = 1$, so $x^2 + 4x^2 = 1$ and therefore $x = \frac{1}{\sqrt{5}}$. The areas of the square $PQRS$ and the triangle $\triangle ABC$ are, respectively,

$$(2x)^2 = \frac{4}{5} \quad \text{and} \quad \frac{1}{2}|AB||CN| = |CN|;$$

these are equal, so $|CN| = \frac{4}{5}$. Since $|ON|^2 + |CN|^2 = |OC|^2 = 1$, we find $|ON| = \frac{3}{5}$.

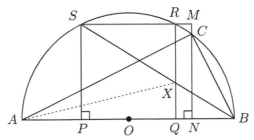

We prove first that SB bisects $\angle ABC$ by showing that the chords AS and CS are equal in length:

$$|AS|^2 = |AP|^2 + |PS|^2 = (|OA| - |OP|)^2 + |PS|^2 = \left(1 - \frac{1}{\sqrt{5}}\right)^2 + \left(\frac{2}{\sqrt{5}}\right)^2 = 2 - \frac{2}{\sqrt{5}}$$

and

$$|CS|^2 = |SM|^2 + |CM|^2$$
$$= (|PO| + |ON|)^2 + (|SP| - |CN|)^2 = \left(\frac{1}{\sqrt{5}} + \frac{3}{5}\right)^2 + \left(\frac{2}{\sqrt{5}} - \frac{4}{5}\right)^2 = 2 - \frac{2}{\sqrt{5}}.$$

It remains to prove that AX bisects $\angle CAB$. Let $t = \tan\left(\frac{1}{2}\angle CAB\right)$. From the double–angle formula for tangent (see Section 3.7), we have

$$\tan \angle CAB = \frac{2t}{1 - t^2};$$

from the right–angled triangle $\triangle CAN$, we have

$$\tan \angle CAB = \frac{|CN|}{|AN|} = \frac{1}{2};$$

equating these two expressions gives $t^2 + 4t - 1 = 0$ and, hence,

$$\tan\left(\frac{1}{2}\angle CAB\right) = t = -2 + \sqrt{5}.$$

On the other hand, by looking at the similar triangles $\triangle BXQ$ and $\triangle BSP$, we have $|QX|/|PS| = |BQ|/|BP|$ and so

$$\tan \angle XAB = \frac{|QX|}{|QA|} = \frac{|PS|}{|QA|}\frac{|BQ|}{|BP|} = \frac{\frac{2}{\sqrt{5}}\left(1 - \frac{1}{\sqrt{5}}\right)}{\left(1 + \frac{1}{\sqrt{5}}\right)^2} = -2 + \sqrt{5} = \tan\left(\frac{1}{2}\angle CAB\right),$$

confirming that AX bisects $\angle CAB$ and completing the verification that X is the intersection of the angle–bisectors of $\triangle ABC$.

Q119 Because of the equality of angles subtending the same arc and the fact that $\triangle ABC$ is isosceles, the angles

$$\angle ABC, \quad \angle ACB, \quad \angle APB, \quad \angle APC$$

are all equal; therefore, the triangles $\triangle AXB$ and $\triangle ABP$ are similar, as are $\triangle AXC$ and $\triangle ACP$. Hence,

$$\frac{|BX|}{|AX|} = \frac{|PB|}{|AB|} \quad \text{and} \quad \frac{|CX|}{|AX|} = \frac{|PC|}{|AC|}.$$

These equations lead to

$$\frac{|BX| + |CX|}{|AX|} = \frac{|PB|}{|AB|} + \frac{|PC|}{|AC|} = \frac{|PB| + |PC|}{|AB|} = 2$$

and so $|BC| = 2|AX|$ as required.

Q120 Choose a point O in the plane which is not equidistant from any two of the points P_k: that is, a point not lying on any of the perpendicular bisectors of $P_i P_j$. Relabel the points so that the lengths of $OP_1, OP_2, \ldots, OP_n$ are in increasing order. For each k, draw a circle C_k with centre O and passing through P_k. Divide C_k into n equal arcs by points

$Q_{k,1}, Q_{k,2}, \ldots, Q_{k,n}$ in anticlockwise order around C_k, with P_k lying in the arc anticlockwise from $Q_{k,k}$.

We shall call a curve from S to T "good" if the distance from O to P increases steadily as P describes the curve. Obviously, such a curve lies in the annulus between circles with centre O passing through S and T.

Now join O to $Q_{1,1}$ by a straight line segment, join $Q_{1,1}$ to $Q_{2,1}$ by a good arc, join $Q_{2,1}$ to $Q_{3,1}$ by another good arc, and so on until we reach $Q_{n,1}$; then draw a ray from $Q_{n,1}$ heading directly away from O. Denote by γ_1 the curve just drawn from O to infinity: see the diagram for an illustration with $n = 4$. For $k = 2, 3, \ldots, n$, let γ_k be the curve obtained by rotating γ_{k-1} anticlockwise about O through an angle of $(360/n)$ degrees. Finally, denote by R_1 the region in the plane between γ_1 and γ_2, by R_2 the region between γ_2 and γ_3, and so on. Since these regions have been constructed by (equal) rotations of the plane, they are all congruent. Also, by construction, the point P_k lies on the arc from $Q_{k,k}$ to $Q_{k,k+1}$ on C_k, which in turn lies in R_k for each k. Therefore, each of the n congruent regions we have constructed contains one of the given points, and we are finished.

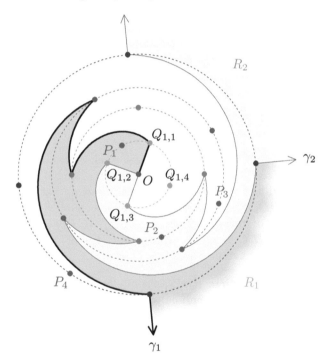

Q121 Let $\triangle ABC$ be a triangle with sides a, b, c. Construct the circumcircle with centre O, and draw the diameter BD. The angles $\angle BAC$ and $\angle BDC$ are subtended by the same chord and therefore are equal. Since BD is a diameter, $\angle BCD$ is a right angle. Therefore,

$$\sin \angle BAC = \sin \angle BDC = \frac{|BC|}{|BD|} = \frac{a}{2R},$$

and so

$$4TR = 4\frac{bc \sin \angle BAC}{2} \frac{a}{2 \sin \angle BAC} = abc,$$

as required.

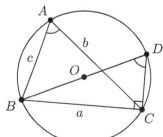

Q122 Let each of A, B, C, D lie at a distance d from the circle with centre O and radius r. The distance from any point P to the circle is the length $|PQ|$, where Q is the intersection of the ray OP with the circle. Hence, if A, B, C, D are all outside the circle, then they all lie on the circle with centre O and radius $r + d$. Since $ABCD$ is not cyclic, this is impossible.

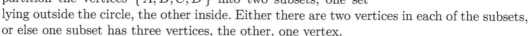

For similar reasons, it is impossible that A, B, C, D all lie inside the circle; hence, the circumference of the circle must partition the vertices $\{A, B, C, D\}$ into two subsets, one set lying outside the circle, the other inside. Either there are two vertices in each of the subsets, or else one subset has three vertices, the other, one vertex.

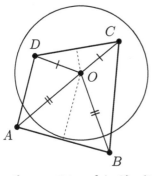

Consider the partition $\{A, B\}, \{C, D\}$: that is, A, B lie outside the circle and C, D inside, or *vice versa*. Then A, B have the same distance from O; and C, D have the same distance from O; so O must lie on both the perpendicular bisectors of AB and CD. Since AB and CD are not parallel, this gives one and only one possibility for O, and r must be the arithmetic mean of the lengths OA and OC. (If r is larger than this, then the distance from A to the circle will decrease, the distance from C to the circle will increase, and the two will no longer be equal; likewise if r is smaller.) Therefore, there is one possible circle in this case; and there is one corresponding to the partition $\{A, C\}, \{B, D\}$, and another corresponding to $\{A, D\}, \{B, C\}$.

Now consider a partition of the four vertices such as $\{A, B, C\}, \{D\}$. The centre O must be the circumcentre of triangle ABC, and the radius r must be the arithmetic mean of the lengths $|OA|$ and $|OD|$. Similarly, there will be one circle corresponding to each of the partitions $\{A, B, D\}, \{C\}$ and $\{A, C, D\}, \{B\}$ and $\{B, C, D\}, \{A\}$. Altogether, there are seven different circles equidistant from A, B, C and D.

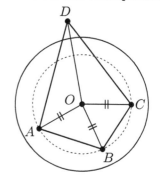

Q123 Consider any thirteen points $P_0, P_1, \ldots, P_{12}$ equally spaced around the circle. By the pigeonhole principle, there are five of these points having the same colour. It is not possible to choose 5 of the vertices of a regular 13–gon without some subset of three of them forming the vertices of an isosceles triangle.

To justify the previous sentence, observe that of our five chosen points, there must be two that are adjacent vertices of the 13–gon, or have just one (not chosen) vertex between them. In the first case, we may assume by relabelling points that two of the chosen five are P_0 and P_1. If there are no three that form an isosceles triangle, then our chosen points cannot include P_2, P_7 or P_{12}, as each of these forms an isosceles triangle with P_0 and P_1. The choice of three further vertices from the remaining eight is governed by the restrictions shown.

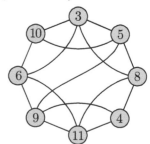

In this diagram, whenever two numbers m, n are joined by a line, it is not possible to choose both P_m and P_n. For instance, we cannot have both P_3 and P_5 since they form an isosceles triangle with P_1. In particular, we see from the diagram that

- our chosen vertices cannot include more than one of P_3, P_6 and P_{10};
- they cannot include more than one of P_4, P_8 and P_{11}; and
- they can only include one of P_5 and P_9.

We must therefore choose exactly one vertex from each of the three groups just mentioned. Starting with the last group and consulting the diagram, it is not hard to confirm that the only choices not forming an isosceles triangle with either P_0 or P_1 are P_5, P_6, P_4 and P_9, P_8, P_{10}; but each of these triples forms an isosceles triangle in itself, and therefore, neither is admissible.

Finally, we consider the case when there are no two adjacent vertices among the chosen five: we may assume that we have P_0 and P_2; and if there are no three forming an isosceles triangle, we cannot choose P_1, P_3, P_4, P_{11} or P_{12}. We then cannot choose P_5 or P_{10}, as either of them leaves only one possible vertex for our remaining two choices; therefore, we must choose three of the vertices P_6, P_7, P_8, P_9, with no two adjacent, which is clearly impossible.

To sum up: the vertices of a regular 13–gon must include 5 of the same colour; and of these 5, there must be three which are the vertices of an isosceles triangle. This completes the proof.

Q124 Let θ be the angle through which the slab has turned. If h_1 is the perpendicular distance from the projecting corner A of the slab to the nearest edge of the path (see the first diagram), then the perimeter of the overhanging triangle is

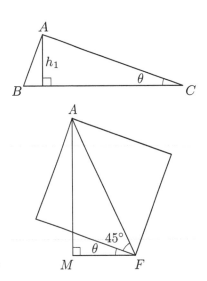

$$|AB| + |AC| + |BC| = h_1\left(\frac{1}{\cos\theta} + \frac{1}{\sin\theta} + \frac{1}{\sin\theta\cos\theta}\right).$$

The perimeter of the overhanging triangle on the other side is the same, except with a perpendicular distance h_2 instead of h_1, and so the sum of the perimeters is

$$S = (h_1 + h_2)\frac{\sin\theta + \cos\theta + 1}{\sin\theta\cos\theta}.$$

Now, the diagonal AF of the slab has length $\sqrt{2}$ metres; from the second diagram, we see that the angle $\angle AFM$ is $45° + \theta$; and so

$$|AM| = \sqrt{2}\sin(45° + \theta) = \sqrt{2}\left(\sin 45°\cos\theta + \cos 45°\sin\theta\right) = \sin\theta + \cos\theta.$$

On the other hand, $|AM|$ is equal to the sum of the two overhangs plus the width of the path, and so

$$h_1 + h_2 = |AM| - 1 = \sin\theta + \cos\theta - 1.$$

Substituting back into the above expression for S gives the sum of perimeters

$$S = \frac{\left((\sin\theta + \cos\theta) - 1\right)\left((\sin\theta + \cos\theta) + 1\right)}{\sin\theta\cos\theta} = \frac{(\sin\theta + \cos\theta)^2 - 1}{\sin\theta\cos\theta}$$
$$= \frac{\sin^2\theta + 2\sin\theta\cos\theta + \cos^2\theta - 1}{\sin\theta\cos\theta} = 2.$$

Q125 We shall express this problem in the terminology of graph theory (Section 3.4). Draw a graph with 36 vertices, representing the 36 members of the club, and an edge joining each pair of vertices. (That is, we have the *complete graph* on 36 vertices.) Colour the edge joining two vertices red if the corresponding members are enemies, and green if they are friends. The question can now be restated as, "How many different one–colour triangles are there in this graph?"

Let v be any one of the vertices of the graph. Denote by R the set of points (13 in number) joined to v by a red edge, and by G the set of 22 points joined to v by a green edge. Of the $C(13, 2)$ edges with both endpoints in R, let r be the number which are red; similarly, let g be the number of green edges having both endpoints in G. Then the number of one–colour triangles having v as a vertex is $r + g$.

For each of the 13 vertices in R, list the 13 red edges incident on this point. Note that each of the r red edges joining vertices in R will be listed twice, once for each of its endpoints. So, this accounts for $2r$ of the 13^2 edges listed; another 13 of them join vertices in R to the vertex v; therefore, the remaining $13^2 - 13 - 2r$ join vertices in R to vertices in G. Similarly, the number of green edges joining vertices in G to vertices in R is $22^2 - 22 - 2g$. But since every one of the 13×22 edges between R and G is either red or green, we have

$$(13^2 - 13 - 2r) + (22^2 - 22 - 2g) = 13 \times 22$$

and hence $r + g = 166$. This is the number of one–colour triangles having v as a vertex; to find the total number we multiply by 36 (the number of vertices) and divide by 3 (because each triangle will have been counted three times, once for each vertex). So the number of one–colour triangles is $36 \times 166 \div 3 = 1992$.

Q126 We use the Remainder Theorem from Section 3.10: if the polynomial $p(x)$ is divided by the linear polynomial $x - a$, then the remainder is $p(a)$. For the present problem, we want $p(0) = 1$, so $p(x)$ must be equal to $x - 0$ times a quotient $p_1(x)$, plus 1. That is,

$$p(x) = 1 + xp_1(x).$$

Now we want $p_1(1) = p(1) - 1 = 2$; therefore,

$$p_1(x) = 2 + (x - 1)p_2(x) \qquad \text{and so} \qquad p(x) = 1 + 2x + x(x - 1)p_2(x).$$

Continuing in the same way,

$$p_2(2) = \frac{p(2) - 5}{2 \times 1} = \frac{4}{2!}$$

$$\Rightarrow \quad p_2(x) = \frac{4}{2!} + (x - 2)p_3(x)$$

$$\Rightarrow \quad p(x) = 1 + 2x + \frac{4}{2!}x(x - 1) + x(x - 1)(x - 2)p_3(x)$$

and

$$p_3(3) = \frac{p(3) - 19}{3 \times 2 \times 1} = \frac{8}{3!}$$

$$\Rightarrow \quad p_3(x) = \frac{8}{3!} + (x - 3)p_4(x)$$

$$\Rightarrow \quad p(x) = 1 + 2x + \frac{4}{2!}x(x - 1) + \frac{8}{3!}x(x - 1)(x - 2)$$

$$+ x(x - 1)(x - 2)(x - 3)p_4(x).$$

Recognising the pattern, we consider the polynomial

$$q(x) = 1 + 2\,\frac{x}{1!} + (2^2)\frac{x(x-1)}{2!} + (2^3)\frac{x(x-1)(x-2)}{3!} + \cdots$$
$$+ (2^n)\frac{x(x-1)\cdots(x-(n-1))}{n!}.$$

This is a polynomial of degree n, and for $k = 0, 1, 2, \ldots, n$ we have

$$q(k) = 1 + 2\,\frac{k}{1!} + (2^2)\frac{k(k-1)}{2!} + (2^3)\frac{k(k-1)(k-2)}{3!} + \cdots$$
$$+ (2^k)\frac{k(k-1)\cdots 1}{k!} + 0 + \cdots + 0,$$

the zeros resulting from the factor $x - k$ in all subsequent terms. By the Binomial Theorem,

$$q(k) = 1 + \binom{k}{1}2 + \binom{k}{2}2^2 + \binom{k}{3}2^3 + \cdots + \binom{k}{k}2^k = (1+2)^k = 3^k,$$

and so $q(x)$ is identical with $p(x)$. The expression for $q(n+1)$ consists of a binomial expansion with the last term missing, and so we have

$$p(n+1) = q(n+1)$$
$$= 1 + \binom{n+1}{1}2 + \binom{n+1}{2}2^2 + \binom{n+1}{3}2^3 + \cdots + \binom{n+1}{n}2^n$$
$$= (1+2)^{n+1} - 2^{n+1}$$
$$= 3^{n+1} - 2^{n+1}.$$

Q127 Note that for any $n \geq 0$, we have

$$\frac{n}{n+1} < \frac{n+1}{n+2}.$$

Writing P for the given product, we have

$$P^2 = \frac{1000}{1001}\,\frac{1000}{1001}\,\frac{1002}{1003}\,\frac{1002}{1003} \cdots \frac{1992}{1993}\,\frac{1992}{1993}$$
$$< \frac{1000}{1001}\,\frac{1001}{1002}\,\frac{1002}{1003}\,\frac{1003}{1004} \cdots \frac{1992}{1993}\,\frac{1993}{1994}$$
$$= \frac{1000}{1994}$$

since almost everything cancels. The job can now be finished by calculator, or more elegantly,

$$P^2 < \frac{1000}{1994} = \frac{500}{997} < \frac{529}{961} = \left(\frac{23}{31}\right)^2.$$

Similarly,

$$P^2 > \frac{999}{1000}\,\frac{1000}{1001}\,\frac{1001}{1002}\,\frac{1002}{1003} \cdots \frac{1991}{1992}\,\frac{1992}{1993} = \frac{999}{1993} > \frac{961}{2025} = \left(\frac{31}{45}\right)^2.$$

Hence,

$$\frac{31}{45} < P < \frac{23}{31}.$$

Q128 Let n be the number of days difference between each birthday and the previous one. Since there are six birthdays, with one in August and no two in consecutive months, there must be one in October and one in December. Thus, a gap of $2n$ days must be at least from August 8 to December 1 (115 days), and at most from August 8 to December 31 (145 days). Hence, remembering that n is an integer, we have

$$58 \leq n \leq 72.$$

As n is prime, we have the four possibilities $n = 59, 61, 67, 71$. Now we need to calculate intervals of n days on each side of August 8 and see which n give admissible answers. A convenient way to do this is to count the days of all months from January to July,

$$31 + 28 + 31 + 30 + 31 + 30 + 31 = 212,$$

to see that August 8 is the 220th day of the year. So if $n = 59$, we consider days 43, 102, 161, 220, 279, 338; by a similar method to above, these days are

February 12, April 12, June 10, August 8, October 6, December 4.

But this is not allowed, since a birthday would fall on the same date in both February and April. Trying $n = 61, 67, 71$ gives, respectively,

February 6, April 8, June 8, August 8, October 8, December 8;
January 19, March 27, June 2, August 8, October 14, December 20;
January 7, March 19, May 29, August 8, October 18, December 28;

of which the first must be discarded. So, two solutions?... well, if you read the question [and the comment preceding it] carefully, I said [in 1993] that all this happened *last year*, so we should have counted not 28 but 29 days for February! Repeating the working, we find the values of n unchanged, and the birthdays corresponding to $n = 59, 61, 67, 71$ are

February 13, April 12, June 10, August 8, October 6, December 4;
February 7, April 8, June 8, August 8, October 8, December 8;
January 20, March 27, June 2, August 8, October 14, December 20;
January 8, March 19, May 29, August 8, October 18, December 28.

Only the first is a solution to the problem.

Q129 Suppose that Simon's statement is correct. Then Alexander came higher up the list than Simon and therefore must also have spoken correctly. But this is impossible since it would mean that Alexander, Esther and Jacinda each occupy one of the top two places. Thus, Simon must have been wrong. This means that Jacinda's first statement is true, and so her second statement must be true too. Thus, Esther came ahead of Jacinda. To summarise what we know so far, (part of) the order of marks is

... Esther... Jacinda... Simon...

and Jacinda made two correct statements, Simon one incorrect statement. Since three of the first five statements are true, we see that of Alexander's and David's remarks, one is true and one false. If David was correct and Alexander incorrect, then David came below Jacinda, and so did Alexander (since his statement was false); thus, Jacinda came second and Alexander's statement was true after all. Thus, David must have made a false statement and finished last, while Alexander made a true statement and came third. So the order (top down) was

Esther, Jacinda, Alexander, Simon, David .

Q130 The kth and $(k+1)$th terms in the expansion are equal when

$$\binom{1993}{k}1993^k m^{1993-k} = \binom{1993}{k+1}1993^{k+1}m^{1993-(k+1)},$$

that is,

$$\frac{1993!}{k!\,(1993-k)!}1993^k m^{1993-k} = \frac{1993!}{(k+1)!\,(1992-k)!}1993^{k+1}m^{1992-k}.$$

After much cancellation, this becomes

$$\frac{m}{1993-k} = \frac{1993}{k+1}$$

which can be rearranged to give

$$(m+1993)k = 1993^2 - m;$$

hence,

$$(m+1993)(k+1) = 1993^2 + 1993 = 1994 \times 1993 = 2 \times 997 \times 1993,$$

where 2, 997 and 1993 are all prime. So $m+1993$ is a factor of $2 \times 997 \times 1993$, and clearly $m+1993 \geq 1994$. There are eight factors, of which 1, 2, 997 and 1993 are rejected as too small, leaving four solutions

$$m+1993 = 2 \times 997,\ 2 \times 1993,\ 997 \times 1993 \quad \text{or} \quad 2 \times 997 \times 1993,$$

that is,

$$m = 1,\ 1993,\ 1985028 \quad \text{or} \quad 3972049.$$

Q131 It is well known how to create a single map that requires four colours.

A bit of experimentation, inspired by this configuration, leads to one possible pair of maps as shown.

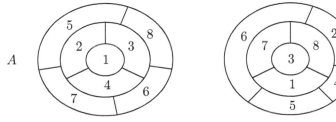

The numbers represent the different nations. It is easy to check that any one of the eight countries has a border with each of the other seven. For example, country 7 shares borders with 2, 4, 5 and 6 on planet A, and with 1, 3, 6 (again) and 8 on planet B. Therefore, no two countries may use the same colour, and eight colours are needed.

Comment. There are no doubt many other examples of pairs of maps which require eight colours. In fact, there are maps which require 9 or more colours – see if you can find one! It has been proved that there are no pairs which need more than 12 colours. Whether any map actually needs 12 colours, 11 or even 10 is at present unknown.

Q132 Let d be the divisor, q the quotient. Looking at the first subtraction in the calculation, d has a multiple of the form EEEO, which must be at least 2001; therefore, $9d \geq 2001$. Likewise, d has a multiple EOE which is not equal to d itself and is at most 898; so $2d \leq 898$. Hence,

$$223 \leq d \leq 449;$$

but since d has the form OEO, we can improve this to

$$301 \leq d \leq 389. \tag{2.12}$$

Now observe that in the last and second last subtractions, two different multiples of d were subtracted from two (possibly different) numbers EOE. But such a number is at most 898, which is less than $3d$, so the last two digits of q must be 12. This allows us to restrict the values of d still further. From (2.12), we have $602 \leq 2d \leq 778$; but $2d$ has digits EOE, so $610 \leq 2d \leq 698$ and

$$305 \leq d \leq 349.$$

To improve this once more, we notice that the second subtraction is $\text{EOE} - d = \text{EO}$, where EO are the first two digits of $2d$; therefore

$$d = \text{EOE} - \text{EO} \geq 410 - 69 = 341.$$

Now we can determine the first digit of q: it is odd, and when multiplied by d gives a result EEEO. But obviously, $d, 3d, 5d < 2000$; and $3069 \leq 9d \leq 3141$; so, to give a four–digit product beginning with an even number, the required digit must be 7. Hence, $q = 712$ and we have five possibilities for d. If d is 341 or 343, then $2d$ is EEE, which is not so; if $d = 345$, then $7d$ is EEOO; while if $d = 349$, then we find that dq is EEEEEE, whereas it should be EEOEEE. Hence, the only valid solution is given by $d = 347$ and $q = 712$.

$$
\begin{array}{r}
7\ 1\ 2 \\
\hline
3\ 4\ 7\ |\ 2\ 4\ 7\ 0\ 6\ 4 \\
2\ 4\ 2\ 9 \\
\hline
4\ 1\ 6 \\
3\ 4\ 7 \\
\hline
6\ 9\ 4 \\
6\ 9\ 4 \\
\hline
\end{array}
$$

Q133 Label the chessboard with the numbers $1, 2, 3$ as shown in the first diagram.

1	2	3	1	2	3	1	2
2	3	1	2	3	1	2	3
3	1	2	3	1	2	3	1
1	2	3	1	2	3	1	2
2	3	1	2	3	1	2	3
3	1	2	3	1	2	3	1
1	2	3	1	2	3	1	2
2	3	1	2	3	1	2	3

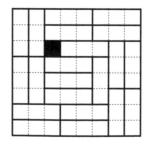

Then any 3 × 1 rectangle placed on the board must cover a 1, a 2 and a 3. The figure 1 occurs twenty–one times: so, by removing all the squares labelled 1, we make it impossible to place a 3 × 1 rectangle on the board. However, removing fewer than 21 squares will not do. To see this, consider the second diagram above. Here, the board contains twenty–one non–overlapping 3 × 1 rectangles; so if only twenty, or fewer, squares are removed, then at least one complete rectangle must remain. So, the fewest possible number of squares that we may remove is 21: one possibility is as shown below.

Q134 Yes, the faces do lie in exactly the same plane.

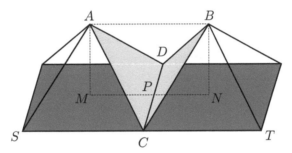

To see this, place *two* pyramids as shown in the diagram: the bases are coplanar and two edges meet. Draw the interval AB joining the summits of the two pyramids. Then it is clear that $ABCD$ is a tetrahedron with one of its faces attached to a face of (say) the left–hand pyramid; it is also clear that faces ASC and ABC (and BCT) are coplanar; what remains is to prove that the tetrahedron $ABCD$ is regular, so that the configuration of one pyramid and the tetrahedron is what was specified in the question. But AC, AD, CD, BC and BD are sides of congruent equilateral triangles (given), while $|AM| = |MN| = 2|MP|$, and $|MP|$ is half of $|DC|$, the side of the base of the pyramid. Hence, $ABCD$ is a regular tetrahedron.

Q135 Label our coins $1, 2, \ldots, 7$. One possible weighing procedure is the following.

(1) Weigh $1, 2$ against $3, 4$.

(2) Weigh $1, 4$ against $5, 6$.

(3) Weigh $1, 3, 5$ against $2, 4, 6$.

We can now work out the results of the weighings for any given pair of heavy coins. If, for example, coins 1 and 5 are heavy, then the first weighing tips to the left, the second balances, the third again tips to the left. Considering in this way all possible pairs, we find that each gives a different set of results for the three weighings, and so these results will determine which two coins are in fact the heavy ones.

To answer the second part of the problem, we observe that each weighing will give one of three results (left, right or balance). Therefore, the total number of results that can arise from a sequence of three weighings is $3 \times 3 \times 3 = 27$. If we have eight coins, then there are $C(8, 2) = 28$ possibilities for the heavy pair. Hence, if we weigh the coins by any procedure at all, then there must be two possibilities which give the same result and therefore cannot be distinguished.

Comment. This is not to say that we can *never* find the two heavy coins from eight. For instance, if we weigh 1 against 2, then 7 against 8, and if both weighings tip to the left, then the heavy coins must be 1 and 7. But this only works if we are lucky. The preceding argument shows that there is no procedure which will find the heavy coins *with certainty* in every situation.

Q136 We label the lines as shown at right. For the configuration given in the question, we can achieve the aim of making all coins heads by turning over the coins on lines $2, 3, 6$ and 7. The lines can be taken in any order, but there is no solution in fewer than four moves, and no other solution in just four.

To answer the second part of the question, note that any allowable move flips two coins in the "outer" ring of five. Thus, any move increases or decreases the number of tails among these coins by two, or (if the affected coins were originally a head and a tail) leaves it unchanged. Since our aim is to reduce the number of tails to zero, we must start with an even number of tails in the outer ring. Likewise, any move affects either two coins in the "inner" ring or none, and we must start also with an even number of tails in the inner ring if we are to have any chance of success.

If the above conditions are met, then does this guarantee that the problem is solvable? Yes, for we can obey the following procedure. If the coin at the intersection of lines 6 and 7 is tails, then flip line 7. If the coin at the intersection of lines 7 and 8 is now tails, then flip line 8. By flipping lines 9 and 10 if necessary, we obtain four heads on the inner ring. Since we started with an even number of tails on the inner ring and changed none or two every move, we cannot have one tail left, and so the fifth coin must be heads too. We can now finish the game by treating lines $2, 3, 4, 5$ in the same way.

Q137 We have

$$\binom{n}{j+1} = a\binom{n}{j}, \quad \binom{n}{j+2} = 23\binom{n}{j},$$

where $\binom{n}{j}$ is the first of the three binomial coefficients. We can write out the binomial coefficients in terms of factorials (page 250), or use a combinatorial proof (Section 3.5.4) to simplify these equations to

$$n - j = a(j+1) \quad \text{and} \quad a(n - j - 1) = 23(j + 2);$$

then taking a times the first of these equations, subtracting the second, and doing a small amount of algebra leads to

$$j = -\frac{a^2 - a - 46}{a^2 - 23}.$$

Now, $j > 0$, so one of the expressions $a^2 - a - 46$ and $a^2 - 23$ must be positive, the other negative. But clearly, $a^2 - 23$ is the larger, so we have

$$a^2 - 23 > 0, \quad a^2 - a - 46 < 0$$

and hence, remembering that a is an integer, $5 \le a \le 7$. Of these, only $a = 5$ gives an integer value for j, namely, $j = 13$; we then find $n = 83$. If you are willing to do a little(?) arithmetic, then you can check that

$$\binom{83}{13} = 528955739755020, \quad \binom{83}{14} = 2644778698775100, \quad \binom{83}{15} = 12165982014365460$$

and that the second and third of these are in fact 5 and 23 times the first.

Q138 This is a multiple-choice question with no information given! Nevertheless, it is possible to deduce the answer. If (d) is true then (b) is true and so (d) is false after all. Since (d) is false, (a) is false. If (c) is true, then (e) is true since we know that (d) is not, and hence (c) is actually false. Since (c) is false, all those following are false: that is, (e) is false. Hence, (b) is true.

Now we have to check that all the true and false specifications are consistent; otherwise, there will be no solution to the problem. This is easily done, and so the answer is (b).

Q139 Noting that $3 < 2 + \sqrt{3} < 4$, the grazing area looks like this,

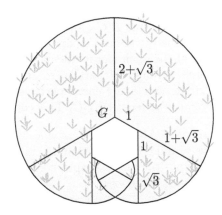

where G is the point to which the goat is tied and we have chosen the hexagon to have side length 1. The reachable region consists of two thirds of a circle with radius $2 + \sqrt{3}$, two sixths of a circle with radius $1 + \sqrt{3}$ and the shaded area below, which we examine more closely.

We can calculate the area as that of two-twelfths of a circle ADE and BDF, plus an equilateral triangle ABD, minus the triangle ABC. Using well–known formulae, the area within which the goat can graze is

$$\frac{2}{3}\pi(2 + \sqrt{3})^2 + \frac{2}{6}\pi(1 + \sqrt{3})^2 + \frac{2}{12}\pi(\sqrt{3})^2 + \frac{3\sqrt{3}}{4} - \frac{\sqrt{3}}{4},$$

which simplifies to

$$\frac{39\pi + 20\pi\sqrt{3} + 3\sqrt{3}}{6}.$$

Q140 Since the points X, Y, Z are given, we can draw through them the circumcircle of $\triangle ABC$. Now, since AY bisects the angle at A, we have equal angles $\angle BAY$ and $\angle CAY$, which are subtended by equal arcs BY and CY. Draw equal arcs (of any length) on each side of Y, cutting the circle at B' and C';

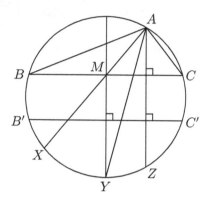

then the line $B'C'$ is parallel to the unknown line BC. From the given information, AZ is perpendicular to BC, and therefore also perpendicular to $B'C'$. Thus, we may draw a line through Z, perpendicular to the known line $B'C'$, and A is the other point of intersection of this line with the circumcircle. Finally, we need to find the line BC, which is bisected by AX. Draw AX, draw a diameter of the circle perpendicular to $B'C'$, call their intersection M, and draw a chord of the circle through M parallel to $B'C'$. Then M is the midpoint of this chord (a diameter of a circle bisects any chord perpendicular to it), and hence AX bisects the chord, which is therefore the required BC. This completes the construction.

Q141 Write c, b, w for the number of ANHP members who live in Canberra, ride bicycles and own whiteboards, respectively, and $\bar{c}, \bar{b}, \bar{w}$ for the numbers who do not. Combinations of properties will be denoted by juxtaposition: for example, $c\bar{w}$ denotes the number of members who live in Canberra and do not own whiteboards.

Suppose that $bw < 8$. Then (c) tells us that $b = 4$; so $b < 9$, and from (b) we have $bw = 8$. This is a contradiction, and hence $bw \geq 8$. But $bw \leq b$ (the number who ride bicycles *and* own whiteboards cannot be more than the number who ride bicycles in the first place), so $b \neq 4$. Therefore, from property (c), we have $c\bar{b} = 12$. We can represent the information we have so far in a Venn diagram.

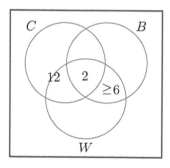

The 12 on a borderline indicates that we do not know exactly how these 12 members are distributed over the two regions. It is clear from the diagram that $\bar{c}b \geq 6 > 3$, so by (a), we have $w = 15$. Since there are altogether 25 members of the party, $\bar{w} = 10 > 7$, and condition

(b) gives $bw = 8$. This leads to an improved Venn diagram.

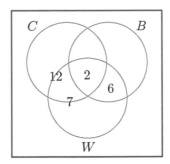

Finally, it is stated that "if you knew the total number of ANHP members who ride bicycles, then you could calculate how many live in Canberra". If there were more who ride bicycles than the 8 we already know about, then we would not know whether or not the additional bike–riding members live in Canberra; so this statement can only be true if $b = 8$. Therefore, we can put $cb\overline{w} = \overline{c}b\overline{w} = 0$ in the above diagram, and we have

$$b = 8, \quad c = 12 + 0 + 2 = 14 \quad \text{and} \quad w = 7 + 2 + 6 = 15.$$

Q142 One way is to print any number of tickets with labels divisible by 6; ten tickets with labels which have remainder 3 when divided by 6; and fifty with remainder 2 when divided by 6. Since 1995 divided by 6 leaves remainder 3, a \$1000 prize can only be won with (at least) one of the second type of ticket, so at most ten such prizes will be won. Similarly, to win a \$100 prize requires one of the fifty tickets of the third type.

Q143 A square can be cut into three pieces, as shown, which can be rearranged to form a 3×4 rectangle,

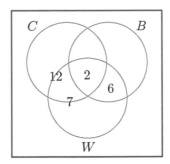

and into four pieces which form a 2×6 rectangle.

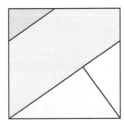
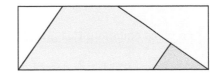

Superimposing the two designs gives a dissection into seven pieces

which will form either rectangle.

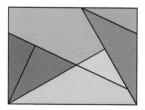   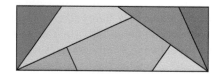

Q144 First note that when three frogs meet, they are transformed into two frogs: so the number of frogs on this island is not constant, but decreases by one whenever three frogs of different colours meet.

At each meeting, the number of frogs of one colour increases by two, while that of each other colour decreases by one. Hence, the difference in numbers between any two colours can only change by 3 (if at all). So, the three colours of frogs which eventually have no members must initially have differed by multiples of 3. These three can only be 50, 62 and 68: thus, the green frogs survived.

Now let b be the number of meetings in which two brown frogs were created, and similarly define g, y and r. Then the total number of red frogs lost in the course of all these meetings is $b + g + y - 2r$, and this must equal 68 since no red frogs were left. Adding $3r$ to both sides gives

$$b + g + y + r = 68 + 3r.$$

Since one frog is lost at each meeting, the final number of frogs is

$$50 + 57 + 62 + 68 - (b + g + y + r) = 169 - 3r, \tag{2.13}$$

which cannot be more than 169. To check that 169 is in fact possible, we first consider the number of yellow and brown frogs lost. As above, we find

$$b + g + y + r = 62 + 3y = 50 + 3b;$$

together with a previous equation, this gives

$$y = r + 2, \quad b = r + 6 \quad \text{and} \quad g = 60.$$

In order that 169 green frogs remain and no others, we must have $r = 0$, $y = 2$, $b = 6$ and $g = 60$. It is now easy to confirm that starting with 50, 57, 62, 68 brown, green, yellow and red frogs, we can create two pairs of yellow frogs to give 48, 55, 66, 66; then six pairs of brown frogs to give 60, 49, 60, 60; and, finally, sixty pairs of green frogs to give 0, 169, 0, 0.

Comment. We can also find the *minimum* possible number of green frogs remaining when the other three colours are extinct. After the final meeting, there must be at least 2 frogs on the island; so from (2.13) we have $169 - 3r \geq 2$ and hence $r \leq 55$. We can take the numbers of various meetings as

$$r = 55, \quad y = 57, \quad b = 61, \quad g = 60$$

to end up with just 4 green frogs, and no others, on the island; though you will have to arrange the meetings carefully in order to avoid having negative frogs at some intermediate stage.

Q145 If n is odd, then all of its factors are odd. Thus, to write n as a sum of factors, an odd number of terms are required, as an even number of odd terms add up to an even sum. If we write $n = de$, then both d and e are odd; so we are seeking a sum

$$n = d_1 + d_2 + \cdots + d_k = \frac{n}{e_1} + \frac{n}{e_2} + \cdots + \frac{n}{e_k},$$

that is,

$$1 = \frac{1}{e_1} + \frac{1}{e_2} + \cdots + \frac{1}{e_k},$$

where $e_1, e_2, \ldots, e_k$ are odd and all different, and k is also odd. This will take rather a lot of computation, but here goes...

First, taking a sum of five fractions is not sufficient, because it will give a maximum sum of

$$\frac{1}{3} + \frac{1}{5} + \frac{1}{7} + \frac{1}{9} + \frac{1}{11} = \frac{3043}{3465},$$

which is less than 1.

Next, consider the possibility of using seven terms. The fractions $\frac{1}{3}, \frac{1}{5}, \frac{1}{7}, \frac{1}{9}$ and $\frac{1}{11}$ must all be used, for if even the least of them is missing, then the maximum attainable is

$$\frac{1}{3} + \frac{1}{5} + \frac{1}{7} + \frac{1}{9} + \frac{1}{13} + \frac{1}{15} + \frac{1}{17} = \frac{68899}{69615} < 1.$$

Also, either $\frac{1}{13}$ or $\frac{1}{15}$ must be used; otherwise the sum cannot exceed

$$\frac{1}{3} + \frac{1}{5} + \frac{1}{7} + \frac{1}{9} + \frac{1}{11} + \frac{1}{17} + \frac{1}{19} = \frac{1107629}{1119195} < 1.$$

Hence, we need either

$$\frac{1}{3} + \frac{1}{5} + \frac{1}{7} + \frac{1}{9} + \frac{1}{11} + \frac{1}{13} + \frac{1}{m} = 1$$

or

$$\frac{1}{3} + \frac{1}{5} + \frac{1}{7} + \frac{1}{9} + \frac{1}{11} + \frac{1}{15} + \frac{1}{m} = 1 \, ;$$

and a final burst of arithmetic gives $m = 22\frac{583}{2021}$ and $m = 18\frac{27}{191}$, respectively. These are both impossible, as m must be an integer. So, the problem cannot be solved with seven factors, and we need at least nine.

Comment. If you've not had enough arithmetic yet, then you could investigate the case of nine factors: it turns out that there are five solutions, one of them being

$$3465 = 1155 + 693 + 495 + 385 + 315 + 231 + 99 + 77 + 15.$$

Q146 First recall two basic facts: (a) in a triangle, each side is shorter than the sum of the other two sides; (b) in a non–acute (that is, right–angled or obtuse) triangle with sides $x \geq y \geq z$, we have $x^2 \geq y^2 + z^2$. Now let the five given segments, in decreasing order of length, be a, b, c, d and e. Then either the triangle a, b, c or the triangle c, d, e is acute–angled. For if not, then

$$a^2 \geq b^2 + c^2 \quad \text{and} \quad c^2 \geq d^2 + e^2;$$

and then

$$a^2 \geq 2c^2 \geq 2(d^2 + e^2) = (d+e)^2 + (d-e)^2 \geq (d+e)^2.$$

But since a, d, e form a triangle, $a < d + e$, and so the above inequality is impossible. Thus, either a, b, c or c, d, e form an acute–angled triangle.

Comment. An example in which *only* one of the triangles is acute is given by five segments of lengths

$$1, \ 1, \ 1, \ \sqrt{2}, \ \sqrt{3}.$$

Here, it is the smallest triangle ("c, d, e" in the above notation) which is acute: in fact, it is impossible to find an example in which *only* the largest triangle a, b, c is acute.

Q147 Divide the 3×4 rectangle into five parts as shown.

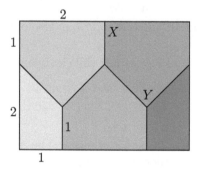

If six points are located in the rectangle, then at least two must lie in (or on the boundary of) the same subdivision. But it is not hard to see that the maximum distance between any two points in the same section is $\sqrt{5}$ – for instance, this is the distance from X to Y. Thus, two of the six points are separated by $\sqrt{5}$ or less.

Q148 Consider the following diagram (in which the 1° angles have been magnified for clarity). The fire must be located within $ABCD$.

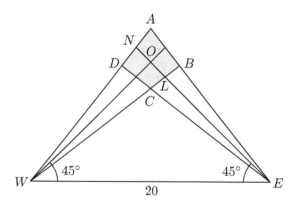

First an approximate calculation of the green area. Since $\angle DAB = 88°$ and $\angle BCD = 92°$ and $\angle ABC = \angle CDA = 90°$, the quadrilateral $ABCD$ is very nearly a square. Also, NL is almost a $2°$ arc of a circle with radius $|WO| = 10\sqrt{2}$. Hence,

$$\text{area}(ABCD) \approx \left(\frac{2}{360} \times 2\pi \times 10\sqrt{2}\right)^2 = \frac{2\pi^2}{81}.$$

To obtain an exact answer, note that if we take the triangle $\triangle WAE$ and subtract both $\triangle WBE$ and $\triangle WDE$ (which are congruent), then we obtain $ABCD$, having removed $\triangle EBC$ and $\triangle WCD$; however, we will have included $\triangle WCE$ once and then subtracted it twice. Therefore,

$$\text{area}(ABCD) = \text{area}(WAE) - 2 \times \text{area}(WBE) + \text{area}(WCE).$$

Now we can find the area of an isosceles triangle with base b and base angles θ; and that of a right–angled triangle with hypotenuse b and angle θ; as

$$\frac{1}{4} b^2 \tan \theta \quad \text{and} \quad \frac{1}{4} b^2 \sin 2\theta$$

respectively. Also, formulae from Section 3.7 can be used to show that

$$\tan \theta + \tan(90° - \theta) = \frac{2}{\sin 2\theta}.$$

Hence, we find

$$\text{area}(ABCD) = \frac{1}{4} 20^2 \tan 46° - \frac{2}{4} 20^2 \sin 88° + \frac{1}{4} 20^2 \tan 44°$$

$$= \frac{200}{\sin 88°} - 200 \sin 88° = 200 \frac{1 - \sin^2 88°}{\sin 88°}$$

$$= 200 \frac{\sin^2 2°}{\cos 2°}.$$

As a check, we may calculate this to six decimal places, giving an area of $0.243743 \, \text{km}^2$; while our first approximation was $0.243694 \, \text{km}^2$.

Q149 The x–axis is tangent to the curve at any x value for which y and dy/dx are simultaneously zero; that is,

$$x - \sin x - (1 - \cos x) \tan \alpha = 0 \tag{2.14}$$

$$1 - \cos x - \sin x \tan \alpha = 0 \tag{2.15}$$

It is easy to check that both equations are true when $x = 0$, and this answers (a). Now suppose that they are true for some $x \neq 0$. Then we also have $\sin x \neq 0$, as otherwise (2.15) gives $1 - \cos x = 0$ and then (2.14) gives $x = 0$ after all. We can write (2.15) as

$$\cos x = 1 - \sin x \tan \alpha, \tag{2.16}$$

square both sides and use $\cos^2 x = 1 - \sin^2 x$ to obtain

$$\sin^2 x = 2 \sin x \tan \alpha - \sin^2 x \tan^2 \alpha;$$

since $\sin x$ is not zero, we may cancel it, and we then solve to find

$$\sin x = 2 \tan \alpha \cos^2 \alpha = \sin 2\alpha.$$

Here, we have also used some trigonometric identities which may be found in Section 3.7. Substituting back into (2.16) yields

$$\cos x = 1 - 2\tan^2\alpha\cos^2\alpha = \cos 2\alpha,$$

and the only way in which these last two equations are both true is

$$x = 2\alpha + 2n\pi$$

for integer n. On the other hand, from (2.14) we now have

$$x = \sin x + (1 - \cos x)\tan\alpha = 2\tan\alpha\cos^2\alpha + 2\tan^3\alpha\cos^2\alpha = 2\tan\alpha;$$

so $2\tan\alpha = 2\alpha + 2n\pi$, and $\tan\alpha - \alpha$ is equal to $n\pi$, a multiple of π, as we wished to show.

Conversely, if α is such that $\tan\alpha - \alpha = n\pi$, then choose

$$x = 2\tan\alpha = 2\alpha + 2n\pi.$$

Since α is not a multiple of π, we have $x \neq 0$. Moreover,

$$\sin x = \sin 2\alpha = 2\sin\alpha\cos\alpha, \quad \cos x = \cos 2\alpha = 1 - 2\sin^2\alpha;$$

from these it is easy to check that (2.14) and (2.15) hold, and so the x–axis is tangent to the curve at the point $x = 2\tan\alpha$.

To illustrate what we have found in this problem, we give the graph of the curve

$$y = x - \sin x - (1 - \cos x)\tan\alpha$$

for two values of α. The green graph has $\alpha = 2$, for which $\tan\alpha - \alpha = -4.1850$ is not a multiple of π; the red graph has $\alpha = 1.3518$, which is (approximately) a solution of $\tan\alpha - \alpha = \pi$. We see that while both curves have the x–axis as a tangent at the origin, only the red curve has the axis as a tangent elsewhere.

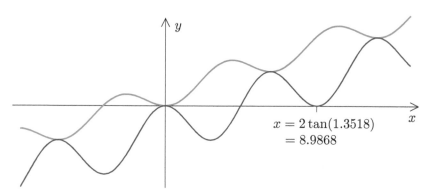

$$x = 2\tan(1.3518)$$
$$= 8.9868$$

Q150 Before we can determine whether statement 5 is true or false, we must consider what happens when the first statement is deleted: that is, consider statements 2, 3, 4 and 5 only, in that order. Be careful to note that any references to "this list" now refer to those four statements, and not to the original five.

Suppose that statement 4 is true. Since it is the third statement in this list, 2 is also true. Now 4 says (truly) that the list contains (at least) two false statements, so 3 and 5 are false; but then the list does not contain two consecutive false statements, so 3 is true after all. This is impossible; so statement 4, which we assumed to be true, must in fact be false. This being so, the list contains only one false statement (namely, statement 4 itself) and three true statements.

Now we return to the complete list of five statements. The fifth says that this list contains four or five true statements. If this is true, then there is only one false statement, or none at all, in the list; so 4 must be false and all the others true. But this is impossible as it would make statement 1 simultaneously true and false. Thus, statement 5 must, after all, be false, and it follows immediately that 1 is true. If statement 4 is false, then (since we know 5 is false) there are at least two false statements; and so 4 is in fact true. (Summary: at this stage, we know that 5 is false while 1 and 4 are true.) Finally, if statement 3 is true, then so is 2; but this is impossible as there would be four true statements, making 5 true. Therefore, statement 3 is false, and to get two consecutive false statements in the list, 2 must also be false.

Answer: statements 1 and 4 are true; statements 2, 3 and 5 are false.

Comment. If we want a logically scrupulous and absolutely watertight solution, then we are not yet finished. For the last statement in our list of four says something about the list of three, which in turn says something about a list of two and then about a list containing just one statement. This last says something about a list with no statements – which we can, at any rate, be certain has no true statements! The point is that we have shown there is only one possible solution to the problem; if the extra considerations we have mentioned in the present paragraph are inconsistent, then there would be no solution at all. We leave it to readers to confirm that everything is consistent, and that truth values are as follows: in the list of three, 3 true, 4 false, 5 true; in the list of two, 4 false, 5 true; in the list of one, 5 false.

Q151

(a) Join each vertex to the circumcentre of the triangle.

(b) One (or both) of the other angles must be greater than 45°. Draw a perpendicular from such a vertex to its opposite side, creating two right–angled triangles. One of these is isosceles; divide the other by bisecting its hypotenuse.

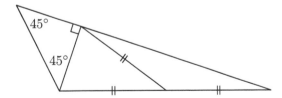

(c) As shown below, where the angle at A is five times that at B.

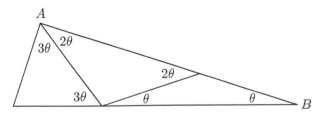

(d) The angle at A is six times that at B.

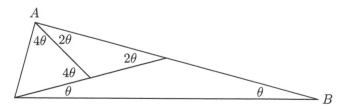

(e) The angle at A is seven times that at B.

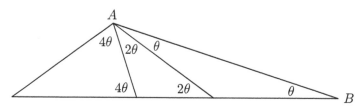

Q152 Draw an $m \times n$ grid; in the kth column (from the left), shade in the bottom a_k squares. We give an example in which $m = 9$ and the first m integers a_k are $2, 3, 3, 5, 7, 7, 7, 11, 12$. Note that the shaded columns always increase in height from left to right; also, that the rightmost column is the full height of the grid, because the height is n, and it was specified in the problem that this is equal to a_m. Now it is clear that the total area of the grid is mn; and also that the total blue area is $a_1 + a_2 + \cdots + a_m$.

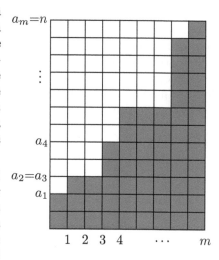

What is the unshaded area in the grid? We shall add it up row by row. The square in column j, row k will be unshaded as long as $a_j < k$. But this says that the number of unshaded squares in row k is equal to the number of $a_1, a_2, a_3, \ldots$ which are less than k, and this is precisely what we mean by b_k!

Therefore, the total unshaded area is $b_1 + b_2 + \cdots + b_n$, and we have

$$(a_1 + a_2 + \cdots + a_m) + (b_1 + b_2 + \cdots + b_n) = \langle \text{total area} \rangle = mn.$$

Q153 Clearly, the greatest distance will be attained if we choose two directions, say north and east, and always drive as far as possible in these directions and as little as possible in the others. So, if we let N, S, E, W be the total distance travelled (in the positive sense) in each of the four directions, then we will want

$$N + E = (2n + 1) + (2n + 2) + (2n + 3) + \cdots + (4n)$$
$$= n(6n + 1)$$
$$S + W = 1 + 2 + 3 + \cdots + (2n)$$
$$= n(2n + 1).$$

Now let V be the overall distance travelled north and H the overall distance travelled east; that is, $V = N - S$ and $H = E - W$. Then

$$V + H = (N + E) - (S + W) = 4n^2,$$

and the square of my eventual distance from home is

$$V^2 + H^2 = V^2 + (4n^2 - V)^2 = 2V^2 - 8n^2 V + 16n^4.$$

Now think of this expression as a function of V. Its graph will be a parabola opening upwards; therefore, there will be no maximum turning point, and the expression will attain its maximum value for either the maximum or minimum value of V. The maximum value of V is

$$V = N - S = ((3n + 1) + (3n + 2) + \cdots + 4n) - (1 + 2 + \cdots + n) = 3n^2,$$

so $H = n^2$ and $V^2 + H^2 = 10n^4$. Since $V + H$ is a fixed number, the minimum value of V corresponds to the maximum value of H; exactly the same working leads to exactly the same result. Hence, the maximum possible distance from home after $4n$ segments is

$$\sqrt{V^2 + H^2} = \sqrt{10}\, n^2.$$

Can I end up at home? If $n = 1$, then it is easy to see that this is impossible. If $n = 2$, then I may travel along 8 segments of distances

$$1, 2, 3, 4, 8, 7, 6, 5,$$

in that order; then the overall distance travelled east is $1 - 3 + 8 - 6 = 0$, and the overall distance travelled north is likewise zero. If $n = 3$, then the sequence

$$1, 2, 3, 4, 8, 6, 7, 5, 12, 10, 11, 9$$

achieves the same object. If I get home after $4n$ segments, then, after having done so, I may continue with segments of lengths

$$4n + 1, 4n + 2, 4n + 3, 4n + 4, 4n + 8, 4n + 7, 4n + 6, 4n + 5,$$

which adds no extra overall distance in either the north–south or east–west directions and therefore lands me back home after $4(n+2)$ segments. Hence, I can always end up at home, provided that $n > 1$.

Q154 We prove the equation for all n by mathematical induction. For each $n = 1, 2, \ldots$, define

$$S_n = -\sum_{k=1}^{n} (-1)^k \frac{1}{k} \binom{n}{k} = \binom{n}{1} - \frac{1}{2}\binom{n}{2} - \cdots - (-1)^n \frac{1}{n}\binom{n}{n}.$$

For $n = 1$, the equation to be proved says

$$S_1 = \binom{1}{1} = 1$$

which is certainly true. Now suppose that the equation is known to be true for some specific $n \geq 1$. We wish to prove that

$$S_{n+1} = 1 + \frac{1}{2} + \cdots + \frac{1}{n+1}.$$

First note that if $1 \leq k \leq n$, then

$$\frac{1}{k}\binom{n+1}{k} = \frac{1}{k}\frac{(n+1)!}{k!\,(n+1-k)!}$$

$$= \frac{n+1}{k(n+1-k)}\frac{n!}{k!\,(n-k)!}$$

$$= \left(\frac{1}{k} + \frac{1}{n+1-k}\right)\frac{n!}{k!\,(n-k)!}$$

$$= \frac{1}{k}\binom{n}{k} + \frac{1}{n+1}\binom{n+1}{k}.$$

Therefore,

$$S_{n+1} = -\sum_{k=1}^{n+1}(-1)^k \frac{1}{k}\binom{n+1}{k}$$

$$= -\sum_{k=1}^{n+1}(-1)^k \left(\frac{1}{k}\binom{n}{k} + \frac{1}{n+1}\binom{n+1}{k}\right)$$

$$= S_n + \frac{1}{n+1}\left(\binom{n+1}{1} - \binom{n+1}{2} - (-1)^{n+1}\binom{n+1}{n+1}\right).$$

By the Binomial Theorem,

$$(x-y)^{n+1} = x^{n+1} - \binom{n+1}{1}x^n y + \cdots + (-1)^{n+1}\binom{n+1}{n+1}y^{n+1},$$

so, letting $x = y = 1$, we obtain

$$0 = 1 - \binom{n+1}{1} + \binom{n+1}{2} - \cdots - (-1)^{n+1}\binom{n+1}{n+1}.$$

Therefore,

$$\binom{n+1}{1} - \binom{n+1}{2} - \cdots - (-1)^{n+1}\binom{n+1}{n+1} = 1,$$

so, from above and the inductive assumption, we have

$$S_{n+1} = S_n + \frac{1}{n+1} = 1 + \frac{1}{2} + \cdots + \frac{1}{n} + \frac{1}{n+1}.$$

The proof now follows by induction.

Q155 Notice that the four given numbers are the last two digits of squares, for example

$$11^2 = 121,\ 18^2 = 324,\ 25^2 = 625,\ 27^2 = 729.$$

Take any two numbers with this property, say x and y, where $100u + x$ and $100v + y$ are squares. Then $(100u + x)(100v + y)$ is also a square. But

$$(100u + x)(100v + y) = 100(100uv + uy + vx) + xy$$

and so the last two digits of xy also form the last two digits of a square. Thus, the only numbers which may ever occur in the set are those which are the last two digits of a square. However, 99 is not such a number. For if $(10m + n)^2$ ends in a 9, then n must be 3 or 7; but

$$(10m + 3)^2 = 100m^2 + 60m + 9$$
$$(10m + 7)^2 = 100m^2 + 140m + 49$$

and, in each case, the second last digit is even. Thus, the set can never contain 99.

Q156 The area of a semicircle is proportional to the area of the square drawn on its diameter. Hence, by Pythagoras' Theorem, the area of the semicircle on the hypotenuse equals the sum of the areas of the semicircles on the other two sides. From the diagram, the shaded region consists of the two smaller semicircular regions, plus the triangle itself, minus the semicircular region on the hypotenuse. Thus, the total area of the shaded crescents equals the area of the triangle, that is, half the product of the two shorter sides.

Q157 Each of the 10^{18} real numbers $r + s\sqrt{2} + t\sqrt{3}$, where $r, s, t \in \{0, 1, \dots, 10^6 - 1\}$, is between 0 and $(1 + \sqrt{2} + \sqrt{3})10^6$. If we partition the interval

$$\{x : 0 \le x < (1 + \sqrt{2} + \sqrt{3})10^6\}$$

into $10^{18} - 1$ equal subintervals, then two of the above numbers $r_1 + s_1\sqrt{2} + t_1\sqrt{3}$ and $r_2 + s_2\sqrt{2} + t_2\sqrt{3}$ must fall in the same subinterval. Set

$$a = r_1 - r_2\,;$$
$$b = s_1 - s_2\,;$$
$$c = t_1 - t_2\,.$$

Then a, b and c lie between -10^6 and 10^6 and are not all zero, and

$$\begin{aligned} |a + b\sqrt{2} + c\sqrt{3}| &= |(r_1 + s_1\sqrt{2} + t_1\sqrt{3}) - (r_2 + s_2\sqrt{2} + t_2\sqrt{3})| \\ &< \frac{(1 + \sqrt{2} + \sqrt{3})10^6}{10^{18} - 1} \\ &< \frac{5 \times 10^6}{5 \times 10^{17}} \\ &= 10^{-11}. \end{aligned}$$

Q158 For $x > 0$, the expression $\sqrt{x + 1} + \sqrt{x}$ defines a strictly increasing function of x. Thus, each integer $n = \sqrt{x + 1} + \sqrt{x}$ is the function value of exactly one x value, and we may therefore count the integers n instead of the real numbers x. For each positive integer n, we have

$$n = \sqrt{x + 1} + \sqrt{x} \quad \Rightarrow \quad \frac{1}{n} = \sqrt{x + 1} - \sqrt{x};$$

subtracting these equations and squaring leads to

$$x = \frac{1}{4}\left(n - \frac{1}{n}\right)^2.$$

We check that, conversely, if x is given by this expression, then $n = \sqrt{x + 1} + \sqrt{x}$. Therefore, we need to find the number of integers n such that

$$0 < \frac{1}{4}\left(n - \frac{1}{n}\right)^2 \le 1997. \tag{2.17}$$

It is easy to see that $n = 1$ does not work. For $n = 2, 3, \ldots, 89$ we have

$$\frac{1}{4}\left(n - \frac{1}{n}\right)^2 \leq \frac{89^2}{4} < 1997,$$

whereas for $n \geq 90$,

$$\frac{1}{4}\left(n - \frac{1}{n}\right)^2 \geq \frac{(89\frac{1}{2})^2}{4} > 1997.$$

There are therefore 88 positive integers n satisfying the inequalities (2.17), and thus 88 positive real numbers x for which $n = \sqrt{x+1} + \sqrt{x}$ is an integer.

Q159 Write $S(k)$ and $P(k)$ for the sum and product of the digits of an integer k, respectively. We must find the smallest positive integer n for which $S(P(S(n))) \geq 10$. Then $P(S(n))$ must be one of the numbers

$$19, \, 28, \, 29, \, 37, \, 38, \, 39, \, 46, \, 47, \, 48, \, 49, \, 55, \, 56, \, 57, \, 58, \, 59, \, 64, \, \ldots.$$

Now, $P(S(n))$ is the product of the digits of $S(n)$ and each of those digits must be non-zero and smaller than 10. The prime factors of $P(S(n))$ can therefore only be 2, 3, 5 or 7. Therefore, $P(S(n))$ cannot be 19 or 39, for instance, since 19 is prime and 39 has 13 as a prime factor. The possible values for $P(S(n))$ are then

$$28, \, 48, \, 49, \, 56, \, 64, \, \ldots, \, .$$

Since these numbers are greater than 9, $S(n)$ is not a single–digit number. The smallest possible two–digit number for $S(n)$ is then 47 since $4 \times 7 = 28$, and this gives the smallest possible value for n, namely 299999.

Q160 Since the problem was posed as a competition to find as many solutions as possible, no specific solution was published. One very impressive student won the competition, as an editorial of a later issue of *Parabola* announced.

> "*Congratulations [go] to Rui–Jue Tan [...] of Strathfield Girls' High School, who managed to find over 300 answers to problem 1000 of our problems section! A book prize has been sent to her.*"

Q161 Yes, it is. One way to do this is as shown in the diagram below. Keep the circle fixed, and consider an isosceles triangle circumscribed around the circle, with a base tangent to the bottom point of the circle. If the height of the triangle is very large, then the area of the triangle must also be very large, as its base can never be less than the diameter of the circle. If the height begins to decrease, then the area will also begin to decrease; however, at a certain point, the area must reach a minimum value and then begin to increase, because if the height is very small – just slightly bigger than the diameter of the circle – then the base will be large and the area will again be very large. As we vary the height, the area of the triangle decreases to a minimum and then increases again, so there must be many pairs of triangles with different heights and the same areas, and these triangles are not congruent since their base angles differ.

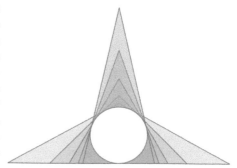

Q162 Let L be the point in the middle of line segment CD and let O be the middle of the line segment ML; that is, the centre of the rectangle.

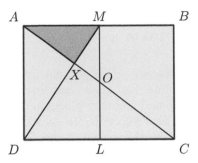

Let a be the area of the rectangle $ABCD$ and note that $\text{area}(ADX) = 4\,\text{area}(MOX)$ since $\triangle ADX$ is similar to $\triangle MOX$ and $|AD| = 2|MO|$. Therefore,

$$
\begin{aligned}
\text{area}(AMX) &= \text{area}(ADM) - \text{area}(ADX) \\
&= \text{area}(ADM) - 4\,\text{area}(MOX) \\
&= \frac{1}{4}\,a - 4\big(\,\text{area}(AMO) - \text{area}(AMX)\big) \\
&= \frac{1}{4}\,a - 4\Big(\frac{1}{8}\,a - \text{area}(AMX)\Big) \\
&= 4\,\text{area}(AMX) - \frac{1}{4}\,a\,.
\end{aligned}
$$

It follows that $\text{area}(AMX) = \dfrac{1}{12}\,a.$

Q163 Pick a player other than Pete in the tournament; let's call that player X_0. We will show that Pete must have beaten X_0 directly.

Since X_0 did not receive a prize, there must have been a player that X_0 could not beat either directly or indirectly; let's call that player X_1. Since X_0 could not beat X_1 directly, X_1 must have beaten X_0 directly. Also, if X_0 directly beat any player Y other than X_1, then X_1 must also have directly beaten Y; otherwise, Y would have directly beaten X_1, and X_0 would have indirectly beaten X_1.

Thus, X_1 has directly beaten everyone that X_0 has directly beaten, and has also directly beaten X_0. If X_1 is Pete, then we are done, so suppose that X_1 is not Pete. By repeating the above argument, we may find a player X_2 who has directly beaten everyone that X_1 has beaten (including X_0), and has also directly beaten X_1. If X_2 is Pete, then we are done; otherwise, we find a player X_3 who has directly beaten everyone that X_2 has beaten (including X_0), and has also directly beaten X_2.

Continuing in this fashion, we must eventually reach Pete. (We cannot cycle in an infinite loop, because each player X_{n+1} has directly beaten more players than the previous player X_n). Thus, Pete has directly beaten X_0.

Q164 The answer is 1001. Note that if the first ticket we buy happens to be 000001, then the first lucky ticket we can get is 001001. Therefore, we need to buy at least 1001 tickets to be sure of getting a lucky ticket.

We now only need to show that 1001 tickets are always sufficient. Write the six–digit number of the first bought ticket as AB where A represents the number formed by the first three digits and B the number formed by the last three. If $A \geq B$, then we can buy

$A - B \leq 1000$ tickets and obtain the lucky ticket AA. If $A < B$, then the purchase of $1001 - B$ tickets leads us to the ticket $A'B'$, where $A' = A + 1$ and $B' = 0$. Then we buy an additional $A + 1$ tickets and obtain the lucky ticket $A'A'$. So, we have our lucky ticket and we have bought $1002 - (B - A)$ tickets. Since $B - A$ is positive, we conclude that 1001 is indeed a sufficient number of tickets to buy.

Q165 It is easy to describe a strategy which will ensure that the first player gains ten points. She makes her first move in the central square of the board and then places her counter in the square symmetric (with respect to the centre of the board) to the square filled by the second player in their previous move. This strategy guarantees that the central row and the central column belong to the first player. Further, all the other rows can be split into pairs of symmetric rows, and we see that each player gets one point from these two rows. The same is true for columns, so the first player gets exactly ten points.

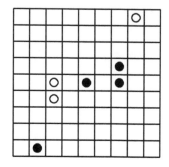

Now we have to prove that the second player is able to play so that he earns at least eight points (since the total number of rows and columns on the board is 18). The strategy is for the second player to achieve a symmetric filling of the board, which, as we have seen, leaves him with the desired eight points. If the first player follows the preceding strategy, then the actions of the second player are of no importance, but, if the first player makes a non–symmetric move, then her opponent should begin to support symmetry. If, at the beginning, the first player makes her first move in a square other than the central square, then the second player can still support the necessary symmetry, and since the last move is made by the first player, she will be compelled to complete the symmetric filling of the board. Thus, we have proved that the answer is ten.

Q166 The cube has six faces and, since the path does not intersect itself, it can include at most six face diagonals. Since the four long diagonals meet at the centre of the cube, the path contains at most one long diagonal. The path must therefore contain at least $8 - 6 - 1 = 1$ edge.

Alternative solution. Paint all the vertices of the cube in two colours, say green and gold, so that each edge joins a green point to a gold point. (Check that this can be done!). The path passing through all vertices must pass through four green vertices and four gold vertices. Since the path is closed, it must therefore pass at least twice either from a green vertex to a gold vertex or from a gold vertex to a green vertex. The path must in other words contain at least two segments joining a green point to a gold point. Such segments are either edges or long diagonals. Since the path does not intersect itself, it contains at most one long diagonal and must therefore contain at least one edge.

Q167 Suppose that one student earned 29 points. Then that student would have exactly 29 numbers in the same position as the leader's. The 30th number would then also be in the same position! Hence, no one can earn 29 points, so each of the other 30 scores must occur. In particular, someone earned 30 points and their sequence was the same as the leader's.

Q168 Assume that no player has two cards with equal numbers. Clearly, the cards numbered 25 do not move – they are fixed forever. After at most two moves, the cards with the number 24 will be fixed as well. Continuing this reasoning, we deduce that the 24 cards with the numbers $25, 24, 23, \ldots, 14$ will all be fixed and one of the cards with a 13 on it will be fixed as well while the other card with a 13 on it will continue to circulate. After at most 24 more moves, the two 13s will be with the same person.

Alternative solution. Let S be the sum of all players' larger numbers. Then S is a non–decreasing function which certainly cannot exceed 650, the sum of all players' numbers. At a certain moment, S must therefore attain its maximum value.

What does this mean? It means that some 25 cards are fixed and the others are moving around the circle. Since 25 is odd, we deduce that there is a number that is written on a fixed card as well as a moving one. Thus, sooner or later, the game will come to an end – that is, one player will have two cards with the same number.

Q169 Imagine that every morning, each pair of people living in the same apartment shake hands. We compare the number of handshakes – that is, the number of pairs of people living in the same apartment – before and after the split–up of an intolerable apartment. Suppose that in the split–up, the n residents of an intolerable apartment move off to n different apartments where the current numbers of residents are $a_1, a_2, \ldots, a_n$. Note that we have $n \geq 15$ and $a_1 + a_2 + \cdots + a_n \leq 119 - n \leq 104$. Then the number of pairs decreases by $C(n, 2) = n(n-1)/2 \geq 105$ and increases by $a_1 + a_2 + \cdots + a_n \leq 104$, so the overall number of pairs decreases. This shows that, whenever there is an intolerable apartment which splits up, the number of pairs of people will decrease by at least 1. Clearly, this cannot go on for ever, so, eventually, there will be no remaining intolerable apartments.

Comment. If just one more person moves into the block, so that there are now 120 people, then it would be possible that the split–up of an intolerable apartment *always* creates a new intolerable apartment. In Problem 324, you are invited to construct a scenario in which this happens.

Q170 Label points as shown in the diagram; the point O is the centre of the star and of both pentagons. It is clear that the area of the star is five times the area of the quadrilateral $ACOD$.

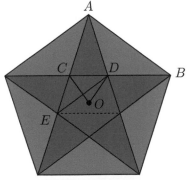

Since each angle of a regular pentagon is 108°, looking at the two lines intersecting at D shows that $\angle ADB = 108°$; by symmetry, $|DA| = |DB|$, so $\angle DAB = \angle DBA = 36°$. A similar argument in $\triangle CDE$ shows that $\angle DEC = 36°$. Furthermore, $\angle ACD = \angle ADC = 72°$ and so $\angle CAD = 36°$. This means that the triangles $\triangle DAB$ and $\triangle DAE$ have the same angles; they also have a common side DA, so they are congruent and have the same area. However, the triangles $\triangle CDO$ and $\triangle CDE$ have the same base CD; it is clear that E is further from CD than O is; so, the latter has the greater area. Adding $\triangle ACD$ to both shows that $\triangle ADE$ has greater area than $ACOD$.

Putting all this back together, the large pentagon consists of five regions congruent to $ACOD$ and five congruent to $\triangle DAB$; so, its area is equal to five times that of $ACOD$ and five times that of $\triangle DAE$, which is larger than ten times that of $ACOD$. So, the area of the pentagon is more than twice that of the star.

Q171 The standard approach to solving a "functional equation" problem is to begin by determining possible values of $f(n)$ for small values of n, and proving some simple properties. Once enough is known about f, it may be possible to guess some more specific properties of f and then to prove them, often by using induction.

Step 1: $f(1) > 1$. Proof: suppose that $f(1) = 1$; then $3 = f(f(1)) = f(1) = 1$, a contradiction.

Step 2: $f(n) > n$ for all n. Proof by induction: we know already that this is true for $n = 1$. Suppose it is true for some specific n: then $f(n) > n$; since $f(n)$ and n are both integers, we have in fact $f(n) \geq n+1$. From the given information, $f(n+1) > n+1$. If needed, then more information about *mathematical induction* can be found in Section 3.8.

Step 3: $f(1) = 2$. Proof: if $f(1) = n \geq 3$, then $3 = f(f(1)) = f(n) > n \geq 3$, a contradiction. Hence, $f(1) = 2$.

We can now calculate a few values of $f(n)$,

$$f(2) = f(f(1)) = 3, \qquad f(3) = f(f(2)) = 6,$$
$$f(6) = f(f(3)) = 9, \qquad f(9) = f(f(6)) = 18,$$
$$f(18) = f(f(9)) = 27, \qquad f(27) = f(f(18)) = 54,$$

and we notice that there seems to be a pattern in the values of $f(n)$ when n is a power of 3 ($n = 1, 3, 9, 27$) or twice a power of 3 ($n = 2, 6, 18$). We guess that

$$f(3^m) = 2 \times 3^m \quad \text{and} \quad f(2 \times 3^m) = 3 \times 3^m = 3^{n+1}.$$

Although this is only a guess, we'll treat it as true and use it as a basis for further guesses. Once we are convinced that we have the right answer, we could prove all our guesses by induction. In fact, this turns out to be quite routine: we shall omit the details and leave them to the diligent reader.

To continue, recall that $f(n)$ always increases as n increases. In particular, the values of $f(3^m), f(3^m + 1), \ldots, f(2 \times 3^m - 1)$ must be chosen in increasing order from the values $2 \times 3^m, 2 \times 3^m + 1, \ldots, 3 \times 3^m - 1$. But since

$$2 \times 3^m - 3^m = 3^m = f(2 \times 3^m) - f(3^m),$$

the two sets of values just mentioned have the same number of elements, and so there is only one possible choice:

$$f(3^m + k) = 2 \times 3^m + k \quad \text{for} \quad k = 0, 1, \ldots, 3^m - 1. \tag{2.18}$$

To illustrate, take $m = 2$. We have guessed that $f(9) = 18$ and $f(18) = 27$; so the nine values $f(9), f(10), \ldots, f(17)$ must be chosen, in increasing order, from the nine values $18, 19, \ldots, 26$: there is only one option.

Unfortunately, the same idea does not work for remaining values of n: for instance, the nine values $f(18), f(19), \ldots, f(26)$ will come from the twenty-seven numbers $27, 28, \ldots, 53$, and it is not immediately clear which are the correct values. However, we note that $18, 19, \ldots, 26$ were found as certain values of $f(n)$ in the previous paragraph, and we have, for example,

$$19 = f(10) \quad \Rightarrow \quad f(19) = f(f(10)) = 3 \times 10 = 30.$$

In general, from equation (2.18) – even though it is still a guess! – we make the further guess that, if $k = 0, 1, \ldots, 3^m - 1$, then

$$f(2 \times 3^m + k) = f(f(3^m + k)) = 3^{m+1} + 3k.$$

A little thought should make it clear that we have now determined – at present, without proof – the value of $f(n)$ for every positive integer n. For any integer $m \geq 0$ and any $k = 0, 1, \ldots, 3^m - 1$, our guess is specified by the two equations

$$f(3^m + k) = 2 \times 3^m + k \quad \text{and} \quad f(2 \times 3^m + k) = 3^{m+1} + 3k. \tag{2.19}$$

We leave the (straightforward) proof by induction as an exercise.

We may also note that these formulae can be expressed very neatly by writing values of n in base 3 – perhaps not a great surprise, in view of the prominence of powers of 3 in the solution. If $3^m \leq n < 3^{m+1}$, then n can be represented as a base 3 numeral with $m + 1$ digits, $n = [d_m d_{m-1} \cdots d_1 d_0]_3$; the two formulae of (2.19) give the value of $f(n)$ in the two cases $d_m = 1$, $d_m = 2$, and can be written

$$f([1d_{m-1} \cdots d_1 d_0]_3) = [2d_{m-1} \cdots d_1 d_0]_3,$$
$$f([2d_{m-1} \cdots d_1 d_0]_3) = [1d_{m-1} \cdots d_1 d_0 0]_3.$$

To answer the question posed, we convert 2001 into base 3, yielding the result

$$f(2001) = f([2202010]_3) = [12020100]_3 = 3816.$$

Q172 We can used the standard factorisations

$$x^3 - 1 = (x - 1)(x^2 + x + 1) \quad \text{and} \quad x^3 + 1 = (x + 1)(x^2 - x + 1)$$

to write the required product P as

$$\frac{(2 - 1)(2^2 + 2 + 1)}{(2 + 1)(2^2 - 2 + 1)} \frac{(3 - 1)(3^2 + 3 + 1)}{(3 + 1)(3^2 - 3 + 1)} \cdots \frac{(2002 - 1)(2002^2 + 2002 + 1)}{(2002 + 1)(2002^2 - 2002 + 1)}.$$

Since
$$x^2 + x + 1 = (x + 1)^2 - (x + 1) + 1,$$

each quadratic in the numerator cancels with the one in the denominator of the following fraction, leaving only the last quadratic in the numerator, the first in the denominator, and the linear terms,

$$P = \frac{(2 - 1)(3 - 1)(4 - 1) \cdots (2001 - 1)(2002 - 1)}{(2 + 1)(3 + 1)(4 + 1) \cdots (2001 + 1)(2002 + 1)} \frac{2002^2 + 2002 + 1}{2^2 - 2 + 1}.$$

It is now clear that $2 + 1$ cancels with $4 - 1$ and so on, and a little calculation gives the final result

$$P = \frac{(2 - 1)(3 - 1)}{(2001 + 1)(2002 + 1)} \frac{2002^2 + 2002 + 1}{2^2 - 2 + 1} = \frac{1336669}{2005003}.$$

Q173 The first inequality, and its method of proof, are moderately well known. For any positive x, y, we have $(x + y)^2 = (x - y)^2 + 4xy \geq 4xy$; dividing both sides by $x + y$ and doing a little algebra,

$$\frac{1}{x} + \frac{1}{y} = \frac{x + y}{xy} \geq \frac{1}{xy} \frac{4xy}{x + y} = \frac{4}{x + y}.$$

For the second, let $a_1, a_2, \ldots, a_n$ be positive, and introduce another positive number x. Using the first inequality repeatedly, we have

$$\frac{1}{x} + \frac{1}{a_1} \geq \frac{4}{x + a_1}$$

$$\frac{1}{x} + \frac{1}{a_1} + \frac{4}{a_2} \geq 4\left(\frac{1}{x + a_1} + \frac{1}{a_2}\right) \geq \frac{4^2}{x + a_1 + a_2}$$

$$\frac{1}{x} + \frac{1}{a_1} + \frac{4}{a_2} + \frac{4^2}{a_3} \geq 4^2\left(\frac{1}{x + a_1 + a_2} + \frac{1}{a_3}\right) \geq \frac{4^3}{x + a_1 + a_2 + a_3}$$

and so on; eventually we reach

$$\frac{1}{x} + \frac{1}{a_1} + \frac{4}{a_2} + \frac{4^2}{a_3} + \cdots + \frac{4^{n-1}}{a_n} \geq \frac{4^n}{x + a_1 + a_2 + a_3 + \cdots + a_n};$$

that is,

$$\frac{1}{a_1} + \frac{4}{a_2} + \frac{4^2}{a_3} + \cdots + \frac{4^{n-1}}{a_n} \geq \frac{4^n}{x + a_1 + a_2 + a_3 + \cdots + a_n} - \frac{1}{x}.$$

This is true for any positive x, so we choose $x = a_1 + a_2 + \cdots + a_n$, giving

$$\frac{1}{a_1} + \frac{4}{a_2} + \frac{4^2}{a_3} + \cdots + \frac{4^{n-1}}{a_n} \geq \frac{4^n}{2(a_1 + a_2 + a_3 + \cdots + a_n)} - \frac{1}{a_1 + a_2 + \cdots + a_n};$$

and the right–hand side simplifies to yield the required result,

$$\frac{1}{a_1} + \frac{4}{a_2} + \frac{4^2}{a_3} + \cdots + \frac{4^{n-1}}{a_n} \geq \frac{2 \times 4^{n-1} - 1}{a_1 + a_2 + a_3 + \cdots + a_n}.$$

Q174 Since there was an outright winner, the final round of the competition must have consisted of one match, and no contestants with byes: three players in all. One of these three was the overall winner; we are told that this player played only one match, and therefore must have had a bye in every previous round. Suppose that some round, not the first, involved n players (including those with byes). Then these players must have been either $n - 2$ winners and 2 byes from the previous round, or $n - 1$ winners and 1 bye: hence, the previous round must have involved $3(n - 2) + 2 = 3n - 4$ or $3(n - 1) + 1 = 3n - 2$ players. So, we can calculate the possible numbers of players in each round by working backwards:

- in the last round, 3 players for certain;
- in the second–last round, either 5 players or 7;
- in the third–last, $11, 13, 17$ or 19;
- before that, $29, 31, 35, 37, 47, 49, 53$ or 55;
- before that, $83, 85, 89, 91, \ldots$

…and we stop here since we were told that there were fewer than 100 entrants. So the greatest possible number of players in the tournament was 91.

Q175 Our expression S is the sum of 668 fractions; since 668 is even, we can add up these fractions in 334 pairs: the first with the last, the second with the second last and so on. The sum of the kth and the kth last fraction is

$$\frac{1}{(3k)(3k - 1)(3k - 2)} + \frac{1}{(2007 - 3k)(2006 - 3k)(2005 - 3k)}$$

$$= \frac{(2007 - 3k)(2006 - 3k)(2005 - 3k) + (3k)(3k - 1)(3k - 2)}{(3k)(3k - 1)(3k - 2)(2007 - 3k)(2006 - 3k)(2005 - 3k)}.$$

Now consider the numerator N of this fraction modulo 2005. (For an introduction to *modular arithmetic*, see Section 3.3.) We have

$$\begin{aligned}
N &= (2007 - 3k)(2006 - 3k)(2005 - 3k) + (3k)(3k - 1)(3k - 2) \\
&\equiv (2 - 3k)(1 - 3k)(-3k) + (3k)(3k - 1)(3k - 2) \quad (\text{mod } 2005) \\
&\equiv -(3k - 2)(3k - 1)(3k) + (3k)(3k - 1)(3k - 2) \quad (\text{mod } 2005) \\
&\equiv 0 \quad (\text{mod } 2005),
\end{aligned}$$

which means that the numerator is a multiple of 2005. If we multiply the fraction by 2004!, then all the factors in the denominator will cancel and we shall get an integer; this integer will be a multiple of the previous numerator and hence a multiple of 2005. Finally, 2004! S is then the sum of 334 integers, each a multiple of 2005, and therefore is itself a multiple of 2005.

Q176 We can rewrite the equation as

$$x \left(\frac{1}{2004} - \frac{1}{2003} + \frac{1}{2002} - \frac{1}{2001} + \cdots + \frac{1}{2} - \frac{1}{1} \right)$$
$$= \frac{1}{2004} - \frac{2}{2003} + \frac{3}{2002} - \frac{4}{2001} + \cdots + \frac{2003}{2} - \frac{2004}{1}.$$

Performing the subtraction of each pair of fractions yields

$$x \left(-\frac{1}{2004 \times 2003} - \frac{1}{2002 \times 2001} - \cdots - \frac{1}{2 \times 1} \right)$$
$$= -\frac{2005}{2004 \times 2003} - \frac{2005}{2002 \times 2001} - \cdots - \frac{2005}{2 \times 1},$$

and so $x = 2005$.

   Alternative solution. A more insightful solution runs as follows. By looking at the equation and *not* actually using algebra to simplify it, but envisaging what would happen *if* we did, we can see that despite its complicated appearance, this is just a linear equation of the form $ax = b$ with $a \neq 0$. Therefore, it will have exactly one solution. If we spot that substituting $x = 2005$ will make each side the sum of 1002 terms, each equal to 1, then we are finished: the only solution is $x = 2005$.

   Of course, it is much easier to find this argument with hindsight as an "alternative" solution than it is to see it in the first place. Moral of the story: if a problem has an intricate solution ending up with a simple answer, then see if you can find a simple way of seeing the answer without going through the complications!

Q177 First, without trigonometry. Let A and B be Annie's position and Brian's position, respectively. Annie spots the sunset when to her eyes the sun is at P, while Brian spots the sunset at Q. The time difference is the time it takes the earth to spin through the arc PQ.

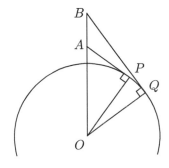

Since the distances from their positions to the surface of the earth are very much smaller than the radius R of the earth, the length of this arc can be approximated by

$$|PQ| \approx |BQ| - |AP|.$$

In units of kilometres, we have $R = 40000/2\pi \approx 6366.1977$; the straight–line distances from A and B to their horizons are

$$|AP| = \sqrt{OA^2 - OP^2} = \sqrt{\left(R + \frac{1}{100}\right)^2 - R^2},$$

$$|BQ| = \sqrt{OB^2 - OQ^2} = \sqrt{\left(R + \frac{4}{100}\right)^2 - R^2}.$$

Now, if x is very much larger than y, then we have the approximation

$$\sqrt{(x+y)^2 - x^2} = \sqrt{2xy + y^2} \approx \sqrt{2xy}.$$

Applying this to the above, we have

$$|AP| \approx \sqrt{\frac{2R}{100}} \approx 11.283792, \quad |BQ| \approx \sqrt{\frac{8R}{100}} \approx 22.567584;$$

so $|PQ| \approx 11.283792$. The earth spins through 40000km in one day; that is, $24 \times 60 \times 60$ seconds; therefore, it spins through 11.283792km in approximately 24.372991 seconds. This is the delay between Annie's moment of sunset and Brian's.

Comment. Note the "coincidence" that our value for $|PQ|$ is just the same as our value for $|AP|$. Of course, this is really no coincidence; and it is not exactly true anyway. It is a consequence of the approximations we made in calculating square roots; readers are invited to go through the argument carefully and find precisely where this (non–)coincidence came from.

Now we solve the problem exactly. Referring once again to the above diagram, we have

$$\theta = \angle POQ = \angle BOQ - \angle AOP = \arccos \frac{|OB|}{|OQ|} - \arccos \frac{|OP|}{|OA|}$$

$$= \arccos \frac{r}{r + 40} - \arccos \frac{r}{r + 10},$$

with the earth's radius r approximately equal to $40000000/2\pi = 6366198$ (in metres this time, just for a change). Then, the time between the two sightings is $t = \theta/2\pi$ days. Doing the calculations gives $\theta = 0.00177245$ and time

$$\frac{\theta}{2\pi} \times 24 \times 60 \times 60 = 24.372937 \text{ seconds.}$$

A note on approximations. A calculation can only be as good as the data it is based on. Since we have stated the circumference of the earth to one significant figure (and never mind the fact that the circumference is not the same in all directions!), it is really rather ridiculous to give an answer to eight significant figures. Our reason for doing so was to show, firstly, that our two answers really are different; and, secondly, that the approximate answer is very close to the true answer. A sensible response to the question would be that the time between Annie's sunset and Brian's is about 24 seconds.

Q178 The total length of 1000 pieces at 70 cm and 2000 pieces at 50 cm is 170000 cm. Since $170000/740 > 229$, the builder certainly cannot buy fewer than 230 bars. The naive approach of cutting 10 pieces of length 70 cm from some bars and 14 pieces of length 50 cm from others, throwing away the leftovers, requires $100 + 143 = 243$ bars.

Suppose that from each bar we cut x pieces of length 70 cm and y pieces of length 50 cm. To do this without any wastage, x and y must be integers satisfying the equation $7x + 5y = 74$. Section 3.2 explains the technique for solving this kind of problem; we leave the reader to follow the procedure expounded there and thus show that

$$x = 2 + 5t, \quad y = 12 - 7t$$

where t is an integer. Since x and y must be non–negative, we have two possibilities,

$$x = 2,\ y = 12 \qquad \text{or} \qquad x = 7,\ y = 5,$$

and so a bar can be cut in two ways:

- first way: 2 pieces of length 70 cm and 12 pieces of length 50 cm;
- second way: 7 pieces of length 70 cm and 5 pieces of length 50 cm.

Now let u be the number of metal bars to be cut in the first way, and v the number of bars to be cut in the second way. Then we would like

$$2u + 7v = 1000$$
$$12u + 5v = 2000.$$

As it stands, this is impossible, since the system has a non–integer solution $u \approx 121.62$, $v \approx 108.11$. As a first attempt, we will cut 121 bars in the first way and 108 bars in the second way, thereby obtaining 998 pieces of length 70 cm and 1992 pieces of length 50 cm. We still require 2 pieces of length 70 cm and 8 pieces of length 50 cm, total 540 cm, which can clearly be cut from one further bar. By doing this we use $121 + 108 + 1 = 230$ bars, the least number possible.

Comment. We have certainly found a solution which cannot be bettered, but we might ask if there are any other solutions which are equally good. In particular, we note that 200 cm of the last bar will be wasted, and this suggests that perhaps we did not need to be so zealous in avoiding wastage by concentrating on our two ways of using up the whole bar. In fact, another option using 230 bars is to cut 110 bars by the first method and 100 by the second; then to take another 20 bars and cut each into four 70 cm pieces and nine 50 cm pieces, with 10 cm wasted.

Q179 If α is the angle at A, then

$$\frac{\text{area}(AFE)}{\text{area}(ABC)} = \frac{\frac{1}{2}|AE||AF|\sin\alpha}{\frac{1}{2}|AB||AC|\sin\alpha} = \frac{|AE||AF|}{|AB||AC|}; \tag{2.20}$$

and there are similar formulae involving the triangles with vertices at B and C. Now, the Arithmetic–Geometric Mean Inequality, Section 3.11, can be written

$$xy \le \left(\frac{x+y}{2}\right)^2$$

for positive x, y; taking $x = |AF|$ and $y = |BF|$ shows that

$$|AF||BF| \le \left(\frac{|AF| + |BF|}{2}\right)^2 = \frac{(AB)^2}{4};$$

that is,

$$\frac{|AF||BF|}{|AB|^2} \leq \frac{1}{4}.$$

Doing the same for terms involving D and E, and multiplying the three results, we obtain

$$\frac{|AF||BF|}{|AB|^2}\frac{|BD||CD|}{|BC|^2}\frac{|CE||AE|}{|CA|^2} \leq \left(\frac{1}{4}\right)^3.$$

But the left–hand side is just a rearrangement of the area ratio (2.20) and the two similar ratios referred to! Therefore,

$$\frac{|AE||AF|}{|AB||AC|}\frac{|BF||BD|}{|BC||BA|}\frac{|CD||CE|}{|CA||CB|} \leq \left(\frac{1}{4}\right)^3,$$

and so at least one of the three area ratios must be $\frac{1}{4}$ or less.

Q180 We begin by rewriting the given equation as

$$ax^3 + bx^2 + cx = -(x^4 + 1), \tag{2.21}$$

and noting that $x = 0$ is certainly not a solution. Now by the Cauchy–Schwartz Inequality (see Section 3.11), we have

$$(ax^3 + bx^2 + cx)^2 \leq (a^2 + b^2 + c^2)(x^6 + x^4 + x^2).$$

Therefore, if x is a real solution of equation (2.21), then

$$(x^4 + 1)^2 = (ax^3 + bx^2 + cx)^2 \leq (a^2 + b^2 + c^2)(x^6 + x^4 + x^2),$$

and so

$$a^2 + b^2 + c^2 \geq \frac{(x^4 + 1)^2}{x^6 + x^4 + x^2}.$$

On the other hand, by factorising polynomials, we have

$$\begin{aligned}
3(x^4 + 1)^2 - 4(x^6 + x^4 + x^2) &= 3x^8 - 4x^6 + 2x^4 - 4x^2 + 3 \\
&= (x^2 - 1)(3x^6 - x^4 + x^2 - 3) \\
&= (x^2 - 1)^2(3x^4 + 2x^2 + 3) \\
&\geq 0;
\end{aligned}$$

rearranging this inequality gives

$$\frac{(x^4 + 1)^2}{x^6 + x^4 + x^2} \geq \frac{4}{3},$$

and so $a^2 + b^2 + c^2 \geq \frac{4}{3}$, as claimed.

Q181 There are at least two convenient ways to express the fact that three numbers x, y, z are in arithmetic progression:

(a) $x = y - k$ and $z = y + k$ for some constant k;
(b) $x - 2y + z = 0$.

Suppose that the numbers a, b, c are in arithmetic progression and so are $\sin^2 a$, $\sin^2 b$, $\sin^2 c$. Applying (a) to the former and (b) to the latter, we seek to solve

$$\sin^2(b - k) - 2\sin^2 b + \sin^2(b + k) = 0.$$

Now simplify the left–hand side, beginning with the "sine–of–a–sum" formula from Section 3.7. This gives

$$\sin^2(b - k) + \sin^2(b + k) = (\sin b \cos k - \cos b \sin k)^2 + (\sin b \cos k + \cos b \sin k)^2$$
$$= 2\sin^2 b \cos^2 k + 2\cos^2 b \sin^2 k$$

and hence

$$\sin^2(b - k) - 2\sin^2 b + \sin^2(b + k) = 2\sin^2 b(\cos^2 k - 1) + 2\cos^2 b \sin^2 k$$
$$= 2\sin^2 k(\cos^2 b - \sin^2 b).$$

This expression is equal to zero, as required, if and only if

$$\sin k = 0 \quad \text{or} \quad \cos^2 b = \sin^2 b;$$

that is, if and only if

$$k = n\pi \quad \text{or} \quad b = \frac{(2n + 1)\pi}{4}$$

for some integer n. The former gives the arithmetic progression

$$b = a + n\pi, \quad c = a + 2n\pi \tag{2.22}$$

with integer n; the latter gives any arithmetic progression at all, as long as the middle term b has the indicated values.

Comment. The solution (2.22) is not particularly interesting, as it gives the arithmetic progression $\sin^2 a$, $\sin^2 a$, $\sin^2 a$ in which the difference between successive terms is zero. The second solution, on the other hand, is very surprising, as it shows that with the stated value for b, it does not matter what a and c are! – that is, so long as a, b, c are in arithmetic progression.

Q182 It's given that

$$f(0) = c \quad \text{and} \quad f(1) = a + b + c \quad \text{and} \quad f(2) = 4a + 2b + c$$

are all integers. As a consequence, $a + b = (a + b + c) - c$, the difference of two integers, is also an integer, and so is $2a = (4a + 2b + c) - 2(a + b + c) + c$.

If $2a$ is an even integer, then a is an integer, so b is an integer, and so $f(x)$ is an integer for all integers x.

On the other hand, suppose that $2a$ is an odd integer, $2a = 2m + 1$ for some integer m. Write $a + b = n$; we know already that n is an integer. Then we have

$$f(x) = \left(m + \frac{1}{2}\right)x^2 + \left(n - m - \frac{1}{2}\right)x + c = mx^2 + (n - m)x + c + \frac{x(x - 1)}{2}.$$

If x is any integer, then either x or $x - 1$ is even, so $x(x - 1)/2$ is an integer; and therefore, so is $f(x)$.

Question. Could we have reached the same conclusion about $f(x)$ if, instead of being told that $f(x)$ is an integer for $x = 0, 1, 2$, then this was given for three other integers? Or for four or more (specific) integers? These questions are pursued in Problem 323.

Q183 Let A be the starting point, B the point x km due south of A; let C be x km due east of B and let D be x km due north of C. Further, let E be the centre of the circle of latitude through A and D – that is, $1°$ south – and let F be the centre of the circle through B and C. The diagram shows a section through A, B and the north and south poles.

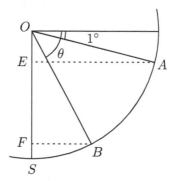

Since the difference in longitude between A and D is equal to that between B and C, and since the arc AD of the circle centred at E is three times the arc BC of the circle centred at F, the radius $|EA|$ is three times the radius $|FB|$. Writing r for the radius of the earth, we see from the diagram that

$$|EA| = r \cos 1°, \quad |FB| = r \cos \theta$$

where θ is the latitude of B, and so

$$\theta = \arccos\left(\frac{1}{3} \cos 1°\right).$$

Now, x is the length of the arc AB, and is given by

$$x = \frac{\theta - 1}{360} 2\pi r.$$

Assuming the radius of the earth to be 6371 kilometres and doing the calculations, we obtain $\theta \approx 70.53°$ and $x \approx 7732$ kilometres.

Comment. A section of the earth through the poles is not actually a circle: the "radius" at the equator is greater than it is at the poles. Taking this into account would have made the problem much harder!

Q184 Suppose that f satisfies the required equation,

$$f\left(\frac{x-3}{1+x}\right) + f\left(\frac{x+3}{1-x}\right) = x \tag{2.23}$$

for $x \neq \pm 1$. To simplify the first term, substitute

$$t = \frac{x-3}{1+x}.$$

Making x the subject of this equation gives $x = (t+3)/(1-t)$ and, hence,

$$f(t) + f\left(\frac{t-3}{1+t}\right) = \frac{t+3}{1-t}.$$

Alternatively, we can simplify the second term by taking $t = (x+3)/(1-x)$; then we get $x = (t-3)/(1+t)$ and

$$f\left(\frac{t+3}{1-t}\right) + f(t) = \frac{t-3}{1+t}.$$

Adding these last two equations,

$$f(t) + f\left(\frac{t-3}{1+t}\right) + f\left(\frac{t+3}{1-t}\right) + f(t) = \frac{t+3}{1-t} + \frac{t-3}{1+t}$$

and, amazingly, the complicated terms on the left–hand side can be entirely eliminated by using the given equation (2.23). Doing so and simplifying the right–hand side gives

$$2f(t) + t = \frac{8t}{1-t^2}$$

and therefore

$$f(t) = \frac{4t}{1-t^2} - \frac{t}{2}.$$

Q185 This solution uses the Sine Rule, the Cosine Rule, and an area formula for triangles, all of which can be found in Section 3.7. Let $a = |BC|$, $b = |AC|$, $c = |AB|$ and $\alpha = \angle BAC$, $\beta = \angle ABC$. Then

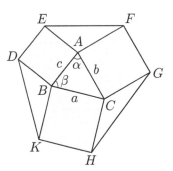

$$\text{area}(ABC) = \frac{1}{2}bc\sin\alpha \quad \text{and} \quad \text{area}(AEF) = \frac{1}{2}bc\sin\alpha,$$

noting that $\angle EAF = \pi - \alpha$. Since $a/\sin\alpha = b/\sin\beta$, we also have

$$\text{area}(BDK) = \frac{1}{2}ac\sin\beta = \frac{1}{2}bc\sin\alpha,$$

and similarly,

$$\text{area}(CHG) = \frac{1}{2}bc\sin\alpha.$$

If we also note that $a^2 = b^2 + c^2 - 2bc\cos\alpha$ by the Cosine Rule, then we find that the area of the hexagon is given by

$$\begin{aligned}
S &= \text{area}(DEFGHK) \\
&= \text{area}(AEDB) + \text{area}(ACGF) + \text{area}(BCHK) \\
&\quad + \text{area}(ABC) + \text{area}(AEF) + \text{area}(BDK) + \text{area}(CHG) \\
&= a^2 + b^2 + c^2 + 2bc\sin\alpha \\
&= 2(b^2 + c^2) + 2bc(\sin\alpha - \cos\alpha) \\
&= 2(b^2 + c^2) + 2bc\sin\left(\alpha - \frac{\pi}{4}\right).
\end{aligned}$$

Since b and c are fixed, S has its maximum value when $\sin(\alpha - \frac{\pi}{4}) = 1$, which is the case when

$$\alpha = \frac{\pi}{2} + \frac{\pi}{4} = \frac{3\pi}{4} = 135°.$$

Comment. It is, perhaps, surprising that this conclusion does not depend on the lengths b and c, or at least on their ratio.

Q186 Let A, B, C be the three angles of a triangle, so that $A + B + C = \pi$. Write

$$S = \cos^2 A + \cos^2 B + \cos^2 C.$$

Using trigonometric formulae from Section 3.7, we have

$$\cos A = \cos(\pi - B - C) = -\cos(B + C);$$

treating $\cos B$ and $\sin C$ similarly, and using the "cosine–of–a–sum" and "sine–of–a–sum" formulae gives

$$\cos A = \sin B \sin C - \cos B \cos C$$
$$\cos B = \sin C \sin A - \cos C \cos A$$
$$\sin C = \sin A \cos B + \cos A . \sin B.$$

Now, take $\cos A$ times the first equation, add $\cos B$ times the second equation, and subtract $\sin C$ times the third. Many terms on the right–hand side cancel, and we obtain

$$\cos^2 A + \cos^2 B - \sin^2 C = -2 \cos A \cos B \cos C;$$

adding 1 to both sides then gives

$$\cos^2 A + \cos^2 B + \cos^2 C = S = 1 - 2 \cos A \cos B \cos C.$$

On the other hand, the Arithmetic–Geometric Mean Inequality (Section 3.11) gives

$$\frac{\cos^2 A + \cos^2 B + \cos^2 C}{3} \geq \left(\cos^2 A \cos^2 B \cos^2 C\right)^{1/3}.$$

If we write $P = (\cos A \cos B \cos C)^{1/3}$, then what we have found can be written as

$$S = 1 - 2P^3 \quad \text{and} \quad S \geq 3P^2;$$

substituting the former into the latter and rearranging leads to

$$2P^3 + 3P^2 - 1 \leq 0.$$

Now, a "randomly chosen" cubic will usually not factorise neatly: but here we are lucky: the last inequality can be written as

$$(P + 1)^2 (2P - 1) \leq 0,$$

and so we have $P \leq \frac{1}{2}$. Substituting back into $S = 1 - 2P^3$ gives $S \geq \frac{3}{4}$, which is what we wished to show.

Q187 By using the Law of Sines from Section 3.7,

$$\frac{a}{\sin \alpha} = \frac{b}{\sin \beta} = 2R,$$

where R is the radius of the circumscribed circle, we have

$$\frac{a + b}{a - b} = \frac{2R(\sin \alpha + \sin \beta)}{2R(\sin \alpha - \sin \beta)} = \frac{\sin \alpha + \sin \beta}{\sin \alpha - \sin \beta}.$$

We can now expand the numerator and denominator of this fraction by means of the sums–to–products formulae for trigonometric functions (Section 3.7) to obtain

$$\frac{a + b}{a - b} = \frac{2 \sin \frac{\alpha+\beta}{2} \cos \frac{\alpha-\beta}{2}}{2 \sin \frac{\alpha-\beta}{2} \cos \frac{\alpha+\beta}{2}} = \frac{\tan \frac{\alpha+\beta}{2}}{\tan \frac{\alpha-\beta}{2}}.$$

Q188 Write
$$a^2 = \tan A, \quad b^2 = \tan B, \quad c^2 = \tan C, \quad d^2 = \tan D,$$

where $0 < A, B, C, D < \frac{\pi}{2}$, and note that

$$\frac{1}{1 + \tan^2 \theta} = \frac{1}{\sec^2 \theta} = \cos^2 \theta.$$

Then the given identity becomes

$$\cos^2 A + \cos^2 B + \cos^2 C + \cos^2 D = 1$$

and therefore
$$\sin^2 A = 1 - \cos^2 A = \cos^2 B + \cos^2 C + \cos^2 D. \tag{2.24}$$

The arithmetic–geometric mean inequality, Theorem 3.15, in the case $n = 3$ states that

$$\frac{x_1 + x_2 + x_3}{3} \geq \sqrt[3]{x_1 x_2 x_3}$$

for any positive x_1, x_2, x_3; applying this to equation (2.24) yields

$$\sin^2 A \geq 3(\cos^2 B \, \cos^2 C \, \cos^2 D)^{1/3}.$$

Similarly, we have

$$\sin^2 B \geq 3(\cos^2 C \, \cos^2 D \, \cos^2 A)^{1/3}$$
$$\sin^2 C \geq 3(\cos^2 D \, \cos^2 A \, \cos^2 B)^{1/3}$$
$$\sin^2 D \geq 3(\cos^2 A \, \cos^2 B \, \cos^2 C)^{1/3}.$$

Multiplying all four inequalities gives

$$\sin^2 A \, \sin^2 B \, \sin^2 C \, \sin^2 D \geq 3^4 \cos^2 A \, \cos^2 B \, \cos^2 C \, \cos^2 D,$$

implying
$$\tan^2 A \, \tan^2 B \, \tan^2 C \, \tan^2 D \geq 3^4$$

and therefore $abcd \geq 3$.

Q189 Let the side opposite the angle γ be c and the other sides a, b. By using an area formula and the Cosine Rule from Section 3.7, we have

$$S = \frac{1}{2} ab \sin \gamma$$

and
$$c^2 = a^2 + b^2 - 2ab \cos \gamma.$$

We can use the area formula to rewrite the latter as

$$c^2 = (a - b)^2 + 2ab(1 - \cos \gamma) = (a - b)^2 + 4S \frac{1 - \cos \gamma}{\sin \gamma}.$$

Since S and γ are fixed, c^2 (and hence c) is minimal when $a = b$; that is, when the triangle is isosceles. In this case, we have

$$c = 2\sqrt{S \frac{1 - \cos \gamma}{\sin \gamma}},$$

and from the area formula,

$$a = b = \sqrt{\frac{2S}{\sin \gamma}}.$$

Q190 We may assume that the five numbers are labelled in such a way that

$$a_1 < a_2 < \cdots < a_5.$$

Let $x > 0$ be such that x^2 is the smallest value of $(a_j - a_i)^2$ with $a_i \neq a_j$. Then for $i = 1, 2, 3, 4$ we have

$$a_{i+1} - a_i \geq x,$$

so that, for $j > i$,

$$a_j - a_i = (a_j - a_{j-1}) + (a_{j-1} - a_{j-2}) + \cdots + (a_{i+1} - a_i) \geq (j - i)x,$$

and hence

$$(a_j - a_i)^2 \geq (j - i)^2 x^2. \tag{2.25}$$

Now let S be the sum of all the expressions $(a_j - a_i)^2$ with $1 \leq i < j \leq 5$. The sum of all $(j - i)^2$ for these values is

$$(1^2 + 2^2 + 3^2 + 4^2) + (1^2 + 2^2 + 3^2) + (1^2 + 2^2) + 1^2 = 50,$$

and so the inequality (2.25) gives

$$S \geq 50x^2.$$

On the other hand, we can evaluate S in the following way. We begin with the expansion

$$(a_j - a_i)^2 = a_j^2 - 2a_i a_j + a_i^2. \tag{2.26}$$

If we add up all these expressions for $1 \leq i < j \leq n$, then we get every possible term $-2a_i a_j$ with $i < j$. We also get all the square terms four times each, because each a_i occurs in a term like (2.26) once with every other a_j. Therefore,

$$S = 4(a_1^2 + \cdots + a_5^2) - 2T = 4 - 2T,$$

where T is the sum of all $a_i a_j$ terms. On the other hand, expanding a square gives

$$(a_1 + \cdots + a_5)^2 = (a_1^2 + \cdots + a_5^2) + 2T = 1 + 2T;$$

combining these last two results gives

$$S = 5 - (a_1 + \cdots + a_5)^2 \leq 5,$$

and so $50x^2 \leq 5$. Therefore, $x^2 \leq \frac{1}{10}$.

Comment. A similar argument will show that, if we have n different numbers $a_1, a_2, \ldots, a_n$ with $a_1^2 + a_2^2 + \cdots + a_n^2 = 1$, then there exist distinct numbers a_j, a_i such that

$$(a_j - a_i)^2 \leq \frac{12}{n(n-1)(n+1)}.$$

Readers may care to supply details for themselves.

Q191 If Alan's statement is true, then Alan is younger than Betty (because he has told the truth) and older than Chris (because the statement is true). If Alan's statement is false, then Alan is older than Betty and younger than Chris. So these three must be, in increasing order of ages, Chris, Alan, Betty or Betty, Alan, Chris. Now, Debbie's statement cannot be true as she would then have told the truth to a younger person. So it is false, which means that Debbie is older than Chris and Chris is not the youngest of the four. This shows that the first of our earlier options must be ruled out, and the order of ages must be Betty, Alan, Chris, Debbie.

Q192 If the side lengths are x, y, z with the 120° angle opposite side z, then the Cosine Rule (Section 3.7) gives
$$z^2 = x^2 + y^2 + xy.$$

The best way to deal with this is to write it as
$$(2z)^2 - (2y + x)^2 = 3x^2;$$

we can find infinitely many solutions by letting x be odd and, noting that an odd number is always the difference of two squares,
$$2n + 1 = (n + 1)^2 - n^2.$$

If $x = 2m + 1$, then we have $3x^2 = 2(6m^2 + 6m + 1) + 1$ and so we use the preceding formula with $n = 6m^2 + 6m + 1$, giving
$$2z = 6m^2 + 6m + 2, \quad 2y + x = 6m^2 + 6m + 1;$$

solving gives
$$x = 2m + 1, \quad y = 3m^2 + 2m, \quad z = 3m^2 + 3m + 1$$

for $m = 1, 2, 3, \ldots$. To show that, with these formulae, x, y, z have no common factor except 1, suppose that a is a common factor of all three. Then a is also a factor of $z - y = m + 1$; and so a is a factor of $2(z - y) - x = 1$. That is, the only common factor of x, y, z is 1.

Comments. Taking $m = 1$, we find that a triangle with sides $3, 5, 7$ includes a 120° angle – not a result which everyone knows! Note that we have found infinitely many possibilities, as the question asked, but we have not found *all* possibilities. For example, you may check that the triangle with sides $7, 8, 13$ includes a 120° angle but does not fit into the above pattern.

Q193

(a) If $a_2 = 1$, then we immediately get $a_3 = 0$ which is not a positive integer.

(b) The proof is by mathematical induction (see Section 3.8): we shall show that the required result is true for $n = 1, 2$, and then that, if it is true for any specific (consecutive) integers $n - 1$ and n, then it is true for the next integer $n + 1$. Firstly, if $a_2 = 2$, then $a_n = n$ for $n = 1, 2$: this proves the basis of the induction. Now suppose that $n \geq 2$ and the result is true both for a_{n-1} and for a_n; then
$$a_{n+1} = \frac{a_n^2 - 1}{a_{n-1}} = \frac{n^2 - 1}{n - 1} = n + 1.$$

By induction, we deduce that $a_n = n$ for all n.

(c) Again we use proof by induction. With $a_2 \geq 3$, it is clearly true that $a_n > a_{n-1} + 1$ for $n = 2$. If the inequality is true for some specific $n \geq 2$, then
$$a_{n+1} = \frac{a_n^2 - 1}{a_{n-1}} = (a_n + 1)\frac{a_n - 1}{a_{n-1}} > a_n + 1.$$

(d) From (b), it is clear that one solution is $a_2 = 2$, $a_{2011} = 2011$; since a_2 can be any positive integer except 1, another solution is $a_2 = 2011$. We shall prove that there are no further solutions. So, suppose that $a_2 \geq 3$ and $n \geq 3$. Then we have

$$a_n a_{n-2} = a_{n-1}^2 - 1 = (a_{n-1} + 1)(a_{n-1} - 1)$$

and so a_n is a factor of $(a_{n-1} + 1)(a_{n-1} - 1)$. But if $a_n = 2011$, which is a prime number, then this means that either

$$a_n \text{ is a factor of } a_{n-1} + 1 \quad \text{or} \quad a_n \text{ is a factor of } a_{n-1} - 1;$$

since $a_n > a_{n-1} + 1$, this is impossible.

Q194 Suppose that the number 2^n consists of a digit a followed by s further digits and 5^n consists of a followed by t further digits. Then we have

$$2^n = (a + x)10^s \quad \text{and} \quad 5^n = (a + y)10^t,$$

where $0 \leq x < 1$ and $0 \leq y < 1$. Multiplying these equations gives

$$10^n = (a + x)(a + y)10^{s+t}$$

and so

$$(a + x)(a + y) = 10^{n-s-t}.$$

However, $n - s - t$ is an integer, and $1 \leq (a + x)(a + y) < 100$, so either

$$(a + x)(a + y) = 1 \quad \text{or} \quad (a + x)(a + y) = 10.$$

In the first case, we have $a = 1$, $x = 0$, $y = 0$, so $2^n = 10^s$; since n is a positive integer, this is impossible. In the second case, we have

$$a^2 \leq (a + x)(a + y) < (a + 1)^2;$$

that is,

$$a^2 \leq 10 < (a + 1)^2,$$

and so $a = 3$ is the only possible digit. And indeed, if $n = 5$, then $2^n = 32$ and $5^n = 3125$, both of which start with a 3.

Comment. In fact, it is possible to show that there are infinitely many such n: the numbers 2^n and 5^n both begin with 3 for $n = 5, 15, 78, 88, 98, 108, 118, \ldots$ (however, the obvious pattern does not continue).

Q195 We have

$$9999999999999999999999999999999991 = 10^{34} - 9$$
$$= \left(10^{17}\right)^2 - 3^2$$
$$= \left(10^{17} - 3\right)\left(10^{17} + 3\right)$$
$$= 99999999999999997 \times 100000000000000003$$

which is composite.

Comment. As it turns out, both of the factors on the right–hand side are prime – this is not a simple calculation, but you don't need to do it in order to answer the question.

Q196 Suppose that two different letters, labelled m and n, are both delivered to the same house. Then $(4m+1)^2 = (4n+1)^2$. This means that either

$$4m+1 = 4n+1, \quad \text{so} \quad m = n,$$

or

$$4m+1 = -(4n+1), \quad \text{so} \quad 4m+4n = -2.$$

But both of these are impossible, the first because different letters have different labels, the second because the left–hand side is a multiple of 4 and the right–hand side is not. So, two letters cannot be delivered to the same house.

Q197 Try a triangle $\triangle ABC$ with $|AB| = 2011$ and $|AM| = 1102$, where M is the point on AC such that BM is perpendicular to AC. This clearly has the required dimensions; since $|BM|^2 = 2011^2 - 1102^2 = 2829717$, we need to find $y = |BC|$ and $x = |MC|$ such that

$$y^2 - x^2 = 2829717,$$

that is,

$$(y+x)(y-x) = 3^2 \times 11 \times 101 \times 283. \tag{2.27}$$

In order that the *longest* side of the triangle be 2011, we also need $y < 2011$ and $x < 909$, and hence $y + x < 2920$. From the factorisations, we have

$$y - x \leq \sqrt{2829717} \quad \text{and} \quad y - x = \frac{2829717}{y+x} > \frac{2829717}{2920}.$$

Doing the calculations and remembering that $y - x$ is an integer, this means that

$$970 \leq y - x \leq 1682. \tag{2.28}$$

Also, $y - x$ must satisfy the factorisation (2.27).

- If 283 is a factor of $y - x$, then $y - x = 283z$ with z a factor of $3^2 \times 11 \times 101$. The inequalities (2.28) imply $4 \leq z \leq 5$, so this is impossible.
- If 283 is not a factor of $y - x$ but 101 is, then $y - x = 101z$ with z a factor of $3^2 \times 11$ and $10 \leq z \leq 16$: the only possibility is $z = 11$.
- If neither 283 nor 101 is a factor of $y - x$, then $y - x \leq 3^2 \times 11$, which is too small.

So, the only possibility is $y - x = 1111$, $y + x = 2547$; this leads to $y = 1829$ and $x = 718$, giving a triangle with sides 2011, 1829 and 1820.

Q198

(a) Up to you! – what did you guess?

(b) We have

$$x = 10^{10^{10^{11}}} / 10^{10^{10}} = 10^{10^{10^{11}} - 10^{10}} = 10^{10^{90000000000}}$$

and so the number of digits in x is $1 + 10^{90000000000}$. Rather more than a million, which is 10^6. Note that the number just given is not x itself, but the number of digits in x.

Q199 Adding all three equations gives

$$1368x + 1368y + 1368z = a + b + c,$$

while subtracting the first from the second gives

$$333x + 333y - 666z = b - a.$$

So,

$$3z = (x + y + z) - (x + y - 2z) = \frac{a + b + c}{1368} - \frac{b - a}{333}$$

This simplifies to

$$z = \frac{189a - 115b + 37c}{151848},$$

and, in the same way, we get

$$x = \frac{189c - 115a + 37b}{151848}, \quad y = \frac{189b - 115c + 37a}{151848}.$$

Q200 Let $a = \lfloor 6x \rfloor$ and $b = \lceil 8x \rceil$ and $\alpha = 6x - a$ and $\beta = b - 8x$. Then a, b are integers and $0 \le \alpha, \beta < 1$. Notice that

$$7x = 9 - a + b, \tag{2.29}$$

so $7x$ is an integer, and

$$\alpha - \beta = 14x - a - b,$$

so $\alpha - \beta$ is also an integer. As α and β both lie between 0 and 1, the only way their difference can be an integer is if they are equal. Thus, we have

$$6x = a + \alpha, \quad 8x = b - \alpha, \tag{2.30}$$

where a, b are integers and $0 \le \alpha < 1$. Now we can eliminate x and b from equations (2.29) and (2.30) to give

$$5a = 54 + 7\alpha.$$

However, 7α is less than 7, so this means that $5a$ is a multiple of 5 lying between 54 and 61. There are two possibilities, $a = 11, 12$; and then from previous equations it is easy to find the respective values of α and x. We have two potential solutions,

$$x = \frac{13}{7} \quad \text{and} \quad x = \frac{15}{7},$$

and substituting back confirms that these both work.

In Problem 205, you are asked to solve another equation involving rounding functions.

Q201 Let $a(n)$ be the number of triples of n-digit integers satisfying the required conditions. Consider the pairs of corresponding digits in x and y. For example, the six–digit solution

$$150050 + 001051 = 151101$$

corresponds to the pairs

$$(1, 0), \ (5, 0), \ (0, 1), \ (0, 0), \ (5, 5), \ (0, 1).$$

The pairs $(1, 1)$, $(1, 5)$ and $(5, 1)$ cannot occur as each would give a digit of 2 or 6 in z (or a 3 or 7 if there is a "carry" from the following digit). The pair $(0, 1)$ can occur but must not be followed by $(5, 5)$ since then the sum $0 + 1$, plus a carry of 1 from the following column, would give a digit 2 in z; the same is true for the pairs $(1, 0)$, $(0, 5)$ and $(5, 0)$. The first pair cannot be $(5, 5)$ as then z would have an extra digit.

Consider numbers having at least two digits, and first consider the case when the second pair of digits is not $(5, 5)$. Then our n–digit numbers contain one of five possibilities for the first pair, followed by $n - 1$ pairs which form an acceptable pair of $(n - 1)$-digit numbers. So, the number of solutions in this case is

$$5a(n - 1).$$

Next, consider the case when the second pair is $(5, 5)$. If the third pair is not $(5, 5)$, then the last $n - 2$ pairs form acceptable numbers; if it is but the fourth is not, then the last $n - 3$ pairs form acceptable numbers; and so on. Therefore, the number of solutions in this case is

$$a(n - 2) + a(n - 3) + \cdots + a(1) + 1,$$

where the final 1 is accounting for the possibility that all pairs except the first are $(5, 5)$. So, the total number of solutions, when $n \geq 2$, is given by

$$a(n) = 5a(n - 1) + a(n - 2) + a(n - 3) + \cdots + a(1) + 1; \tag{2.31}$$

it is easy to see that $a(1) = 5$, so

$$a(2) = 5 \times 5 + 1 = 26, \quad a(3) = 5 \times 26 + 5 + 1 = 136,$$

and so on. In fact, we can make things a bit easier for ourselves by observing that, if $n \geq 3$, then (2.31) holds both for n and for $n - 1$, so

$$a(n) = 5a(n - 1) + a(n - 2) + a(n - 3) + \cdots + a(1) + 1$$
$$a(n - 1) = 5a(n - 2) + a(n - 3) + a(n - 4) + \cdots + a(1) + 1.$$

Subtracting these equations and rearranging gives

$$a(n) = 6a(n - 1) - 4a(n - 2) \quad \text{for} \quad n \geq 3.$$

So, we have

$$a(1) = 5$$
$$a(2) = 26$$
$$a(3) = 6 \times 26 - 4 \times 5 = 136$$
$$a(4) = 6 \times 136 - 4 \times 26 = 712$$

and so on, which eventually gives our answer of $a(10) = 14\,672\,384$.

Q202 First assume that $|x_1| \leq 1$. Then we can write $x_1 = \cos \theta$ and we have

$$x_2 = 1 - \cos^2 \theta = \sin^2 \theta, \quad x_3 = 1 - 2\sin^2 \theta = \cos 2\theta.$$

It is easy to show inductively that

$$x_{2k} = \sin^2(2^{k-1}\theta) \quad \text{and} \quad x_{2k+1} = \cos(2^k\theta), \tag{2.32}$$

and we therefore have a solution provided that

$$2\sin^2(2^{n-1}\theta) + \cos \theta = 1;$$

that is, $\cos \theta = \cos(2^n\theta)$. The solutions

$$\theta = \frac{2m\pi}{2^n - 1} \quad \text{for} \quad m = 0, 1, 2, \ldots, 2^{n-1} - 1 \tag{2.33}$$

and

$$\theta = \frac{2m\pi}{2^n + 1} \quad \text{for} \quad m = 1, 2, 3, \ldots, 2^{n-1} \tag{2.34}$$

are all different and between 0 and π, and therefore give 2^n different values for $x_1 = \cos\theta$. On the other hand, by elementary algebra we have

$$x_2 = 1 - x_1^2, \quad x_3 = -1 + 2x_1^2, \quad x_4 = 4x_1^2 - 4x_1^4$$

and we can show that x_{2k} is a polynomial of degree 2^k in x_1. So, the equation $2x_{2n} + x_1 = 1$ becomes a polynomial of degree 2^n in x_1; since we have already found 2^n possible values of x_1, there are no more to find! Therefore, all solutions of the system are given by the formulae (2.32) with the values of θ given by (2.33) and (2.34).

Alternative solution. Adding 16 times all the quadratic equations, subtracting 8 times the others, and completing squares, gives

$$(4x_1 - 1)^2 + (4x_3 - 1)^2 + \cdots + (4x_{2n-1} - 1)^2 = 9n.$$

So, there is at least one k for which $(4x_{2k-1} - 1)^2 \le 9$ and hence $-1 \le x_{2k-1} \le 1$. But then we see from the original equations that x_{2k-3} satisfies the same inequality; eventually, we have $-1 \le x_1 \le 1$ and we proceed as in the previous solution.

Q203 First, we note that $24186470400000 = 2^{17} \times 3^{10} \times 5^5$. Therefore, the numbers in the "outer" region must include five 5s; the sixth is obviously in the bottom–right square. Suppose that the numbers of 6s, 4s, 3s, 2s and 1s in the outer region are a, b, c, d, e, respectively. Now, this region contains twenty–five numbers (including the five 5s), so

$$a + b + c + d + e = 20.$$

Also, the total of all thirty–six numbers is $6(1 + 2 + 3 + 4 + 5 + 6) = 126$, so the total of those in the outer region is $126 - 26 - 5 = 95$. Once again, this includes the five 5s, so

$$6a + 4b + 3c + 2d + e = 70.$$

The numbers in the outer region must have altogether seventeen factors of 2 and ten factors of 3, so

$$a + 2b + d = 17 \quad \text{and} \quad a + c = 10.$$

Eliminating a, b, c from these four equations gives

$$d + 3e = 9. \tag{2.35}$$

Now look at the bottom row and the right–hand column. Each of these must contain a 6, a 4, a 3, a 2 and a 1; so, $a, b, c, d, e \ge 2$. Therefore, (2.35) gives only one possibility $e = 2$, $d = 3$, and then it is not hard to find $c = 6$, $b = 5$, $a = 4$. Now we know exactly what numbers are in each region of the diagram, and it remains to arrange them correctly.

Consider which numbers go around the edges, and which go in the "indents" of the outer region. There are six 3s and five 4s; only four of each can go around the edges, so the indents must contain two 3s and a 4. Because of the 5 in the bottom right–hand corner, only two more 5s can go around the edges, and the other three must occupy the indents. It is not hard to see that they must be arranged as follows.

		5		3	
				5	
	5	3		4	
					5

Now we know that the central region contains no 5s or 3s, so it is easy to place the rest of these numbers. Also, the central region contains four 1s, and so the two 1s in the outer region must occupy two of the three corners; clearly, they must be in the top–right and bottom–left corners. So, we get

	3		5		1
		5		3	
3				5	
	5	3		4	
5					3
1			3		5

The bottom row and right–hand column must each contain a 6, a 4 and a 2; this leaves two 6s, two 4s and a 2 in the top row and left–hand column. Clearly, the 2 must be in the corner, and it is now not very difficult to fill in the rest of the puzzle.

2	3	4	5	6	1
4	1	5	2	3	6
3	2	1	6	5	4
6	5	3	1	4	2
5	6	2	4	1	3
1	4	6	3	2	5

Q204 Jack's travel time home was twice that of his outward journey; the total travel time, plus the 60 minutes' visit, adds up to 2 hours 15 minutes. So Jack's outward travel time was 25 minutes, and the time when he looked at Jill's clock was 1 hour 25 minutes after looking at his own. Since it then appeared to show the time at which Jack left, double this time plus the 1 hour 25 minute stay must add up to 12 hours (if you have trouble seeing why, then draw pictures of the two clocks). So, the time at which Jack left was 5 : 17 : 30.

Comment. The times involved could have added up to 24 hours instead of 12, in which case Jack's departure time would have been 11 : 17 : 30. However, this does not fit in with the information that he "left home one afternoon".

Q205 Following the solution to Problem 200, let $a = \lfloor x \rfloor$ and $b = \{2x\}$ and $c = \lceil 3x \rceil$. Then a, b, c are integers and we can write

$$x = a + \alpha, \quad 2x = b + \beta, \quad 3x = c - \gamma \tag{2.36}$$

where

$$0 \leq \alpha < 1, \quad -\tfrac{1}{2} \leq \beta < \tfrac{1}{2}, \quad 0 \leq \gamma < 1. \tag{2.37}$$

We have immediately

$$a - b + c = 5,$$

and from (2.36) we obtain

$$2a + 2\alpha = b + \beta \quad \text{and} \quad 3a + 3\alpha = c - \gamma. \tag{2.38}$$

Solving the last three equations to find a, b, c in terms of α, β, γ is easy (exercise!) and gives the results

$$a = \tfrac{1}{2}(5 - \alpha - \beta - \gamma), \quad b = 5 + \alpha - 2\beta - \gamma, \quad c = \tfrac{1}{2}(15 + 3\alpha - 3\beta - \gamma).$$

But now the inequalities (2.37) imply

$$1\tfrac{1}{4} < a \leq 2\tfrac{3}{4}, \quad 3 < b < 7, \quad 6\tfrac{1}{4} < c < 9\tfrac{3}{4},$$

and, since a, b, c are integers, we have

$$a = 2, \quad b = 4 \text{ or } 5 \text{ or } 6, \quad c = 7 \text{ or } 8 \text{ or } 9$$

with $c = b + 3$. Returning to equations (2.38) and solving for α, we obtain

$$\alpha = \tfrac{1}{2}(b - 2a + \beta) \quad \text{and} \quad \alpha = \tfrac{1}{3}(c - 3a - \gamma),$$

and using one last time the inequalities (2.37) together with the known possibilities for a, b, c gives restrictions on the value of α. Not forgetting that we already know $0 \leq \alpha < 1$, there are three possibilities:

- $a = 2, b = 4, c = 7$, so $-\tfrac{1}{4} \leq \alpha < \tfrac{1}{4}$ and $0 < \alpha \leq \tfrac{1}{3}$, so

$$0 < \alpha < \tfrac{1}{4};$$

- $a = 2, b = 5, c = 8$, so $\tfrac{1}{4} \leq \alpha < \tfrac{3}{4}$ and $\tfrac{1}{3} < \alpha \leq \tfrac{2}{3}$, so

$$\tfrac{1}{3} < \alpha \leq \tfrac{2}{3};$$

- $a = 2, b = 6, c = 9$, so $\tfrac{3}{4} \leq \alpha < \tfrac{5}{4}$ and $\tfrac{2}{3} < \alpha \leq 1$, so

$$\tfrac{3}{4} \leq \alpha < 1.$$

Since $x = 2 + \alpha$, the solutions of the equation must satisfy

$$2 < x < 2\tfrac{1}{4} \quad \text{or} \quad 2\tfrac{1}{3} < x \leq 2\tfrac{2}{3} \quad \text{or} \quad 2\tfrac{3}{4} \leq x < 3.$$

There are other conditions which could potentially restrict the solutions still further, so it is necessary to check that all these values really work; however, this is not difficult, and all of these x values are in fact solutions.

Q206 Imagine that we make a video of the whole procedure, and then view the video backwards, so that instead of getting up and leaving, people arrive and sit down. We can then see that the answer to the problem posed is the same as to the following: if people arrive one by one in the empty carriage and choose seats at random, then find the probability that the first k people form no pairs, and person $k+1$ sits next to one of the people already there. The probability is clearly zero if $k > n$, so we suppose that $k \leq n$. Now, the first person clearly does not form a pair; the second avoids a pair with probability $(2n - 2)/(2n - 1)$; the third avoids a pair with probability $(2n - 4)/(2n - 2)$; and so on. The probability that the first k people form no pairs is

$$\frac{2n - 2}{2n - 1} \frac{2n - 4}{2n - 2} \cdots \frac{2n - 2(k - 1)}{2n - (k - 1)}.$$

There are now $2n - k$ unoccupied seats, including k which are next to an occupied seat; so the probability that person $k + 1$ *does* form a pair is $k/(2n - k)$. The final answer is the above product times this fraction; it can be simplified in various ways to get

$$\frac{2n}{2n} \frac{2n - 2}{2n - 1} \frac{2n - 4}{2n - 2} \cdots \frac{2n - 2(k - 1)}{2n - (k - 1)} \frac{k}{2n - k} = k \, 2^k \frac{n!}{(n - k)!} \frac{(2n - k - 1)!}{(2n - k)!}.$$

Q207

(a) In dealing a five–card hand from a 52–card pack, cards are chosen without repetition, and order is not important. Therefore – see Section 3.5.1 – the number of hands is the binomial coefficient $C(52, 5)$. To choose a "four of a kind" hand, we must choose the value for the "four" (13 possibilities), then choose one further card (48 possibilities). So, the probability that my opponent has "four of a kind" is

$$\frac{13 \times 48}{C(52, 5)} = 0.0002400960384.$$

(b) Before doing the calculations, it's interesting to try to work out how the answer should compare with part (a). It is fairly plausible that, if we increase the number of cards in each suit (for example, consider a 100–card pack with 25 different values in each suit), then the likelihood of "four of a kind" should decrease. In the present scenario, an entire four (and one extra card) is ruled out from my opponent's hand; so, more or less, his hand has come from a smaller pack (of 48 cards); so we should expect his probability of "fours" to be greater than in the previous case.

To do this more precisely, since we know that I have "fours", my opponent's hand must be chosen from a 47–card pack consisting of 11 groups of 4 and one group of 3. The number of possible hands is $C(47, 5)$; to choose "fours" we pick a four (11 possibilities) and then one other card (43 possibilities). So the probability that my opponent has "fours" in this case is

$$\frac{11 \times 43}{C(47, 5)} = 0.0003083564601;$$

as expected, this is bigger than the previous answer.

(c) Once again, we try first to understand what we expect to happen. We know from (a) that the chance of my having four of a kind is very small. That is, even before I look at my cards, I feel fairly certain that I don't have "fours"; if I make this *absolutely* certain by looking at my cards, then I have scarcely changed the situation at all. So, we should expect that the probabilities in (a) and (c) should be very similar, and we'll need to calculate in order to find out which is actually larger.

Working out the probability here is harder than in (a), and we'll do it by counting the number of possibilities for both my hand and my opponent's. The probability is equal to the number of deals in which he has "fours" and I do not, divided by the number of deals in which I do not (and he may or may not). For the denominator, I can have any hand ($C(52,5)$ possibilities) except for "fours" (13×48 possibilities); my opponent can then have any five cards chosen from 47, giving $C(47,5)$ possibilities. The numerator is the number of deals in which he has fours and I have anything ($13 \times 48 \, C(47,5)$ possibilities), minus the number in which we both have fours. To choose a deal in this last category, choose a "four" for my opponent (13 possibilities) and then one other card (48 possibilities); then a four for me (11 possibilities – two have been knocked out) and one final card (43 possibilities). Putting all this together, the probability that my opponent has "four of a kind", if it is known that I do not, is

$$\frac{13 \times 48 \, C(47,5) - 13 \times 48 \times 11 \times 43}{(C(52,5) - 13 \times 48) \, C(47,5)} = 0.0002400796454.$$

This is less than the probability in (a), though, as we predicted, the difference is very small.

Q208 The arctangents of the seven numbers are seven different real numbers lying strictly between $-\frac{1}{2}\pi$ and $\frac{1}{2}\pi$. That is, we have now seven numbers in an interval of length π, and so there must be two of them which are separated by less than $\frac{1}{6}\pi$. If $\arctan x$ is the larger of these two numbers and $\arctan y$ is the smaller, then we have

$$0 < \arctan x - \arctan y < \frac{\pi}{6}.$$

Now, tan is an increasing function between 0 and $\frac{1}{6}\pi$ and so

$$0 < \tan\left(\arctan x - \arctan y\right) < \frac{1}{\sqrt{3}};$$

using the "tan of a difference" formula and the fact that $\tan(\arctan x) = x$ gives

$$0 < \frac{x-y}{1+xy} < \frac{1}{\sqrt{3}};$$

finally, since the first inequality shows that we are dealing with positive numbers, we may take reciprocals in the second inequality to give

$$\frac{1+xy}{x-y} > \sqrt{3}$$

as claimed.

Q209 Let O be the intersection of AC and MN. The triangles $\triangle ABC$, $\triangle AON$ and $\triangle MAN$ are all right–angled. The first two have $\angle BAC$ in common, so they are similar; the last two have $\angle ANM$ in common, so they are similar; thus, $\triangle ABC$ and $\triangle MAN$ are similar. By Pythagoras' Theorem, we calculate the ratio of hypotenuse lengths:

$$\frac{|MN|}{|AC|} = \frac{|AD|}{|AC|} = \frac{1}{\sqrt{3}};$$

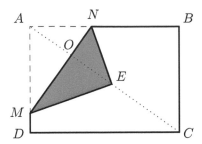

therefore, the ratio of areas is

$$\frac{\text{area}(MAN)}{\text{area}(ABC)} = \frac{1}{3}.$$

This means that $\triangle MAN$ has area which is $\frac{1}{6}$ of the whole sheet $ABCD$; but $\triangle MEN$ has the same area, and so the proportion we are looking for is

$$\frac{1/6}{5/6} = \frac{1}{5}.$$

The reason that an A4 sheet has proportions $1 : \sqrt{2}$ is that then half of the sheet has the same proportions (easy exercise: prove it), which means that two A4 sheets can be reduced and printed on one A4 sheet without being distorted.

Q210 We shall write the superknight's move as $(3,8)$, meaning 3 squares to the right and 8 upwards. The knight can also make moves of $(8,3)$, $(-3,8)$ and so on. We wish either to add up lots of these moves to get a result of $(1,0)$, or to prove that it's not possible to do so. We'll begin by just getting a little closer to the start point: we have

$$(8,-3) + (-8,-3) + (3,8) = (3,2)$$

– that is, in plain language: 8 right 3 down, followed by 8 left 3 down, followed by 3 right 8 up, takes the knight altogether 3 squares right and 2 squares up from the starting point. Repeating the same idea with $(3,2)$ and related moves, we have

$$(3,2) + (-3,2) + (2,-3) = (2,1)$$

– we have now "synthesised" an ordinary knight's move – and then

$$(2,1) + (-2,1) + (1,-2) = (1,0).$$

We have achieved our object, but it could probably do with a bit of clarification. If we take the second equation above and change all upwards moves to downwards and vice versa, then we have the related equation

$$(-3,2) + (3,2) + (-2,-3) = (-2,1);$$

if we now interchange rightwards and upwards moves, then we obtain

$$(2,-3) + (2,3) + (-3,-2) = (1,-2).$$

We can substitute all these into the third equation: moves such as $(3,2)$ and $(-3,-2)$ are just the opposites of each other and can be cancelled. This gives

$$(3,2) + 2(-3,2) + 2(2,-3) = (1,0).$$

Using the first equation in a similar way, we have

$$(-8, -3) + (8, -3) + (-3, 8) = (-3, 2)$$

and

$$(-3, -8) + (-3, 8) + (8, -3) = (2, -3)$$

and so, finally,

$$5(8, -3) + 3(-8, -3) + (-3, -8) + 4(-3, 8) = (1, 0).$$

That is, our "superknight" can get one square to its right in thirteen moves: five of 8 right 3 down, three of 8 left 3 down, one of 3 left 8 down and four of 3 left 8 up.

To generalise this puzzle, try Problem 212.

Q211 Completing the square (in a slightly unusual way), we can write the equation as

$$(x^2 + 2x)^2 - 10(x^2 + 2x) + 13 = 0.$$

If we write $q = x^2 + 2x$, then this is the quadratic equation $q^2 - 10q + 13 = 0$, which has solutions $q = 5 \pm \sqrt{12}$. To complete the problem, we have to solve two further quadratic equations

$$x^2 + 2x = 5 + \sqrt{12} \quad \text{and} \quad x^2 + 2x = 5 - \sqrt{12},$$

and there are altogether four solutions

$$x = -1 + \sqrt{6 + \sqrt{12}}, \quad x = -1 + \sqrt{6 - \sqrt{12}},$$
$$x = -1 - \sqrt{6 + \sqrt{12}}, \quad x = -1 - \sqrt{6 - \sqrt{12}}.$$

You could now try Problem 218.

Q212 For this to be possible, a and b must have no common factor greater than 1: for if they have a common factor $g > 1$, then every move involves either a or b steps to the right or left; so the superknight must always be a multiple of g squares to the right or left of its initial square, and cannot be one square to the right. Furthermore, a and b must not both be odd: for if this is so, then any move by the superknight takes it from one chessboard square to another of the same colour, and it can never reach the square immediately to its right, which has the opposite colour.

There are no further requirements on a and b. To prove this, we shall show that if a and b are positive integers, not both odd and having no common factor, then an (a, b)–superknight can move from one square to that immediately to the right. The proof will imitate the case of the $(3, 8)$–superknight in Problem 210. So, suppose that we have an (a, b)–superknight; since this is the same as a (b, a)–superknight, we may assume that $0 < b < a$. We can make a sequence of moves (see Solution 210 for an explanation of the notation)

$$(a, -b) + (-a, -b) + (b, a) = (b, c), \quad \text{where} \quad c = a - 2b;$$

by turning all the moves "upside down", we may assume that $c \geq 0$. Now if $c = 0$, then $a = 2b$; since a and b have no common factor, this is only possible when $b = 1$; so, we have reached $(1, 0)$ and we are finished. If, on the other hand, $c > 0$, then we have "synthesised" the move of a (b, c)–superknight. Now, b and c have no common factor greater than 1 (if they did, then it would also be a common factor of a and b); and they are not both odd (if they were, then so would be a and b). Moreover, b and c are both less than a. So, b and c satisfy the same conditions as a and b; and the (b, c)–superknight is "smaller" than the original (a, b)–superknight; therefore, we can repeat the procedure, as we did in the solution to Problem 210, obtaining superknights with smaller and smaller moves, until at last we reach the smallest possible superknight, which is an ordinary knight. And we know that we can use this to get from one square to that immediately to the right, so we are finished.

If you wish, then you can write this argument more formally as a proof by mathematical induction; see Section 3.8.

For one more question about superknights, see Problem 216.

Q213 This is a question of "applied" rather than of "pure" geometry, and to answer it, we will of course need to know some basic facts about the solar system – fortunately, not too many. The Earth revolves around the sun once per year, tracing an elliptical path in a plane which also includes the sun – this is known as the *plane of the ecliptic*. At the same time, the earth rotates about its axis, the line passing through the north and south poles. Imagine a plane passing through the earth's equator and extending indefinitely in every direction. This *equatorial plane* rotates with the Earth, and so its position as a whole, relative to the Earth, never changes. As the Earth travels in its orbit around the sun, the polar axis and the equatorial plane move with it, always retaining the same orientation. The sun is directly overhead at the equator whenever it is found to be on the equatorial plane. If the polar axis were perpendicular to the ecliptic, then the sun would always be on the equatorial plane; in reality, however (and this is the last astronomical fact we shall need), the axis is tilted away from the ecliptic, and the equatorial plane is also tilted away from the ecliptic plane. As the earth revolves once about the sun (here you will probably wish to draw some diagrams or wave your hands a bit – imagine the sun as fixed and the equatorial plane moving around it), the equatorial plane will contain the sun twice, once "on each side" of the earth's orbit.

In conclusion, the Sun is directly overhead at the equator twice a year.

Q214 In order to obtain the maximum product, none of the summands (the numbers in the sum making up n) can be 1; for any 1 could be combined with another summand to increase the product. For example, replacing $1 + 2$ by 3 increases the product from 2 to 3. Furthermore, none of the summands can be 5 or more, as 5 could be replaced by $2 + 3$, and 6 could be replaced by $2 + 4$, and so on, in each case increasing the product of summands. So to get the largest possible product, we should write n as a sum of 2s, 3s and 4s only. And indeed, a summand of 4, while permissible, is not needed, because 4 could be replaced by $2 + 2$, leaving the product unchanged; so, we may write n as a sum of 2s and 3s only. Finally, we must not have three or more 2s, since replacing $2 + 2 + 2$ by $3 + 3$ will again give a larger product, 9 instead of 8.

Therefore, we shall use 2s and 3s only, with no more than two 2s. And since this leaves only one way to create a sum adding up to any given number, the problem is finished. Specifically, if we divide n by 3, then we shall obtain a remainder of $0, 1$ or 2:

- if the remainder is 0, then we write $n = 3 + \cdots + 3$, giving a maximal product of $3^{n/3}$;
- if the remainder is 1, then we write $n = 2 + 2 + 3 + \cdots + 3$ with $(n - 4)/3$ summands of 3, giving a product $4 \times 3^{(n-4)/3}$;
- if the remainder is 2, then we write $n = 2 + 3 + \cdots + 3$, giving a product $2 \times 3^{(n-2)/3}$.

Comment. In the case of remainder 1, we could alternatively have taken $n = 4 + 3 + \cdots + 3$, resulting in the same product $4 \times 3^{(n-4)/3}$.

There is an extension of this puzzle in Problem 217.

Q215 If $f(x)$ is a polynomial having coefficients 1 and -1 only and degree m, then $f(1)$ is a sum of $m + 1$ terms, each equal to 1 or -1; that is, it is a sum of $m + 1$ odd numbers.

Thus, $f(1)$ is even if m is odd, and odd if m is even. Similarly, if $g(x)$ has coefficients 1 and -1 only and degree k, then $g(1)$ is even if k is odd and odd if k is even. Now suppose that $f(x)$ and $g(x)$ also satisfy the given equation. Then, by substituting $x = 1$ and also comparing degrees, we have

$$f(1) = g(1) \quad \text{and} \quad m = n + k.$$

However, if n is odd, then these equations are inconsistent with the previous conclusions. For the first equation implies that m and k are both odd or both even, while the second implies that one is odd and the other even. The situation is impossible, and we must conclude that there are no polynomials satisfying the requirements of the question.

Q216 No, it is not. To move the $(3, 8)$–superknight one square to its right, we need

$$w(8, -3) + x(-8, -3) + y(-3, -8) + z(-3, 8) = (1, 0)$$

where w, x, y, z are integers, and the number of moves taken will be $|w| + |x| + |y| + |z|$. (Note that we can't "subtract" moves – a coefficient of -5 would still count as 5 moves.) Separating out the right/left moves and the up/down moves, we have

$$8w - 8x - 3y - 3z = 1 \quad \text{and} \quad -3w - 3x - 8y + 8z = 0.$$

Adding and subtracting these equations yields

$$5(w + z) - 11(x + y) = 1 \quad \text{and} \quad 11(w - z) - 5(x - y) = 1.$$

We solve the first equation by regarding $w+z$ as a single variable, $x+y$ likewise, and treating it as a linear Diophantine equation (Section 3.2); similarly for the second; this gives

$$
\begin{aligned}
w + z &= -2 + 11s, & x + y &= -1 + 5s \ , \\
w - z &= \ \ 1 + 5t \ , & x - y &= \ \ 2 + 11t
\end{aligned}
$$

for some integers s, t; finally, we derive formulae for w, x, y, z by adding and subtracting these:

$$
\begin{aligned}
2w &= -1 + 11s + 5t, & 2x &= \ \ 1 + 5s + 11t, \\
2y &= -3 + 5s - 11t, & 2z &= -3 + 11s - 5t.
\end{aligned}
$$

Now if $|w| + |x| + |y| + |z| < 13$, then $|2w| \le 24$ and so on; hence,

$$
\begin{aligned}
-24 &\le -1 + 11s + \ 5t \le 24 \\
-24 &\le \ \ 1 + \ 5s + 11t \le 24 \\
-24 &\le -3 + \ 5s - 11t \le 24 \\
-24 &\le -3 + 11s - \ 5t \le 24.
\end{aligned}
$$

Adding the first and fourth inequalities, subtracting the second and third, gives

$$-48 \le -4 + 22s \le 48 \quad \text{and} \quad -48 \le 4 + 22t \le 48$$

and so $s = -2, -1, 0, 1, 2$ and $t = -2, -1, 0, 1, 2$. From previous equations, $s+t$ is odd, which leaves twelve pairs to check; the values of $|w| + |x| + |y| + |z|$ calculated in the following table show that the number of moves is never less than 13:

	$s = -2$	$s = -1$	$s = 0$	$s = 1$	$s = 2$
$t = -2$		33		29	
$t = -1$	35		13		29
$t = 0$		19		13	
$t = 1$	37		19		33
$t = 2$		37		35	

Comment ... and one of the 13–move solutions has $s = 1$, $t = 0$, giving $w = 5$, $x = 3$, $y = 1$, $z = 4$, the solution we found in Problem 210.

Q217 The equations

$$8 = 4 + 4, \quad 9 = 4 + 5, \quad 10 = 4 + 6$$

and so on show that we can never obtain the largest possible product if we use summands of 8 or more, because in each case, the product of summands on the right–hand side exceeds the left–hand side. For the same reason, the equations

$$7 + 7 = 4 + 4 + 6, \quad 7 + 6 = 4 + 4 + 5, \quad 7 + 5 = 4 + 4 + 4, \quad 7 + 4 = 5 + 6$$

show that we can never use a 7. Thirdly,

$$6 + 6 = 4 + 4 + 4, \quad 6 + 5 + 5 = 4 + 4 + 4 + 4, \quad 6 + 4 = 5 + 5$$

show that we cannot have two 6s; if we have a 6, then we cannot have two 5s; and if we have a 6, then we cannot have a 4. Finally,

$$5 + 5 + 5 + 5 = 4 + 4 + 4 + 4 + 4$$

shows that we cannot have more than three 5s. So, how can we find suitable numbers adding up to 2013? If we don't use any 4s, then the most we can get is $6 + 5$ or $5 + 5 + 5$, which are both way too small. If we do have 4s, then we cannot have 6s and so we must make up 2013 from 4s and 5s only, with at most three 5s. The only possibility is

$$2013 = 4 + \cdots + 4 + 5$$

with one 5 and all the rest 4s. Therefore, the maximum possible product is 5×4^{502}.

Q218 We need to find coefficients k and l such that

$$x^4 + ax^3 + bx^2 + cx + d = (x^2 + kx)^2 + l(x^2 + kx) + d. \qquad (2.39)$$

Expanding the right hand side and equating coefficients of x^3, x^2 and x, we have

$$a = 2k, \quad b = k^2 + l, \quad c = kl.$$

Solving the first two equations for k and l in terms of a and b, then substituting into the third, gives

$$8c = a(4b - a^2);$$

for the method to succeed, it is necessary that this condition hold. Conversely, if that condition is true and we choose

$$k = \frac{1}{2}a, \quad l = b - \frac{1}{4}a^2,$$

then it is easy to check that (2.39) is true, and we can proceed to solve the quartic as in the earlier problem.

Comment. We could also try looking for something like

$$x^4 + ax^3 + bx^2 + cx + d = (x^2 + kx + m)^2 + l(x^2 + kx + m) + d,$$

but if you think about it, you should be able to see that this will not give us any further possibilities.

Q219 Let $n = \lfloor x \rfloor$. Then $x^2 = 12n - 23$, so that $n \geq 2$; and hence,

$$x = \sqrt{12n - 23},$$

taking the positive root since $x \geq n > 0$. Since $n \leq x < n + 1$, we have

$$n \leq \sqrt{12n - 23} < n + 1;$$

this gives $n^2 - 12n + 23 \leq 0$ and $n^2 - 10n + 24 > 0$. Calculation shows that the first quadratic has a root between 2 and 3 and another between 9 and 10; since n is an integer and the quadratic has a negative value, we have

$$3 \leq n \leq 9.$$

Similarly, the second inequality shows that

$$\text{either} \quad n \leq 3 \quad \text{or} \quad n \geq 7.$$

Combining all the information we have found about n gives four possibilities $n = 3, 7, 8, 9$, and hence the four solutions

$$x = \sqrt{13},\ \sqrt{61},\ \sqrt{73},\ \sqrt{85}.$$

Q220 As in Solution 219, we write $n = \lfloor x \rfloor$, so that $x = \sqrt{2an - b}$. This will give a solution of the equation provided that

$$n \leq \sqrt{2an - b} < n + 1;$$

a bit of algebra leads to

$$n^2 - 2an + b \leq 0 \quad \text{and} \quad n^2 - (2a - 2)n + b + 1 > 0.$$

Therefore, a solution will exist for every value of n which is between the larger root of the second quadratic and the smaller root of the first:

$$a - 1 + \sqrt{a^2 - 2a - b} < n < a + \sqrt{a^2 - b}.$$

Now, there will certainly be k or more integers n which satisfy this inequality if the difference between the left–hand side and right–hand side is more than k. We have

$$\left(a + \sqrt{a^2 - b}\right) - \left(a - 1 + \sqrt{a^2 - 2a - b}\right) = 1 + \sqrt{a^2 - b} - \sqrt{a^2 - 2a - b}$$
$$= 1 + \frac{2a}{\sqrt{a^2 - b} + \sqrt{a^2 - 2a - b}};$$

this will be largest when b has its largest possible value. So, take $b = a^2 - 2a$, giving

$$1 + \sqrt{2a}.$$

This can be made as large as we like by choosing a suitably large, and, therefore, we may obtain as many solutions as desired. In particular, to get more than 2013 solutions, we take $1 + \sqrt{2a} > 2014$; that is, $a > 2013^2/2$. For example, we could take $a = 2026085$ and $b = a^2 - 2a = 4105016375055$. Thus, the equation

$$x^2 - 4052170\lfloor x \rfloor + 4105016375055 = 0$$

has more than 2013 solutions.

Q221 We can write the given equations as

$$xf(x) - 1 = 0$$

for $x = 1, 2, 3, \ldots, 2013$. Since $xf(x) - 1$ is a polynomial with degree 2013 which has 2013 roots, we can use the Factor Theorem (Section 3.10) to write an expression for $xf(x) - 1$ involving one unknown constant:

$$xf(x) - 1 = c(x - 1)(x - 2) \cdots (x - 2012)(x - 2013). \qquad (2.40)$$

By substituting $x = 0$, we obtain

$$-1 = c(-1)(-2) \cdots (-2012)(-2013)$$

and so $c = 1/2013!$. By substituting this value back into equation (2.40) and taking $x = 2014$, we find

$$2014f(2014) - 1 = \frac{1}{2013!}(2013)(2012) \cdots (2)(1) = 1$$

and so

$$f(2014) = \frac{2}{2014} = \frac{1}{1007}.$$

Q222

(a) It is not hard to see that $x = 1$ is a root of the quartic, and by long (or short) division, we have

$$x^4 + 6x^2 - 16x + 9 = (x - 1)(x^3 + x^2 + 7x - 9).$$

And now $x = 1$ is also a root of the cubic, so

$$x^4 + 6x^2 - 16x + 9 = (x - 1)^2(x^2 + 2x + 9).$$

This is the complete factorisation because the quadratic has no roots: to see this, either calculate the discriminant and observe that it is negative, or complete the square to see that

$$x^2 + 2x + 9 = (x + 1)^2 + 8 > 0.$$

Comment. To state this more precisely, the quadratic has no *real* roots. Readers who are familiar with complex numbers could use them to take the factorisation one step further.

(b) From (a), we have

$$x^4 + 6x^2 - 16x + 9 \geq 0$$

for all real values of x. This can be written as

$$(x^2 + 3)^2 \geq 16x$$

and so

$$\frac{x}{(x^2 + 3)^2} \leq \frac{1}{16}.$$

Thus, the expression on the left–hand side can never be greater than $\frac{1}{16}$; and it is equal to $\frac{1}{16}$ when $x = 1$; so, this is the required maximum value.

(c) Let α be the angle shown, let (x, y) be the "top right" vertex of the triangle, and let A be the area of the triangle.

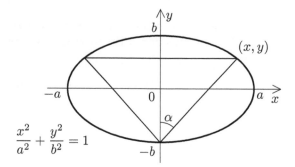

Since the triangle has base length $2x$ and altitude $y + b$, we have

$$A = x(y + b).$$

Now we need to find (x, y), the point in the first quadrant where the triangle meets the ellipse, in terms of α. The line through $(0, -b)$ and (x, y) has gradient $\tan(\frac{\pi}{2} - \alpha) = \cot \alpha$ and so its equation is

$$y = x \cot \alpha - b.$$

Substituting into the equation of the ellipse and simplifying,

$$\frac{x^2}{a^2} + \frac{(x \cot \alpha - b)^2}{b^2} = 1$$

and so

$$b^2 x^2 + a^2(x^2 \cot^2 \alpha - 2bx \cot \alpha + b^2) = a^2 b^2$$

and so

$$(b^2 + a^2 \cot^2 \alpha)x^2 - 2a^2 bx \cot \alpha = 0.$$

Now, this equation has a solution $x = 0$ giving the intersection point $(0, -b)$, which is not the one we want; we need the other solution

$$x = \frac{2a^2 b \cot \alpha}{b^2 + a^2 \cot^2 \alpha}.$$

Therefore,

$$A = x(y + b) = x^2 \cot \alpha = \frac{4a^4 b^2 \cot^3 \alpha}{(b^2 + a^2 \cot^2 \alpha)^2}.$$

Multiplying top and bottom by $\tan^4 \alpha$ gives a slightly simpler expression

$$A = \frac{4a^4 b^2 \tan \alpha}{(b^2 \tan^2 \alpha + a^2)^2},$$

and if we write

$$z = \frac{b}{a}\sqrt{3} \tan \alpha,$$

then this becomes

$$A = 12\sqrt{3}\, ab \, \frac{z}{(z^2 + 3)^2}.$$

But now recall the result of (b). The quotient $z/(z^2+3)^2$ has a maximum value when $z=1$; that is, $\tan\alpha = a/b\sqrt{3}$, and this maximum value is $\frac{1}{16}$. So, the maximum possible area for the triangle is

$$A = \frac{12\sqrt{3}\,ab}{16} = \frac{3\sqrt{3}}{4}\,ab.$$

Alternative solution, for those who know about parametrically defined curves and calculus. Since (x,y) is on the ellipse, we have $x = a\cos\theta$ and $y = b\sin\theta$ for some θ, and then the area of the triangle is

$$A = x(y+b) = ab\cos\theta(1+\sin\theta).$$

Using standard techniques – readers are invited to fill in the details – this has a maximum when $\sin\theta = \frac{1}{2}$, $\cos\theta = \frac{\sqrt{3}}{2}$, and so the greatest possible area is

$$A = ab\,\frac{\sqrt{3}}{2}\frac{3}{2} = \frac{3\sqrt{3}}{4}\,ab$$

as above.

Now try Problem 224.

Q223 Kevin is certainly a truth–teller, as a liar will never admit, "I don't know". If Laura were a truth–teller, then, having heard Kevin's answer, she would know that he was a truth–teller too, and would not give the false answer "One"; so she must be a liar. Mike must be a liar for the same reason. Now suppose that Noela is a liar. Then Laura would have given the right answer: as she is a liar, the only way this could happen is if she didn't know any *definitely* false answer. But this is not the case: she knew that Kevin was a truth–teller and therefore that "None" would have been a false answer. It must therefore be that Noela is a truth–teller (and that Laura knew this, and therefore knew that "One" was a false answer).

For another puzzle like this, try Problem 225.

Q224 The maximum (if it exists) is a positive number; for convenience we call it $1/c$. Therefore, we want

$$\frac{x}{(x^2+a^2)^2} \le \frac{1}{c}$$

for all x, with equality holding for at least one value of x. Multiplying out this inequality gives

$$x^4 + 2a^2x^2 - cx + a^4 \ge 0.$$

Now, if equality holds for some x, say $x = \alpha$, then the left–hand side must have a factor $x - \alpha$; but if the left–hand side is never negative, then $x = \alpha$ must in fact be a double root: that is, $(x - \alpha)^2$ is a factor. Dividing out to obtain a quotient and remainder,

$$x^4 + 2a^2x^2 - cx + a^4 = (x-\alpha)^2(x^2 + 2\alpha x + 3\alpha^2 + 2a^2)$$
$$+ (4\alpha^3 + 4a^2\alpha - c)x - (3\alpha^4 + 2a^2\alpha^2 - a^4),$$

and we require the remainder to be zero. Thus,

$$3\alpha^4 + 2a^2\alpha^2 - a^4 = 0 \tag{2.41}$$

and

$$4\alpha^3 + 4a^2\alpha - c = 0. \tag{2.42}$$

Now take 3α times equation (2.42) minus 4 times (2.41):

$$4a^2\alpha^2 - 3\alpha c + 4a^4 = 0, \tag{2.43}$$

and now a^2 times (2.42) minus α times (2.43):

$$3\alpha^2 c - a^2 c = 0.$$

Since c cannot be zero, we have $\alpha^2 = a^2/3$, giving the location of the double root, and then equation (2.42) yields

$$c = 4\alpha^3 + 4a^2\alpha = \pm\frac{16a^3}{3\sqrt{3}}.$$

As c is positive, we reject the negative root, and so the required maximum is

$$\frac{1}{c} = \frac{3\sqrt{3}}{16a^3}.$$

Alternative solution. Instead of investigating the whole problem from the beginning, we can reduce it to the special case we saw in Problem 222 (solution on page 165). If we take $y = \sqrt{3}\,x/a$, then the expression of interest is

$$\frac{x}{(x^2 + a^2)^2} = \frac{ay/\sqrt{3}}{((a^2 y^2/3) + a^2)^2} = \frac{3\sqrt{3}}{a^3}\frac{y}{(y^2 + 3)^2}.$$

Since we already know that the maximum value of $y/(y^2 + 3)^2$ is $1/16$, the answer to the present problem is

$$\frac{3\sqrt{3}}{16a^3}$$

as above.

Q225 As in the solution to Problem 223, Helen must be a truth–teller (liars don't admit their ignorance).

Now consider Ian's answer. He has heard Helen's reply and knows that she is a truth–teller. Thus, Ian has given a false answer, and he must be a liar.

Suppose that Jacqui is a liar. She has heard the others' statements and she can work out, just as we have done, that Helen is a truth–teller and Ian is a liar. So she knows for sure that "None" and "Three" are false answers. Therefore, as she has given the answer "Two", she must know for sure that that is false too. That is, she knows that George is a truth–teller. But in that case, George has given a false answer (the correct answer is "Two"), which is impossible: truth–tellers either give the right answer or say "I don't know". Therefore, the assumption that Jacqui is a liar is untenable, and she must be a truth–teller. This being the case, and since she knows that Helen is a truth–teller, she must also know that George is a liar. (She could not deduce this from what he said, so she must have known it already.)

So, George and Ian are liars, while Helen and Jacqui are truth–tellers.

Comment. Notice that George, although he is a liar, has given a true answer! It must be that he had no idea at all whether the others were truth–tellers or not, so he gave a random answer which by accident happened to be correct.

Q226

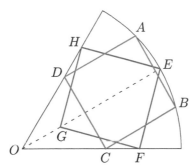

Since OE bisects $\angle O$ and $DC \perp OE$, the triangle $\triangle OCD$ is isosceles; and since it includes a 60° angle, it is equilateral. Therefore, $|CD| = |CO|$, and $|CD| = |CB|$ (sides of the same square); so $|CO| = |CB|$, and $\triangle OCB$ is isosceles. Moreover,

$$\angle OCB = \angle OCD + \angle DCB = 60° + 90° = 150°,$$

and so $\angle CBO = 15°$. Hence,

$$\angle OBD = \angle CBD - \angle CBO = 45° - 15° = 30°.$$

Now consider $\triangle OBD$ and $\triangle EOH$. The former has angles $\angle O = 45°$ and $\angle B = 30°$, with included side length $|OB|$ equal to the radius of the circle; and the other has the same properties. Therefore, the two triangles are congruent, $|HE| = |DO| = |DC|$, and we are finished.

Q227 Let m be Mitchell's probability of winning. There is a $\frac{1}{5}$ chance that he will win on his first throw. For Mitchell to win the game, but not on his first throw, he must fail on his first throw (probability $\frac{4}{5}$), and Dale must fail on his first throw (probability $\frac{5}{6}$). From this point on, it is just as if the game is played again from the start, and so Mitchell's winning probability is again m. Putting all this together gives the equation

$$m = \frac{1}{5} + \left(\frac{4}{5}\right)\left(\frac{5}{6}\right)m,$$

which is easily solved to give $m = \frac{3}{5}$. Dale's winning probability is therefore $d = 1 - m = \frac{2}{5}$, which can be checked by setting up an equation as we did for Mitchell.

Mitchell and Dale's next game is the subject of Problem 228.

Q228 Suppose that the player who throws first uses a die with s sides, and the other uses a die with t sides. If p is the first player's winning probability, then, as in the solution to Problem 227, we have

$$p = \frac{1}{s} + \left(\frac{s-1}{s}\right)\left(\frac{t-1}{t}\right)p;$$

this equation is easily solved (exercise!) to give

$$p = \frac{t}{s+t-1}.$$

For question (a), we find Dale's winning probability when he goes first by taking $s = 6$ and $t = 5$: this gives $p = \frac{5}{10} = \frac{1}{2}$, so he has an even chance of winning. In case (b), we take $s = 12$ and $t = 10$; Dale's winning probability is $\frac{10}{21}$, which is less than $\frac{1}{2}$, so he should not accept the offer.

Q229

(a) Since $\pi > 3$, the two ratios are

$$\frac{\pi a^2}{(2a)^2} = \frac{\pi}{4} > \frac{3}{4} > \frac{2}{3} \quad \text{and} \quad \frac{(2b)^2/2}{\pi b^2} = \frac{2}{\pi} < \frac{2}{3}.$$

(b) We have

$$\frac{\pi a^2 + 2b^2}{4a^2 + \pi b^2} = \frac{2}{3} \quad \Leftrightarrow \quad (3\pi - 8)a^2 = (2\pi - 6)b^2$$

and so

$$\frac{a}{b} = \sqrt{\frac{2\pi - 6}{3\pi - 8}}.$$

(c) To satisfy the given condition, we require

$$\frac{m\pi a^2 + 2na^2}{4ma^2 + n\pi a^2} = \frac{2}{3} \quad \Leftrightarrow \quad 3m\pi + 6n = 8m + 2n\pi.$$

Treating π as a "variable" and "solving" the equation, we get

$$\pi = \frac{8m - 6n}{3m - 2n};$$

But this is impossible since π is an irrational number and cannot be expressed as the ratio of two integers.

Q230 By the Sine Rule we have

$$\frac{\sin A}{a} = \frac{\sin B}{b} = \frac{\sin C}{c};$$

denote the common value by K. Using the "sine of a difference" formula, we have

$$a^3 \sin(B - C) + b^3 \sin(C - A) + c^3 \sin(A - B)$$
$$= a^3(\sin B \cos C - \cos B \sin C) + b^3(\sin C \cos A - \cos C \sin A)$$
$$\quad + c^3(\sin A \cos B - \cos A \sin B)$$
$$= a^3(bK \cos C - cK \cos B) + b^3(cK \cos A - aK \cos C)$$
$$\quad + c^3(aK \cos B - bK \cos A)$$
$$= K\left[(a^2 - b^2)ab \cos C + (b^2 - c^2)bc \cos A + (c^2 - a^2)ca \cos B\right] \ldots$$

...and now we use the Cosine Rule to show that this equals

$$\frac{K}{2}\left[(a^2 - b^2)(a^2 + b^2 - c^2) + (b^2 - c^2)(b^2 + c^2 - a^2) + (c^2 - a^2)(c^2 + a^2 - b^2)\right].$$

If we expand this, then everything cancels and we get zero, as required. For the trigonometric formulae used in this solution, see Section 3.7.

Q231 We follow the method of solution of Problem 227. Let p be the first player's probability of winning overall. Their chance of winning on their first throw is $\frac{1}{6}$. If they do not win on their first throw, then there are two options which will still allow them to win. They could pass the die to the other player who could then make a losing throw: the probability of this is $(\frac{3}{6})(\frac{2}{6}) = \frac{1}{6}$. Alternatively, they could pass the die and the second player could pass it back, probability $(\frac{3}{6})(\frac{3}{6}) = \frac{1}{4}$; the first player is then exactly in the same position as they were at the beginning of the game and their winning probability from here is just p. Putting all this together gives the equation

$$p = \frac{1}{6} + \frac{1}{6} + \frac{1}{4}p,$$

which can be solved to give the winning probability

$$p = \frac{4}{9}.$$

Check: since the first player has a higher chance of losing than winning on the first throw of the game, it makes sense that their overall winning probability should work out as less than $\frac{1}{2}$.

Q232 The kth term on the right–hand side is

$$\frac{2^k}{10^{2^k} + 1}.$$

Now observe that

$$\frac{2^{k+1}}{10^{2^{k+1}} - 1} = \frac{2^k[(10^{2^k} + 1) - (10^{2^k} - 1)]}{(10^{2^k} + 1)(10^{2^k} - 1)} = \frac{2^k}{10^{2^k} - 1} - \frac{2^k}{10^{2^k} + 1};$$

therefore, the kth term can be written alternatively as

$$\frac{2^k}{10^{2^k} - 1} - \frac{2^{k+1}}{10^{2^{k+1}} - 1}.$$

So, the entire sum is equal to

$$\left(\frac{2}{10^2 - 1} - \frac{4}{10^4 - 1}\right) + \left(\frac{4}{10^4 - 1} - \frac{8}{10^8 - 1}\right)$$

$$+ \left(\frac{8}{10^8 - 1} - \frac{16}{10^{16} - 1}\right)$$

$$+ \cdots$$

$$+ \left(\frac{512}{10^{512} - 1} - \frac{1024}{10^{1024} - 1}\right)$$

$$+ \left(\frac{1024}{10^{1024} - 1} - \frac{2048}{10^{2048} - 1}\right);$$

all of the terms except the first and last cancel to give

$$S = \frac{2}{10^2 - 1} - \frac{2048}{10^{2048} - 1} = \frac{2}{99} - \frac{2048}{999 \cdots 999},$$

where there are 2048 nines in the last denominator. Now, if the denominator of a fraction $p/q < 1$ is the number consisting of n digits, all equal to 9, then p/q can be written as a repeating decimal in which the repeating part has length n and contains the digits of p, preceded by a sufficient number of 0s to give that length – to prove this, write down the repeating decimal described and sum a geometric series to get

$$\frac{p}{10^n} + \frac{p}{10^{2n}} + \frac{p}{10^{3n}} + \cdots = \frac{p}{10^n} \frac{1}{1 - \frac{1}{10^n}} = \frac{p}{10^n - 1} = \frac{p}{q}.$$

Using this fact gives the decimals

$$\frac{2}{99} = 0.02020202020202\cdots \quad \text{and} \quad \frac{2048}{10^{2048} - 1} = 0.000\cdots 0002048\cdots,$$

where there are 2044 zeros and the decimal now repeats. Subtracting these gives

$$S = 0.020202\cdots 020202018154\cdots:$$

the pair 02 occurs 1021 times, and the entire part after the decimal point which we have shown now repeats.

Q233 If x is a 10–digit number, then $10^9 \le x < 10^{10}$ and so

$$10^{18} \le x^2 < 10^{20}.$$

If x^2 begins with the digits 2015, then there are two options:

$$2015 \times 10^{15} \le x^2 < 2016 \times 10^{15}$$
$$\text{or} \quad 2015 \times 10^{16} \le x^2 < 2016 \times 10^{16}.$$

Using a calculator (and remembering that x is an integer),

$$1419506957 \le x \le 1419859147$$
$$\text{or} \quad 4488875138 \le x \le 4489988864.$$

But since x ends in the digits 2015, the options are

$$x = 1419512015 \text{ to } 1419852015 \quad \text{or} \quad 4488882015 \text{ to } 4489982015,$$

and the number of possibilities is

$$(141985 - 141950) + (448998 - 448887) = 146.$$

Q234 Let $\lfloor x \rfloor = n$. Then n is an integer, $n \le 2014$, and $x = n + \alpha$, where $0 \le \alpha < 1$. Substituting into the "quadratic" and simplifying, we get

$$n\alpha + \alpha^2 = 20.15.$$

Now, $0 \le \alpha < 1$, so n cannot be negative and we have

$$n + 1 > n\alpha + \alpha^2,$$

so $n \ge 20$. There are 1995 values of n from 20 to 2014; we shall show that, for each of these values, there is exactly one possible value of α and hence one possible value of x.

Consider $n\alpha + \alpha^2$ as a quadratic in α. This quadratic increases continuously from the value 0 (which obviously is less than 20.15) when $\alpha = 0$ to the value $n + 1$ (which by assumption is greater than 20.15) when $\alpha = 1$; so, there is exactly one value of α at which $n\alpha + \alpha^2 = 20.15$. Thus, we have 1995 possible values of n, each with one associated value of α, giving 1995 values of x.

Comment. For example, take $n = 100$. Then $\alpha^2 + 100\alpha = 20.15$, which has the positive solution $\alpha = -50 + \sqrt{2520.15} \approx 0.2011$. So, one solution is $x = 100.2011$, approximately.

Q235 Consider the possible remainders when the expression $x^3 - x^2$, for any integer x, is divided by m. All possible remainders will be found by taking $x = 0, 1, 2, \ldots, m - 1$. This gives m remainders; but they will not all be different, since $x = 0$ and $x = 1$ both give remainder 0. Therefore, there will be an integer c from 0 to $m - 1$ which is never found as a remainder when $x^3 - x^2$ is divided by m; and therefore $p(x) = x^3 - x^2 - c$ never has remainder 0; that is, it is never a multiple of m.

Q236 Let the points on the parabola be $A = (a, a^2)$ and $B = (b, b^2)$ and $C = (c, c^2)$, where $\angle BAC$ is a right angle. Then (1) AB and AC are perpendicular, so the product of their gradients is -1; and (2) AB and AC have equal length; and (3) the point (s, t) is the midpoint of BC. Writing these facts as equations and using the given value $s = \frac{3}{2}$, we have

$$\frac{b^2 - a^2}{b - a} \frac{c^2 - a^2}{c - a} = -1 \tag{2.44}$$

$$(b - a)^2 + (b^2 - a^2)^2 = (c - a)^2 + (c^2 - a^2)^2 \tag{2.45}$$

$$b + c = 3. \tag{2.46}$$

Simplifying (2.44) and factorising (2.45) gives

$$(b + a)(c + a) = -1 \tag{2.47}$$

$$(b - a)^2[1 + (b + a)^2] = (c - a)^2[1 + (c + a)^2]. \tag{2.48}$$

Now multiply both sides of (2.48) by $(b + a)^2$ and substitute from (2.47) to get

$$(b + a)^2(b - a)^2[1 + (b + a)^2] = (c - a)^2[(b + a)^2 + 1];$$

that is,

$$(b^2 - a^2)^2 = (c - a)^2.$$

There are two possibilities: $b^2 - a^2 = c - a$ or $b^2 - a^2 = a - c$. We shall solve the first and shall leave the reader to check that the second gives the same solution with the points B and C interchanged, which makes no difference in terms of the stated question. So, suppose that $b^2 - a^2 = c - a$, and use (2.46) to eliminate c from this equation and from (2.47). We have

$$b^2 - a^2 = 3 - a - b \tag{2.49}$$

$$(b + a)(3 + a - b) = -1. \tag{2.50}$$

Expanding (2.50) gives $3b + 3a + a^2 - b^2 = -1$; adding (2.49) to this equation and simplifying gives

$$a + b = \frac{1}{2}.$$

Substituting back into (2.50) leads to

$$a - b = -5$$

and it is now easy to get

$$a = -\frac{9}{4}, \quad b = \frac{11}{4}, \quad c = \frac{1}{4}.$$

Therefore, three of the vertices of the square are

$$A = \left(-\frac{9}{4}, \frac{81}{16}\right), \quad B = \left(\frac{11}{4}, \frac{121}{16}\right), \quad C = \left(\frac{1}{4}, \frac{1}{16}\right),$$

and the fourth, which does not lie on the parabola, is

$$D = B + C - A = \left(\frac{21}{4}, \frac{41}{16} \right).$$

Q237 Consider

$$N = \left(\sqrt{20} + \sqrt{15} \right)^{2016} + \left(\sqrt{20} - \sqrt{15} \right)^{2016}.$$

Expanding by the Binomial Theorem (see Section 3.6),

$$N = \sum_{k=0}^{2016} \binom{2016}{k} \left(\sqrt{20} \right)^k \left(\sqrt{15} \right)^{2016-k} + \sum_{k=0}^{2016} (-1)^k \binom{2016}{k} \left(\sqrt{20} \right)^k \left(\sqrt{15} \right)^{2016-k}.$$

Now, the terms with even values of k will be integers (since an even power of $\sqrt{a}$ is a whole power of a), while the terms with odd k in the first sum will cancel with the corresponding terms in the second sum. Therefore, N is an integer, and we have

$$\left(\sqrt{20} + \sqrt{15} \right)^{2016} = N - \left(\sqrt{20} - \sqrt{15} \right)^{2016}.$$

Moreover,

$$\sqrt{20} - \sqrt{15} = \frac{20 - 15}{\sqrt{20} + \sqrt{15}} < \frac{5}{8};$$

a (fairly) easy calculation then shows that

$$\left(\sqrt{20} - \sqrt{15} \right)^5 < \frac{3125}{32768} < \frac{1}{10}$$

and so

$$\left(\sqrt{20} - \sqrt{15} \right)^{2016} < \left(\sqrt{20} - \sqrt{15} \right)^{2015} < \frac{1}{10^{403}}.$$

So, the decimal expansion of $\left(\sqrt{20} - \sqrt{15} \right)^{2016}$ begins with at least 403 zeros; when this is subtracted from the integer N, the part after the decimal point begins with at least 403 nines. So, the 400th digit is a 9.

Q238 We'll refer to a town which can be reached from every other town either directly, or with just one intermediate town, as a "central" town. We shall use mathematical induction on n, the number of towns, to prove that there always exists a central town. (See Section 3.8 for a brief explanation of mathematical induction.) The claim that there exists a central town is clearly true if the country has only one town (as there are no other towns to cause any problems!), and if it has two towns (as one of the towns will have a highway from the other, and so the former is central). Assume that the claim is true for a town with n countries, and let x be the central town. Now add another town y, and construct highways between y and every other town. We shall prove that either x or y is a centre of the expanded country.

So, suppose that x is **not** a centre of the expanded country. Since we know that x is reachable in at most two steps from every "old" town, the only possible problem is with y. Therefore, there is no highway $y \to x$, and no pair of highways $y \to ? \to x$.

Now consider any town a other than y.

- If $a = x$, then there is no highway $y \to a$, so there is a highway $a \to y$.
- If there is a highway $a \to x$, then there is a pair $a \to x \to y$.

- If there is a pair $a \to b \to x$, then there is no highway $y \to b$, so there is a highway $b \to y$ and there is a pair $a \to b \to y$.

This shows that, if a country with n towns has a central town, then so does a country with $n + 1$ towns. By mathematical induction, no matter how many towns are in the country, there must be a central town.

See also Problem 240.

Q239 The given expression is the distance from $(0,0)$ to $(8,6)$, going via an unspecified point $(x_1, 1)$ on the line $y - 1$ and then an unspecified point $(x_1 + x_2, 3)$ on the line $y = 3$, as shown in the diagram.

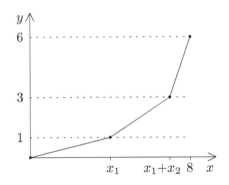

The minimum value is clearly the straight line distance from $(0,0)$ to $(8,6)$, which by Pythagoras' Theorem is 10.

See also Problem 241.

Q240 Draw a schematic diagram where the towns (irrespective of their actual geographic locations) are arranged on a circle. From each town T, build a highway to every other town which is an odd number of steps around the circle from T in a clockwise direction, stopping just before we have gone all the way round the circle once. The first thing to do is to check that we have not infringed the highway–building rules by having a highway in both directions between the same pair of towns. If this were so for towns T_1 and T_2, then we would be able to travel along a highway from T_1 to T_2 and then another from T_2 to T_1; since each highway goes an odd number of steps around the circle in a clockwise direction, we have gone once around the circle from T_1 to T_1 in an even number of steps. But this is impossible as it takes n steps, an odd number, to go once round the circle.

So, we have a legitimate construction. Any town T can be reached directly from each town which is an odd number of steps anticlockwise from T; it can also be reached in two steps from any town an even number of steps anticlockwise from T by first going to the town before T on the circle. Therefore, every town is central.

More on this strange collection of highways in Problem 244.

Q241 To solve (a), draw a diagram like that in the solution to Problem 239: the given expression can be written as $y_1 + y_2 + y_3$, where the distances from $(0,0)$ to (x_1, y_1) and then to $(x_1 + x_2, y_1 + y_2)$ and then to $(x_1 + x_2 + x_3, y_1 + y_2 + y_3)$ are 1 and 2 and 3 respectively. So, we have to find the maximum height y for which $(1, y)$ can be reached from $(0,0)$ in steps of length $1, 2, 3$. The maximum will occur when we make these three steps collinear, forming a right triangle with side 1 and hypotenuse 6. The required maximum is the other side of the triangle; that is, $\sqrt{35}$.

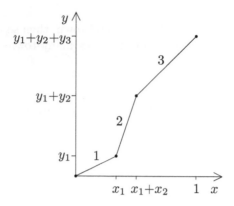

For (b), the given expression represents the distance from $(1, 2)$ to $(3, 4)$ via the parabola $y = x^2$. Since $(1, 2)$ is above the parabola and $(3, 4)$ is below, the minimum length path will again be a straight line, and the distance is $2\sqrt{2}$. The value of x is obtained by finding the point (x, x^2) which is collinear with the given points:

$$\frac{x^2 - 2}{x - 1} = \frac{4 - 2}{3 - 1} = 1,$$

which simplifies to $x^2 - x - 1 = 0$ and, hence,

$$x = \frac{1 + \sqrt{5}}{2} :$$

we have rejected the negative root since it is clear that for the intersection point we must have $1 < x < 3$.

See also Problem 245.

Q242 Let x be a number greater than 1 such that

$$a_1 = x + \frac{1}{x}$$

(we shall confirm later that such a number exists). Then

$$a_2 = \left(x + \frac{1}{x}\right)^2 - 2 = x^2 + \frac{1}{x^2}, \quad a_3 = \left(x^2 + \frac{1}{x^2}\right)^2 - 2 = x^4 + \frac{1}{x^4}$$

and so on: in general,

$$a_n = x^{2^{n-1}} + \frac{1}{x^{2^{n-1}}}.$$

Next, we have

$$\left(x - \frac{1}{x}\right)a_1 = \left(x - \frac{1}{x}\right)\left(x + \frac{1}{x}\right) = x^2 - \frac{1}{x^2}$$

$$\left(x - \frac{1}{x}\right)a_1 a_2 = \left(x^2 - \frac{1}{x^2}\right)\left(x^2 + \frac{1}{x^2}\right) = x^4 - \frac{1}{x^4}$$

and in general

$$\left(x - \frac{1}{x}\right)a_1 a_2 a_3 \cdots a_{n-1} = x^{2^{n-1}} - \frac{1}{x^{2^{n-1}}}.$$

Therefore,

$$\frac{a_n}{a_1 a_2 a_3 \cdots a_{n-1}} = \frac{x^{2^{n-1}} + \frac{1}{x^{2^{n-1}}}}{x^{2^{n-1}} - \frac{1}{x^{2^{n-1}}}} \left(x - \frac{1}{x} \right) = \frac{1 + \frac{1}{x^{2^n}}}{1 - \frac{1}{x^{2^n}}} \left(x - \frac{1}{x} \right);$$

in the limit as $n \to \infty$, we have $1/x^{2^n} \to 0$ and the expression approaches

$$x - \frac{1}{x}.$$

To finish, we have

$$x + \frac{1}{x} = 3 \quad \Leftrightarrow \quad x^2 - 3x + 1 = 0 \quad \Leftrightarrow \quad x = \frac{3 \pm \sqrt{5}}{2};$$

in order to have $x > 1$ we choose the positive root, and hence

$$\frac{a_n}{a_1 a_2 a_3 \cdots a_{n-1}} \to x - \frac{1}{x} = \frac{3 + \sqrt{5}}{2} - \frac{2}{3 + \sqrt{5}} = \sqrt{5}.$$

Q243 Starting at B, the task cannot be achieved. The required path must contain A, C, E, G, J, L, none of which can be accessed directly from any of the others. So these six points must be separated by the remaining five points B, D, F, H, K; but since B is to be the first point on the path, this is impossible.

Q244

(a) With a bit of trial and error we see that the following will work:

from	1	2	3	4
to	$2, 4, 5$	$3, 5, 8$	$1, 4, 6, 7$	$2, 5, 6, 7$

from	5	6	7	8
to	$3, 6, 8$	$1, 2, 7$	$1, 2, 5, 8$	$1, 3, 4, 6$

You may easily check that this is a legitimate arrangement (there is never a highway both from x to y and from y to x) and that every town is accessible from every other town in either one or two steps.

(b) We shall show that if it is possible to arrange highways in a country of n towns (with $n > 1$) in such a way that every town is central, then the same is true in a country of $2n$ towns.

So, suppose that we have a suitable arrangement for a country having towns $T_1, T_2, \ldots, T_n$. Consider a country with $2n$ towns; suppose that it is divided into two states called A and B, and that the towns are called $A_1, A_2, \ldots, A_n$ and $B_1, B_2, \ldots, B_n$. (The people who choose town names in these countries are not very imaginative.) Now construct highways as follows:

- highways in A are arranged as in the smaller country: that is, there is a highway from A_i to A_j if and only if there is a highway from T_i to T_j;
- highways in B are arranged as in the smaller country but with directions reversed: that is, there is a highway from B_i to B_j if and only if there is a highway from T_j to T_i;

- the highway between A_i and B_j will be directed from A_i to B_j if $i = j$, and from B_j to A_i if $i \neq j$.

Since every town in the smaller country is central, it is possible to get from any town in A to any other town in A with at most one intermediate town; this property is not changed by reversing the directions of all highways, so the same is true in B. We have to check that we can get to every town in B from every town in A with at most one intermediate town, and likewise to A from B.

So, consider any town B_j. If $i = j$, then there is a direct connection $A_i \to B_j$. If $i \neq j$ and there is a highway $B_i \to B_j$, then we have $A_i \to B_i \to B_j$. If $i \neq j$ and there is no highway $B_i \to B_j$, then $B_j \to B_i$ (remember that *every* pair of towns is connected one way or the other), so $T_i \to T_j$, so $A_i \to A_j$ and we have $A_i \to A_j \to B_j$. Thus, B_j can be reached from every A_i in at most two steps.

Finally, consider any town A_i. If $j \neq i$, then we have $B_j \to A_i$. If $j = i$, then choose a town T_k having a highway to T_i. (There must be such a town; otherwise, T_i would be inaccessible and hence not central: this is why we required $n > 1$.) Therefore, $A_k \to A_i$, and since $k \neq j$ we have $B_j \to A_k \to A_i$.

This shows that every town is central in the country of $2n$ towns. Since we know that it is possible to make every town central in a country of 8 towns (above) or $1, 3, 5, 7, \ldots$ towns (Problem 240, solution on page 175), doubling a sufficient number of times (but not for $n = 1$) gives, for any number of towns except 2 or 4, an arrangement in which every town is central.

Q245 Draw a diagonal of length $\sqrt{t}$ to form triangles with sides $2, 3, \sqrt{t}$ and $4, 5, \sqrt{t}$. Using Heron's formula (page 262) for the area of a triangle in terms of its sides, the first triangle has area given by

$$4A = \sqrt{\left(2 + 3 + \sqrt{t}\right)\left(2 + 3 - \sqrt{t}\right)\left(2 + \sqrt{t} - 3\right)\left(3 + \sqrt{t} - 2\right)}$$
$$= \sqrt{(25 - t)(t - 1)}$$
$$= \sqrt{-25 + 26t - t^2}$$
$$= \sqrt{12^2 - (t - 13)^2}.$$

Doing the same sort of thing for the other triangle, the total area of the quadrilateral is given by

$$4A = \sqrt{12^2 - (t - 13)^2} + \sqrt{40^2 - (41 - t)^2}. \tag{2.51}$$

To maximise this, we use the technique of Problem 241(a). The right–hand side of (2.51) is the vertical distance gained by the broken line going from $(13, 0)$ to (t, y_1) and then to $(41, y_1 + y_2)$, where the two line segments have lengths 12 and 40. We maximise the height gain by making the two parts collinear, giving a right–angled triangle with hypotenuse 52, horizontal side $41 - 13 = 28$ and therefore vertical side

$$\sqrt{52^2 - 28^2} = \sqrt{80 \times 24} = 8\sqrt{30}.$$

Therefore, the maximum area is $2\sqrt{30}$.

Q246 First we note that there are exactly 11 collections of positive integers which add up to 6, namely,

$$6 = 5 + 1 = 4 + 2 = 4 + 1 + 1 = 3 + 3 = 3 + 2 + 1$$
$$= 3 + 1 + 1 + 1 = 2 + 2 + 2 = 2 + 2 + 1 + 1$$
$$= 2 + 1 + 1 + 1 + 1 = 1 + 1 + 1 + 1 + 1 + 1.$$

Therefore, we must use every one of these once each. Obviously, the 6 must go in a region by itself. There are two options for $2+2+2$; but if we take that in the top left corner, then there is nowhere to obtain $4+2$. The rest can be done by trial and error.

2	2	1	2	1	1	1
2	4	3	1	1	2	2
1	1	2	1	1	2	1
5	1	3	1	6	1	1
1	3	3	1	1	1	4

Q247 Imagine that instead of the wedge remaining fixed and the ball bouncing, the ball keeps going and the wedge is reflected; and that this happens every time the ball hits a wall. The scenario looks like this:

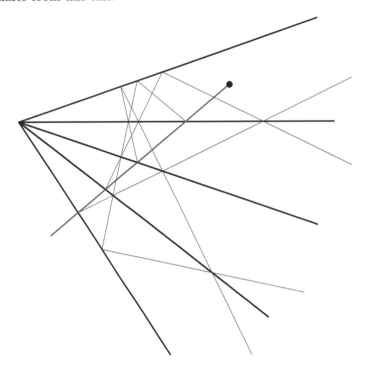

Since for the bouncing ball (ignoring any spin) the angle of incidence equals the angle of reflection, the successive segments of the ball's path form a straight line (shown in red in the diagram). So, the answer we are looking for is the shortest distance from this line to the vertex. If we treat the vertex as the origin in the Cartesian plane, then the line goes through (x_0, y_0) and has gradient $\tan\theta$; by a well–known formula, the distance from the line to the origin is

$$|x_0 \sin\theta - y_0 \cos\theta|.$$

Another similar problem is suggested in Problem 251.

Q248 To assist with the solution, give the grid a chessboard colouring as shown.

START

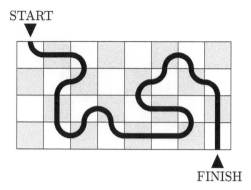

FINISH

The connections are on light and dark squares alternately; the path both starts and finishes on a light square; so there must be an odd number of connections overall. Each quarter–circle connection changes the direction of the path from horizontal to vertical, or *vice versa*; the path both starts and finishes vertically; therefore there must be an even number of quarter–circles. Hence, there are an odd number of straight connections.

Q249 There are many ways to solve the equation

$$\sqrt{x+20} - \sqrt{x} = 17, \tag{2.52}$$

but perhaps this is the simplest: using a difference of two squares we have

$$\left(\sqrt{x+20} + \sqrt{x}\right)\left(\sqrt{x+20} - \sqrt{x}\right) = (x+20) - x = 20.$$

Dividing this by equation (2.52), we get

$$\sqrt{x+20} - \sqrt{x} = \frac{20}{17},$$

and now subtracting this from (2.52) gives

$$2\sqrt{x} = 17 - \frac{20}{17}.$$

Therefore,

$$x = \left(\frac{1}{2}\left(17 - \frac{20}{17}\right)\right)^2 = \left(\frac{269}{34}\right)^2.$$

Q250 Let $T = (x, y)$ be any point on the circle such that OT is perpendicular to PT. By the "gradients of perpendicular lines" theorem, we have

$$\frac{y-b}{x-a}\frac{y-q}{x-p} = -1,$$

which can be rewritten

$$(x-a)(x-p) + (y-b)(y-q) = 0;$$

and since T is on the circle we have

$$(x-a)^2 + (y-b)^2 = r^2.$$

Subtract the previous equation from this one and do a little algebra to get

$$(p-a)(x-a) + (q-b)(y-b) = r^2. \tag{2.53}$$

Now, observe that (i) this is the equation of a line, since it can be simplified to take the form $Ax + By = C$ for certain constants A, B, C with A, B not both zero; and (ii) both P and Q lie on this line, since both $T = P$ and $T = Q$ satisfy the conditions OT on the circle and $OT \perp PT$ ("tangent perpendicular to radius" theorem). So (2.53) is our answer!

Comment. The "gradients of perpendicular lines" equation does not make sense if OT is horizontal and PT is vertical, or *vice versa*; but the following equation is still correct in these cases, as you may easily check, and so our solution is still correct.

Q251 As in the solution to Problem 247, we imagine that instead of the wedge remaining fixed and the ball being reflected, the wedge is reflected and the ball keeps going in a straight line. The successive angles (in degrees) made by the ball as it "exits" a wedge are $180 - \theta$, $180 - \theta - \alpha$, $180 - \theta - 2\alpha$ and so on, as shown in the diagram.

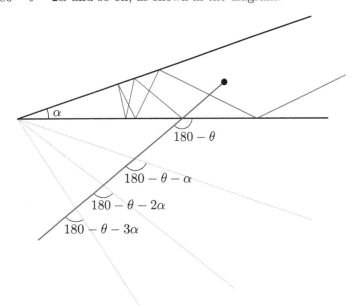

Now let k be a positive integer. The "exit angle" after the ball hits the wedge for the kth time is $180 - \theta - (k - 1)\alpha$; and the ball will never hit the wedge again if this is less than or equal to the angle α of the wedge. So, the number of hits is the smallest k such that

$$180 - \theta - (k - 1)\alpha \le \alpha;$$

that is, the smallest k such that

$$k \ge \frac{180 - \theta}{\alpha};$$

that is, $(180 - \theta)/\alpha$, rounded to the nearest integer upwards.

Q252 First note that, if we write

$$D = a^2b + a^2c + b^2c + b^2a + c^2a + c^2b,$$

then

$$(a + b + c)^3 = (a^3 + b^3 + c^3) + 3D + 6abc$$

and

$$(a^2 + b^2 + c^2)(a + b + c) = (a^3 + b^3 + c^3) + D.$$

Eliminating D from these equations and making abc the subject yields

$$abc = \frac{(a+b+c)^3 + 2(a^3+b^3+c^3) - 3(a^2+b^2+c^2)(a+b+c)}{6} = 1.$$

Moreover, the identity

$$(a+b+c)^2 - (a^2+b^2+c^2) = 2ab + 2ac + 2bc$$

yields

$$ab + ac + bc = \frac{(a+b+c)^2 - (a^2+b^2+c^2)}{2} = 5.$$

We now have $a+b+c = 5$ and $ab+ac+bc = 4$ and $abc = 1$; from Section 3.10, we see that a, b, c are the roots of the cubic $x^3 - 5x^2 + 5x - 1$. Factorising,

$$x^3 - 5x^2 + 5x - 1 = (x-1)(x^2 - 4x + 1) = (x-1)(x - (2 - \sqrt{3}))(x - (2 + \sqrt{3}));$$

and since we are given that $a < b < c$, we have the only solution

$$a = 2 - \sqrt{3}, \quad b = 1, \quad c = 2 + \sqrt{3}.$$

Q253 Since the radius of the cone must be greater than the radius of the sphere, let the cone's radius be ar, such that $a > 1$ and ar is the volume–minimising radius of the cone. In addition, let $x = |AD|$.

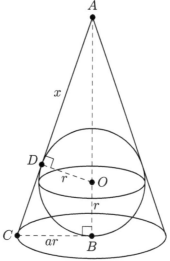

Using the fact that $\triangle ADO$ is similar to $\triangle ABC$, we can write the equation

$$\frac{x}{r} = \frac{\sqrt{x^2 + r^2} + r}{ar}. \tag{2.54}$$

We solve for x in terms of a and r, first getting $(ax - r)^2 = x^2 + r^2$ and thus

$$x = \frac{2ar}{a^2 - 1}.$$

From (2.54), the height of the cone is

$$\sqrt{x^2 + r^2} + r = ax = \frac{2a^2 r}{a^2 - 1}, \tag{2.55}$$

and so the volume of the cone, expressed as a function $f(a)$ of a, is

$$f(a) = \frac{1}{3}\pi(ar)^2(ax) = \frac{2\pi}{3}\frac{a^4 r^3}{a^2 - 1}.$$

Because the cone must have minimum volume, we need to find the value of a that minimises our function, so we calculate the derivative:

$$f'(a) = \frac{2\pi r^3}{3}\frac{a^3(a^2 - 2)}{(a^2 - 1)^2};$$

setting this derivative equal to 0 and solving for a, we find that $a = 0, \pm\sqrt{2}$. Since $a > 1$, the relevant solution is $a = \sqrt{2}$. Thus, the radius of the cone is $r\sqrt{2}$ and by (2.55), its height is $4r$. Therefore, the volume of the cone is

$$\frac{1}{3}\pi\left(r\sqrt{2}\right)^2(4r) = \frac{8\pi r^3}{3},$$

which is exactly double the volume of the sphere, as claimed.

Comment. The minimisation problem can be solved without the use of calculus by substituting $a = \csc\theta$ with $0 < \theta \leq \frac{\pi}{2}$ and then using trigonometric identities (including some of those reviewed in Section 3.7) to obtain

$$f(a) = \frac{8\pi r^3}{3}\frac{1}{\sin^2 2\theta},$$

which has a minimum when $\theta = \frac{\pi}{4}$.

Q254 Let a, b and c be the average amounts required before exiting if the driver starts at roundabout A, B or C respectively. Starting from A, the driver has a $\frac{1}{3}$ chance of taking the road to B; in this case, they will pay \$1 for taking that road, plus, on average, another \$$b$ to get out. There is also a $\frac{1}{3}$ chance of looping back to A; this costs \$9 plus, on average, another \$$a$. Finally, there is a $\frac{1}{3}$ chance of going to C, which costs \$7 immediately plus, on average, another \$$c$. The total average cost is

$$a = \frac{1}{3}(1 + b) + \frac{1}{3}(9 + a) + \frac{1}{3}(7 + c).$$

Applying similar ideas to drivers who start from B or from C gives the equations

$$b = \frac{1}{3}(8 + a) + \frac{1}{3}(2) + \frac{1}{3}(3 + c)$$

and

$$c = \frac{1}{3}(5 + a) + \frac{1}{3}(6 + a) + \frac{1}{3}(4).$$

Note that the last term in the last equation corresponds to choosing the exit from C, in which case the driver pays the \$4 toll but has nothing further to pay. Simplifying these equations yields

$$
\begin{aligned}
2a - b - c &= 17 \\
-a + 3b - c &= 13 \\
-2a \quad\quad + 3c &= 15
\end{aligned}
$$

which can be solved (exercise!) to give $a = 36$, $b = 26$, $c = 29$. Since the first roundabout the driver reaches is A, the average cost to get out of MessConnex is \$36.

Q255 It is well known that a leap year contains 1 month of 29 days, 4 months of 30 days and 7 months of 31 days, a total of 366 days. That is,

$$(29 \times 1) + (30 \times 4) + (31 \times 7) = 366,$$

and so a solution is $x = 1$, $y = 4$, $z = 7$. To show that there is no other solution, we consider two possibilities: $x = 1$ or $x \geq 2$.

If $x = 1$, then we have

$$30y + 31z = 337.$$

Since we know one solution $y = 4$, $z = 7$, we can use the technique explained in Section 3.2 to find all possible solutions

$$y = 4 - 31t, \quad z = 7 + 30t$$

where t is an integer. But if $t > 0$, then y is negative, which is not allowed; if $t < 0$, then z is negative, which is not allowed; and if $t = 0$, then the solution is the one we have already found.

Finally, consider the case $x \geq 2$. We have $y > x$ and so $y \geq 3$. Also, our original equation can be written as

$$z - x = 6(61 + 5x - 5y - 5z);$$

so $z - x$ is a multiple of 6; and hence $z \geq 8$. Therefore, we have

$$29x + 30y + 31z \geq (29 \times 2) + (30 \times 3) + (31 \times 8) = 396,$$

and so $29x + 30y + 31z$ cannot possibly equal 366. Thus, the solution $x = 1$, $y = 4$, $z = 7$ found in (a) is the only possibility.

Q256

(a) Taking (positive) square roots of the given inequality, we have

$$n < \sqrt{a} < \sqrt{b} < n + 1,$$

and hence

$$0 < \sqrt{b} - \sqrt{a} < 1.$$

Squaring both sides,

$$0 < a + b - 2\sqrt{ab} < 1,$$

and rearranging this inequality yields

$$a + b - 1 < 2\sqrt{ab} < a + b.$$

But $a + b - 1$ and $a + b$ are consecutive integers; so $2\sqrt{ab}$, which lies between them, cannot be an integer; so, $\sqrt{ab}$ cannot be an integer; that is, ab cannot be a perfect square.

(b) If $n = 2$, then

$$(n + 1)^{3/2} - n^{3/2} = \sqrt{27} - \sqrt{8} > \sqrt{25} - \sqrt{9} = 2;$$

since the graph of $y = x^{3/2}$ is concave upwards, values of the left–hand side for $n > 2$ are even larger than this. That is, for any integer $n \geq 2$ we have $(n+1)^{3/2} - n^{3/2} > 2$; so there are integers p, q such that

$$n^{3/2} < p < q < (n + 1)^{3/2};$$

consequently,
$$n^3 < p^2 < pq < q^2 < (n+1)^3.$$

So, the integers $a = p^2$, $b = pq$, $c = q^2$ satisfy the required inequality, and their product is
$$abc = (p^2)(pq)(q^2) = (pq)^3,$$

a cube.

Q257 First, suppose that $1 \leq \frac{m}{n} \leq 3$. Consider the following polyomino which contains n dark squares, $n-1$ light squares, and $2n+2$ further squares (marked with crosses) which are optional, and will be light if they are used.

Since $n \leq m \leq 3n$, we have $1 \leq m - n + 1 \leq 2n + 1$; so we can choose $m - n + 1$ of the "optional" squares to be light, and omit the rest. Then the total number of light squares is m, and the ratio is $\frac{m}{n}$ as required.

In the case $\frac{1}{3} \leq \frac{m}{n} \leq 1$, we can obtain a ratio of $\frac{n}{m}$ as above; then interchange light and dark squares.

Q258 Replace the dots by numerical labels as shown.

```
0  1  2  0  1  2  0  1  2  0  1
1  2  0  1  2  0  1  2  0
2  0  1  2  0  1  2
0  1  2  0  1
1  2
```

In this pattern, any three dots which may be removed together (that is, one of them is midway between the other two) either will have the labels $0, 1, 2$, or will have the same label three times. In each case, the sum of the labels removed will be a multiple of 3. Now, the total of all the labels is
$$(11 \times 0) + (12 \times 1) + (11 \times 2) = 34;$$

if we keep on decreasing this total by multiples of 3 until only one label remains, then it must be 1. Now label the dots in a similar but different way.

```
0  1  2  0  1  2  0  1  2  0  1
2  0  1  2  0  1  2  0  1
1  2  0  1  2  0  1
0  1  2  0  1
2  0
```

The total of these labels is 32; as above, whenever we remove three permitted dots, the total must diminish by a multiple of 3; so the remaining label at the end must be 2.

Therefore, the dots which could remain are those which have the label 1 in the first diagram and 2 in the second: these are the first, fourth and seventh dots in the second row, and the first dot in the fifth row.

To confirm that it is possible to leave a dot in the place marked by the red dot in the fifth row, we can remove the groups of three dots marked A, B, ..., K in the following diagram; to confirm that each of the other red dots can also be the survivor is left for you as an exercise.

A A A B B B C C C J K
D F F F G G G J K
D E H H H J K
D E I I I
● E

Q259 The coefficient of n^4 suggests that we try something like this:

$$m^4 + 3n^4 = (n^2 + a)^2 + (n^2 + b)^2 + (n^2 + c)^2,$$

where a, b, c are expressions in terms of m, n. Expanding, we get

$$m^4 + 3n^4 = 3n^4 + 2(a + b + c)n^2 + a^2 + b^2 + c^2.$$

Now somehow, we have to get an m^4 term on the right–hand side. It clearly can't come from the terms containing n, so it looks like one of the terms a^2, b^2, c^2 should be m^4. It obviously doesn't matter which one we go for, so let's say $c^2 = m^4$. Taking $c = m^2$ doesn't seem to lead anywhere (try it!), so we explore $c = -m^2$. This gives

$$m^4 + 3n^4 = m^4 + 3n^4 + 2(a + b - m^2)n^2 + a^2 + b^2.$$

If we remove the a and b from the middle term by choosing $b = -a$, then we have

$$m^4 + 3n^4 = m^4 + 3n^4 - 2m^2n^2 + 2a^2,$$

and it is now clear that we get what we want by taking $a = mn$. So, to sum up, we have found that

$$m^4 + 3n^4 = (n^2 + mn)^2 + (n^2 - mn)^2 + (n^2 - m^2)^2;$$

it is easy to check this by multiplying out the right–hand side, and since m, n are unequal positive integers, it is clear that each bracketed term on the right–hand side is a non–zero integer.

Q260 Eric's second roll cannot have been 1, as it was bigger than his first. There is one way that the second roll could have been 2 (he threw $1, 2$); and he then has four winning possibilities on the third roll $(3, 4, 5, 6)$. There are two ways that the second roll could have been 3 (he threw $1, 3$ or $2, 3$); and in each of these two cases, he has three winning possibilities on the third roll $(4, 5, 6)$. Continuing to think in this way gives the following.

second roll	2	3	4	5	6
number of possibilities	1	2	3	4	5
number of winning third rolls	4	3	2	1	0

So, the number of possibilities for the first two rolls is $1+2+3+4+5 = 15$, and the number for all three is $15 \times 6 = 90$. The number of winning possibilities is

$$(1 \times 4) + (2 \times 3) + (3 \times 2) + (4 \times 1) = 20,$$

and so Eric's chance of winning is $\frac{20}{90} = \frac{2}{9}$.

Q261 We solve a more general case with angles $\angle XAY = \angle XAZ = a$ and $\angle BAY = b$. Draw a perpendicular from B to AY extended, meeting AY at C; extend XA to meet BC at D.

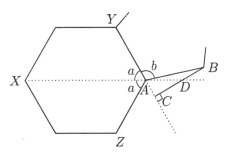

Since the pentagon $YAB\cdots$ is rotated around YC, the perpendicular line BC will always remain directly above its initial position. When the point B meets the other point B (from the diagram in the question), they will be directly above the line XA; and also, as already shown, directly above BC; so they will be directly above the point D. We will denote the location of the points B when this happens by B'; so the problem is to find the angle $\angle XAB'$.

Since $\angle ADB'$ is a right angle, we have

$$\cos \angle DAB' = \frac{|AD|}{|AB'|}.$$

Now, triangles $\triangle ACB$ and $\triangle ACD$ are both right–angled, so

$$|AC| = |AB| \cos(\pi - b) \quad \text{and} \quad |AC| = |AD| \cos a;$$

moreover, the intervals AB' and AB have the same length as one is a rotation of the other. So

$$\cos \angle DAB' = \frac{|AD|}{|AB|} = \frac{|AC|/|AB|}{|AC|/|AD|} = \frac{\cos(\pi - b)}{\cos a},$$

and the required angle is the supplement of this,

$$\angle XAB' = \pi - \arccos \frac{\cos(\pi - b)}{\cos a} = \arccos \frac{\cos b}{\cos a}.$$

In the specific case asked, we have $a = 60°$, $b = 108°$ and so

$$\angle XAB' = \arccos \frac{\cos 108°}{\cos 60°} = 128.2°.$$

Q262 Because of the limit of 2018 on their size, the numbers we have are

$$2^0 5^0 = 1, \ 2^1 5^0 = 2, \ 2^2 5^0 = 4, \ 2^3 5^0 = 8, \ldots, \ 2^{10} 5^0 = 1024,$$
$$2^0 5^1 = 5, \ 2^1 5^1 = 10, \ 2^2 5^1 = 20, \ldots, \ 2^8 5^1 = 1280,$$
$$2^0 5^2 = 25, \ 2^1 5^2 = 50, \ldots, \ 2^6 5^2 = 1600,$$
$$2^0 5^3 = 125, \ldots, \ 2^4 5^3 = 2000,$$
$$2^0 5^4 = 625, \ 2^1 5^4 = 1250.$$

Let a be the sum of all the exponents on powers of 2 among these numbers, and b the sum of all the exponents on powers of 5. We calculate

$$a = (5 \times 0) + (5 \times 1) + (4 \times 2) + (4 \times 3) + (4 \times 4)$$
$$+ (3 \times 5) + (3 \times 6) + (2 \times 7) + (2 \times 8) + 9 + 10 = 123$$
$$b = (11 \times 0) + (9 \times 1) + (7 \times 2) + (5 \times 3) + (2 \times 4) = 46.$$

Now, suppose that we have a set of three numbers in geometric progression, in which the middle term is $2^p 5^q$ and the common ratio is $2^s 5^t$. The numbers in the progression are

$$2^{p-s} 5^{q-t}, \quad 2^p 5^q, \quad 2^{p+s} 5^{q+t};$$

the sum of the exponents on powers of 2 is $3p$, and on powers of 5 is $3q$. So, if we place these three numbers in a set and from now on ignore them, then the sum of the exponents on powers of 2 has decreased by a multiple of 3; and likewise for the sum of the exponents on powers of 5. Since 123 has remainder 0 when divided by 3 and 46 has remainder 1 and these remainders will never change, our leftover term $2^x 5^y$ must have an x value with remainder 0 and a y value with remainder 1 when divided by 3. Within the set of numbers we began with, the possibilities are

$$x = 0, \ y = 1 \ \text{ or } \ x = 0, \ y = 4 \ \text{ or } \ x = 3, \ y = 1 \ \text{ or } \ x = 6, \ y = 1$$

and so the remaining number is

$$5 \quad \text{or} \quad 625 \quad \text{or} \quad 40 \quad \text{or} \quad 320.$$

Alternative solution. If we write the numbers in an array as shown at the beginning of the previous solution,

1	2	4	8	16	32	64	128	256	512	1024
5	10	20	40	80	160	320	640	1280		
25	50	100	200	400	800	1600				
125	250	500	1000	2000						
625	1250									

we see that three numbers in a geometric progression must lie on a straight line, with one of the numbers exactly halfway between the other two. So, this problem is really just Problem 258 in disguise! Using our previous solution (page 185), the leftover number must be the first, fourth or seventh in the second row, or the first in the fifth row: that is, 5, 40, 320 or 625.

Q263 We have

$$\frac{a+b+c}{c} = \frac{a+b}{c} + 1 = 2019, \qquad \frac{a+b+c}{a} = \frac{b+c}{a} + 1 = 2020$$

and so

$$\frac{b}{a+b+c} = \frac{a+b+c}{a+b+c} - \frac{a}{a+b+c} - \frac{c}{a+b+c} = 1 - \frac{1}{2019} - \frac{1}{2020}.$$

Therefore,

$$\frac{a+c}{b} = \frac{a+b+c}{b} - 1 = \frac{1}{1 - \frac{1}{2019} - \frac{1}{2020}} - 1 = \frac{4039}{4074341}.$$

Q264 First, we note that using the hint, $10^{20} = 61s + 13$ for some integer s, and therefore

$$10^{40} = (61s + 13)^2$$
$$= 61^2 s^2 + 2 \times 13 \times 61s + 13^2$$
$$= 61(61s^2 + 26s + 2) + 47$$

which leaves remainder 47 when divided by 61; and

$$10^{60} = (61s + 13)^3$$
$$= 61^3 s^3 + 3 \times 13 \times 61^2 s^2 + 3 \times 13^2 \times 61s + 13^3$$
$$= 61(61^2 s^3 + 3 \times 13 \times 61s^2 + 3 \times 13^2 s + 36) + 1$$

which leaves remainder 1 when divided by 61. (In fact, the latter can be found without calculation from a very important result of number theory known as *Fermat's Little Theorem*, see page 243.)

Now, multiplying $\frac{n}{61}$ by 10^{20} will shift the 21st digit after the decimal point to first place after the decimal point; since this digit is 1, we have

$$\frac{10^{20}n}{61} = a + x,$$

where a is an integer and $0.1 < x < 0.2$. Similarly,

$$\frac{10^{40}n}{61} = b + y,$$

where b is an integer and $0.9 < y < 1$. Multiplying both equations by 61 we find

$$10^{20}n = 61a + X, \qquad 10^{40}n = 61b + Y, \tag{2.56}$$

where $X = 61x$ and $Y = 61y$. Now X is an integer since it is a difference of two integers, $X = 10^{20}n - 61a$; and $6.1 < X < 12.2$. Similar reasoning for Y gives the possibilities

$$X = 7, 8, 9, 10, 11, 12 \quad \text{and} \quad Y = 55, 56, 57, 58, 59, 60.$$

Using the first of equations (2.56) once again, we have

$$10^{60}n = 10^{40} \times 61a + 10^{40}X;$$

taking remainders after division by 61 and using facts mentioned above, the remainder when n is divided by 61 is the same as that for $47X$. As the first option for X is 7, the first value of $47X$ is 329 and the remainder is 24. The full set of options for n is

$$24, 10, 57, 43, 29, 15.$$

Similarly, the second equation from (2.56) shows that the remainder when n is divided by 61 is one of

$$44, 57, 9, 22, 35, 48.$$

The only possibility common to both lists is 57. Therefore, n has remainder 57 when divided by 61, and the smallest possible (positive) value for n is 57. We can check this result by calculating

$$\frac{57}{61} = 0.93442622950819672131147540983606557377049180 3 \cdots.$$

Q265 Suppose that in the first guess, a and b are correct and correctly placed, while c is part of the code but not correctly placed. This means that a is the first letter in the code, b is the second, and c is not the third so it must be the fourth. The third letter must be the one in the correct place in your second guess, so it is h. Working carefully through all the possibilities gives the following potential answers:

```
1 2 4      1 2 ×      1 4 3      1 × 3
× × ×      × × 3      × × ×      × × 2
× 3 ×      × × 4      2 × ×      × × 4

1 3 ×      1 × 2      4 2 3      × 2 3
× × 4      × × 4      1 × ×      × × 1
2 × ×      × 3 ×      × × ×      × × 4

3 2 ×      × 2 1      2 × 3      × 1 3
1 × 4      × × 4      1 × 4      × × 4
× × ×      × 3 ×      × × ×      2 × ×
```

Now, the first of these is impossible as the line from b to h crosses an unused dot (represented by a cross) at e; the third, fifth, sixth, tenth, eleventh and twelfth are impossible for similar reasons; which leaves five possible patterns:

```
1 2 ×      1 × 3      4 2 3      × 2 3      3 2 ×
× × 3      × × 2      1 × ×      × × 1      1 × 4
× × 4      × × 4      × × ×      × × 4      × × ×
```

For our next guess, we try $dbaf$ (other solutions may be possible). The five remaining patterns will give a response of m correctly placed, n incorrectly placed, where m, n are

$$1,\ 2; \qquad 0,\ 2; \qquad 2,\ 1; \qquad 1,\ 1; \qquad 4,\ 0$$

respectively. Since these are all different we can definitely pick the right answer on the fourth guess (unless we were lucky enough to have already picked it on the third).

Q266 Let y be the area of the middle rectangle in the top row. The ratio of areas of two rectangles having the same height is the same as the ratio of their widths. Therefore,

$$\frac{x+1}{y} = \frac{\text{width of column 3}}{\text{width of column 2}} = \frac{x+5}{x+4}.$$

For the same reason,

$$\frac{y}{x} = \frac{\text{width of column 2}}{\text{width of column 1}} = \frac{x+3}{x+2};$$

multiplying these equations, the y cancels and we have

$$\frac{x+1}{x} = \frac{x+5}{x+4}\frac{x+3}{x+2}.$$

Multiplying out, we get a cubic on each side, but the x^3 terms cancel, leaving the quadratic $x^2 + x - 8 = 0$. Solving for x and rejecting the negative root gives

$$x = \frac{-1+\sqrt{33}}{2}.$$

Q267 A polynomial $f(x)$ has a point of inflexion at $x = a$, with tangent $y = mx + c$ at that point, if

$$f(x) - (mx + c)$$

has a factor $(x - a)^3$. In this case, trying to do as little work as possible, we want

$$x^4 - 2x^3 + \cdots = (x - a)^3(x - b) = x^4 - (3a + b)x^3 + \cdots,$$

and so $3a + b = 2$. Using this and completing the factorisation, we have

$$x^4 - 2x^3 - 36x^2 + (28 - m)x + (99 - c)$$
$$= (x - a)^3(x + 3a - 2)$$
$$= x^4 - 2x^3 - (6a^2 - 6a)x^2 + (8a^3 - 6a^2)x - (3a^4 - 2a^3).$$

Equating coefficients,

$$6a^2 - 6a = 36, \quad 28 - m = 8a^3 - 6a^2, \quad 99 - c = -3a^4 + 2a^3.$$

The first equation is easily solved to find possible values of a, then the others give corresponding values of m and c. Therefore, the curve has

- an inflection point at $x = 3$, with tangent $y = -134x + 288$;
- an inflection point at $x = -2$, with tangent $y = 116x + 163$.

Q268 Consider, for example, the case where the answer to A is wrong and B and C are right. Then the correct answers to the questions are

$$a = 2, \quad b = 1, \quad c = 2;$$

since A is wrong, the answers given must satisfy

$$a \neq 2, \quad b = 1, \quad c = 2$$

and so the options are

$$(a, b, c) = (0, 1, 2) \text{ or } (1, 1, 2) \text{ or } (3, 1, 2).$$

Continue in this way and tabulate the options for all eight possible combinations of right and wrong answers. You will find that certain options for the given answers come up once only: $(0, 1, 2)$ is an example of this. So, if the given answers were in fact $(0, 1, 2)$, then we could deduce that A was wrong and the others right: that is, two correct and one incorrect: which is inconsistent with the given information. So, this case must be ruled out. Then again, some options come up twice: an example is $(3, 1, 2)$, which would mean that *either* A is wrong and the others are right, *or* all three answers are wrong. From these two options, we could say that either A is wrong or A is wrong, so A is definitely wrong; and either B is wrong or B is right, so the status of B is unknown; and either C is wrong or C is right, so C is unknown. This gives one incorrect and two unknown, which again is inconsistent with the given information.

Looking through all results, we find that the only satisfactory possibility is $(a, b, c) = (2, 0, 1)$, which occurs when either B is wrong and the others right, or C is right and the others are wrong. From this, we find that A is unknown, B is definitely incorrect and C is definitely correct.

Q269 We'll – sort of – complete the square to get the x^4, x^3, x and constant terms right, then see what happens with the x^2 term. We have

$$(x^2 + px + q)^2 = x^4 + 2px^3 + (p^2 + 2q)x^2 + 2pqx + q^2,$$

and so we choose

$$p = \frac{a}{2} \quad \text{and} \quad q = \frac{c}{a}.$$

This gives

$$(x^2 + px + q)^2 = x^4 + ax^3 + \left(\frac{a^2}{4} + \frac{2c}{a}\right)x^2 + cx + d;$$

so, we need

$$b = \frac{a^2}{4} + \frac{2c}{a}.$$

This answers the first part of the question. For the second, we note that, if $a = 2$ and $c = 6$ and $d = 9$, then $d = c^2/a^2$, so, the first part applies with $p = 1$ and $q = 3$ and $b = 7$. Therefore,

$$x^4 + 2x^3 - 20x^2 + 6x + 9 = (x^2 + x + 3)^2 - 27x^2$$
$$= \left(x^2 + (1 - 3\sqrt{3})x + 3\right)\left(x^2 + (1 + 3\sqrt{3})x + 3\right)$$

by using the difference of two squares. We can now use the quadratic formula to find the solutions

$$x = \frac{-1 + 3\sqrt{3} \pm \sqrt{16 - 6\sqrt{3}}}{2} \quad \text{and} \quad x = \frac{-1 - 3\sqrt{3} \pm \sqrt{16 + 6\sqrt{3}}}{2}.$$

Q270 To get rid of the large term on the right–hand side of the equation, we substitute $a_n = 2^{2^n+1}b_n$. This yields

$$2^{2^n+1}b_n = \left(2^{2^{n-1}+1}b_{n-1}\right)^2 - 2^{2^n+1} = 2^{2^n+2}b_{n-1}^2 - 2^{2^n+1}$$

and so

$$b_n = 2b_{n-1}^2 - 1.$$

This is reminiscent of the double–angle formula for cosine; see Section 3.7: if we can write $b_{n-1} = \cos\theta$ for some θ, then we have

$$b_n = 2\cos^2\theta - 1 = \cos(2\theta).$$

Now, $b_0 = \frac{1}{4} = \cos\theta$, where $\theta = \arccos\frac{1}{4}$; so by the double–angle formula

$$b_1 = \cos(2\theta), \quad b_2 = \cos(4\theta), \quad b_3 = \cos(8\theta)$$

and so on. In general, $b_n = \cos(2^n\theta)$, and therefore

$$a_n = 2^{2^n+1}\cos\left(2^n \arccos\frac{1}{4}\right).$$

Q271 The line $y = mx + c$ is tangent to $y = f(x)$ at the point $x = a$ if and only if $(x - a)^2$ is a factor of $f(x) - (mx + c)$. In the present problem, we want this to occur for two points $x = a$ and $x = b$, so we need

$$f(x) - (mx + c) = (x - a)^2(x - b)^2,$$

that is,

$$x^4 + 2x^3 - 7x^2 - mx + (11 - c) = [(x - a)(x - b)]^2.$$

Since the quartic is the square of a quadratic, Problem 269 shows that we need

$$11 - c = \frac{m^2}{4} \quad \text{and} \quad -7 = 1 - m.$$

We easily find $m = 8$, $c = -5$, so the double tangent is $y = 8x - 5$. We may also calculate

$$f(x) - (mx + c) = x^4 + 2x^3 - 7x^2 - 8x + 16 = (x^2 + x - 4)^2,$$

and so the points of tangency are

$$x = \frac{-1 \pm \sqrt{17}}{2}.$$

Q272 First, observe that each time a red counter moves, the number of red counters on white squares increases or decreases by 1.

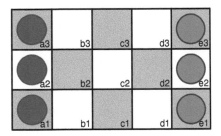

Since there must be one red counter on a white square both at the beginning and at the end, red must make an even number of moves; and the same applies to blue. It's clear that each counter must make 4 horizontal moves, a total of 24. Moreover in each row, at least one of the two original counters must make a vertical move in order to get out of the way of the other; this is at least 3 more moves, 27 altogether; but the total number of moves must be even, and therefore we need at least 28. For all we know at this stage, we might require even more moves than this; but trial and error gives various 28–move solutions, for example,

$$
\begin{array}{ccccccc}
a1{-}b1, & e1{-}d1, & b1{-}b2, & d1{-}c1, & b2{-}b3, & c1{-}b1, & c2{-}c1, \\
e3{-}d3, & a2{-}b2, & d3{-}c3, & b2{-}c2, & c3{-}b3, & c1{-}d1, & b3{-}b2, \\
a3{-}b3, & b2{-}a2, & b3{-}c3, & e2{-}d2, & c3{-}d3, & b1{-}a1, & d3{-}e3, \\
d2{-}d3, & d1{-}e1, & d3{-}c3, & c2{-}d2, & c3{-}b3, & d2{-}e2, & b3{-}a3.
\end{array}
$$

Another question. For a 4×3 board, a similar argument shows that we need at least 22 moves. Is a solution possible in 22 moves, or do you need more?

Q273 First observe that we can ignore the "missing chunk" in the top right corner, because if this chunk is filled in, then the perimeter of the figure remains the same. Now suppose that n unit squares have been used, that a $k \times k$ square has been finished and the next column on the right has been started.

- If the next row on the top has *not* been started, then the number of tiles satisfies $k^2 < n \le k^2 + k$, and the perimeter $P(n)$ is that of a k by $k + 1$ rectangle.
- If the next row on the top *has* been started, then the number of tiles satisfies $k^2 + k < n \le (k + 1)^2$, and the perimeter $P(n)$ is that of a $k + 1$ by $k + 1$ rectangle.

Therefore, we have

$$P(n) = \begin{cases} 4k+2 & k^2 < n \le k(k+1) \\ 4k+4 & \text{if } k(k+1) < n \le (k+1)^2. \end{cases}$$

For the case of 2019 tiles, we have

$$44 \times 45 < 2019 \le 45^2$$

and so the perimeter is

$$P(2019) = 4 \times 45 = 180.$$

Q274 If d is a common factor of m and n, then it is also a factor of

$$3m - n = 3 \times 2^{20} - 2^{19} = 5 \times 2^{19}.$$

But m and n are clearly odd, so all their factors are odd; so $d = 1$ or $d = 5$. However, d cannot be 5 since m is not a multiple of 5: we may prove this by using arithmetic modulo 5 (see Section 3.3) to show that

$$2^{20} + 3^{19} \equiv 2^{20} + (-2)^{19} \equiv 2^{20} - 2^{19} = 2^{19} \not\equiv 0 \pmod 5.$$

Hence, the only common factor, and therefore also the highest common factor, of $2^{20} + 3^{19}$ and $2^{19} + 3^{20}$ is $d = 1$.

Alternative. To show that m is not a multiple of 5 without using modular arithmetic, consider the following table, which gives the last digit of $2^{k+1} + 3^k$ for $k = 1, 2, 3, \ldots$.

k	1	2	3	4	5	6	7 …
2^{k+1}	4	8	6	2	4	8	6 …
3^k	3	9	7	1	3	9	7 …
$2^{k+1} + 3^k$	7	7	3	3	7	7	3 …

This table is easy to compute: in row 2, each entry is the last digit of twice the previous entry; similarly for row 3; and row 4 is the last digit of the sum of the two rows above it. It is also clear that the first four columns will repeat indefinitely; so the last digit of $2^{k+1} + 3^k$ is never 0 or 5, and this expression cannot be a multiple of 5.

Comment. The exponents 20 and 19 were chosen purely because this problem appeared in *Parabola* in 2019. In fact, the highest common factor of $2^{k+1} + 3^k$ and $2^k + 3^{k+1}$ is always equal to 1.

Q275 Label the squares a to p as shown.

a	b	c	d
e	f	g	h
i	j	k	l
m	n	o	p

We begin by listing all possible moves:

$$ad, am, ap; \quad bn; \quad co; \quad db, dj, dl; \quad ea, eb, ef, ei, ej;$$
$$fh, fn, fp; \quad ge, gm, go; \quad he; \quad il; \quad jb, jd, jl; \quad ka, kc, ki;$$
$$li; \quad mi, mj, mn; \quad ni, nj, nk, nm, no; \quad oe, og, om; \quad pk, pl, po.$$

For convenience, here are the same moves, but ordered according to the destination rather than the origin:

$$ea, ka; \quad db, eb, jb; \quad kc; \quad ad, jd; \quad ge, he, oe; \quad ef; \quad og; \quad fh;$$
$$ei, ki, li, mi, ni; \quad dj, ej, mj, nj; \quad nk, pk; \quad dl, il, jl, pl;$$
$$am, gm, nm, om; \quad bn, fn, mn; \quad co, go, no, po; \quad ap, fp.$$

We need to list the letters a to p in a sequence such that every pair of adjacent letters in the sequence comes from the above pairs.

- First note that i can go only to l, and l can go only to i; so the end of the sequence must be either il or li.

- Since h is not the last letter it must be followed by something, and the only possibility is e. If h does not start the sequence, then it must be preceded by f; and if f does not start the sequence, then it must be preceded by e. So the start of the sequence must be he or fhe.

- We now know that every letter other than f, h, i, l must have another letter both before and after it. This means we must have the subsequences bn and $kcogm$.

- Now a can only be preceded by e, so no letter can precede f. So, the sequence begins $fheap$.

- Therefore, p is the 5th letter in the sequence and l is the 15th or 16th; hence, pl is impossible; so is po; so, the beginning is $fheapkcogm...$

- ...which means that m is not followed by i; it is not followed by n as we already have bn; hence we get $fheapkcogmjd$.

- There is now no letter to follow n except for i, and the whole sequence is $fheapkcogmjdbnil$.

Q276 Imagine the power written out as a product of 10 factors,

$$\left(1 + x^2 + \frac{1}{x}\right)\left(1 + x^2 + \frac{1}{x}\right) \cdots \left(1 + x^2 + \frac{1}{x}\right).$$

If we multiply out and *do not* collect terms, then every product will have coefficient 1. Noting that

$$x^3 = (x^2)^m \left(\frac{1}{x}\right)^{2m-3},$$

we will obtain a term in x^3 by choosing m of the 10 factors to supply an x^2 and $2m - 3$ of the remaining $10 - m$ factors to supply a $1/x$; all the other factors must supply a 1. Both choices must be made without choosing the same factor more than once, and with the order of factors unimportant, so – see Section 3.5.1 if required – the number of ways of doing this is

$$\binom{10}{m}\binom{10 - m}{2m - 3}.$$

To obtain a final result of x^3, clearly at least two factors of x^2 must be included; so $m \geq 2$. The total number of terms specified above is $3m - 3$, which is at most 10; so $m \leq 4$. Hence, the relevant values of m are $2, 3, 4$, and the coefficient of x^3 is

$$\binom{10}{2}\binom{8}{1} + \binom{10}{3}\binom{7}{3} + \binom{10}{4}\binom{6}{5} = 360 + 4200 + 1260 = 5820.$$

Q277 Let $2m$ be an **even number** which contains the given digits consecutively and in order. (So if the last of the digits is even we can just take the given digits; if it is odd, then we would take, for example, the given digits followed by a zero.) Let k be the number of digits in $2m$. Take

$$n = m10^k + 1$$

and consider

$$n^2 = m^2\,10^{2k} + 2m\,10^k + 1.$$

Now, the first term here ends in $2k$ zeros. The second consists of the digits of $2m$ (there are k of these) followed by another k zeros. This is $2k$ digits altogether, so if we add the first two terms, then the digits of $2m$ are added to zeros and do not change. And the final 1 is then added to the last of the k zeros, and does not affect the digits of $2m$. Therefore, n^2 contains the digits of $2m$, which contain the given digits. In more detail, if the digits of $2m$ are $d_1 d_2 \cdots d_k$ and the digits of m^2 are $c_1 c_2 \cdots c_{2k}$, then we have

$$n^2 = c_1 c_2 \cdots c_{2k} \overbrace{00\cdots 000\cdots 0}^{2k\ \text{digits}} + \overbrace{d_1 d_2 \cdots d_k 00\cdots 0}^{2k\ \text{digits}} + 1$$
$$= c_1 c_2 \cdots c_{2k} d_1 d_2 \cdots d_k 00\cdots 1$$

which contains the required digits.

Q278 Let the side of the square be 1, so that we are looking for three regions of area $\frac{1}{3}$ each. Place the point O on a diagonal of the square, very close to one corner, with one of the rays extending to the opposite corner (and the others spaced at angles of $120°$ as required) – see the first diagram.

Then the shaded area is very small, certainly less than $\frac{1}{3}$. Now, gradually move the point O up the diagonal until it reaches the point shown in the second diagram: at this stage, the shaded area is clearly bigger than $\frac{1}{2}$. So, somewhere in between, this area must be exactly $\frac{1}{3}$. The combined area of the other two regions is then $\frac{2}{3}$; and these two regions are clearly congruent, so they also have area $\frac{1}{3}$ each. This solves the problem.

Comments.

- It is unnecessary to find the actual location of O in order to answer the question! However, if you want to do so, then this is not too hard and is left as an exercise. You should find that the distance from O to the bottom left corner of the square is

$$\frac{\sqrt{3 + \sqrt{3}}}{3}.$$

- Another way to answer this question is to place O on the "mid-line" of the square, near the bottom, and then gradually move it vertically upwards. The argument is very similar to the one we have given.

For part (b), the three rays starting at O will meet the square in three points; so, there must be a side of the square which does not meet any of the rays, except perhaps at a corner. If we draw this side as the top of the square and place O at the centre, then it looks like this.

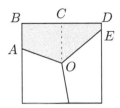

The shaded area is split into two trapezoids by the dotted line; since $BC = CD = OC = \frac{1}{2}$, its area is

$$\Delta = \frac{1}{2}\frac{|AB| + |OC|}{2} + \frac{1}{2}\frac{|OC| + |ED|}{2} = \frac{|AB| + 1 + |ED|}{4}. \qquad (2.57)$$

Now, let $\angle AOC = 60° + \theta$; then $\angle EOC = 60° - \theta$, and since A and E are on the vertical sides of the square we have $-15° \leq \theta \leq 15°$. Then

$$|AB| = \tfrac{1}{2} - \tfrac{1}{2}\tan(30° - \theta), \quad |DE| = \tfrac{1}{2} - \tfrac{1}{2}\tan(30° + \theta);$$

so, by using the tangent–of–a–sum formula (see Section 3.7) and the value $\tan 30° = \frac{1}{\sqrt{3}}$, substituting back into (2.57) and simplifying gives

$$\Delta = \frac{1}{4}\left(2 - \sqrt{3}\frac{1 + \tan^2 \theta}{3 - \tan^2 \theta}\right).$$

Multiplying top and bottom of the fraction in this formula by $\cos^2 \theta$; then, using the identities $\cos^2 \theta + \sin^2 \theta = 1$ and $\cos^2 \theta - \sin^2 \theta = \cos 2\theta$, we have

$$\frac{1 + \tan^2 \theta}{3 - \tan^2 \theta} = \frac{\cos^2 \theta + \sin^2 \theta}{3\cos^2 \theta - \sin^2 \theta} = \frac{1}{1 + 2\cos 2\theta}$$

and so

$$\Delta = \frac{1}{4}\left(2 - \frac{\sqrt{3}}{1 + 2\cos 2\theta}\right).$$

Now, the minimum value of the area occurs for the minimum value of $\cos 2\theta$; this is when $\theta = \pm 15°$, and the minimum area is

$$\Delta_{\min} = \frac{1}{4}\left(2 - \frac{\sqrt{3}}{1 + \sqrt{3}}\right) = \frac{\sqrt{3} + 1}{8}.$$

Finally, noting that $3\sqrt{3} = \sqrt{27} > \sqrt{25} = 5$, the area of the shaded region satisfies

$$\Delta \geq \Delta_{\min} = \frac{3\sqrt{3} + 3}{24} > \frac{5 + 3}{24} = \frac{1}{3}.$$

Therefore, the shaded area can never be $\frac{1}{3}$, and it is impossible to divide the square into three regions of area $\frac{1}{3}$ by means of three equally spaced rays meeting at the centre of the square.

Comment. Problems 279 and 284 ask further questions about dividing up a square using equally spaced rays meeting at a point.

Q279 Let the area of the square be 1, and suppose that such an arrangement exists, as shown in the diagram.

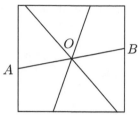

There are three regions of area $\frac{1}{6}$ on one side of the line AB, total area $\frac{1}{2}$, and the same on the other; so AB must pass through the centre of the square. The same holds for the other two lines in the "star", and so O must lie at the centre of the square. But then by combining pairs of adjacent regions, we obtain an arrangement of three equally spaced rays meeting at the centre of the square and dividing the square into regions of area $\frac{2}{6} = \frac{1}{3}$, and we know from the solution to Problem 278 that this is impossible. Therefore, six rays cannot be arranged as desired.

Comment. This topic is pursued still further in Problem 284.

Q280 Multiply both sides of the equation by 9 to get

$$9S = 9 + 99 + 999 + 9999 + \cdots + \overbrace{99\cdots99}^{\text{999 digits}}.$$

Now add 1 for each term on the right–hand side; since there are 999 terms, this gives

$$9S + 999 = 10 + 100 + 1000 + 10000 + \cdots + 1\overbrace{00\cdots00}^{\text{999 zeros}}$$
$$= \underbrace{11\cdots11}_{\text{999 ones}}0;$$

which can be written with the 1s split into blocks of nine digits,

$$9S + 999 = \overbrace{(111111111)\cdots(111111111)}^{\text{111 blocks}}0.$$

Now, 111111111 is exactly divisible by 9, so when we divide both sides by 9, there will be no "carries" between blocks, and we get

$$S + 111 = \overbrace{(012345679)\cdots(012345679)}^{\text{111 blocks}}0.$$

The sum of the digits on the right–hand side is

$$111 \times (0 + 1 + 2 + 3 + 4 + 5 + 6 + 7 + 9) = 4107.$$

Subtracting 111 from the right–hand side changes the last three digits 790 to 679, and therefore increases the sum of the digits by 6. Hence, the sum of the digits of S is 4113.

Q281 To get a multiple of 4321 in which the last digit is 4, we simply multiply by 4:

$$4321 \times 4 = 17284.$$

We want to add another multiple to this to make the second–last digit 3, without changing the last. This means we need to add 5 in the second–last place and 0 in the last; so we add 4321×50. Continue in the same way: as only the last four digits are relevant, we omit the rest.

$$4321 \times 4 = \cdots 7284$$
$$4321 \times 50 = \cdots 6050$$
$$\therefore \quad 4321 \times 54 = \cdots 3334$$
$$4321 \times 900 = \cdots 8900$$
$$\therefore \quad 4321 \times 954 = \cdots 2234$$
$$4321 \times 9000 = \cdots 9000$$
$$\therefore \quad 4321 \times 9954 = \cdots 1234.$$

Thus, 4321×9954 ends in the digits 1234. Is this the smallest? We want a multiple of 4321 which equals a multiple of 10000 plus 1234; that is,

$$4321x = 10000y + 1234,$$

where x and y are integers. The Bézout property (Theorem 3.1, Section 3.2) shows that the general solution for x is

$$x = 9954 + 10000t$$

where t is an integer. Now if $t > 0$, then this gives a value larger than the one we have already, while if $t < 0$, then we have $x < 0$, which is not a valid solution; so, the smallest possibility is the one we have already. That is, the smallest multiple of 4321 ending in the digits 1234 is

$$4321 \times 9954 = 43011234.$$

Q282 We have added some extra lines to the diagram; and extracted and labelled the important section.

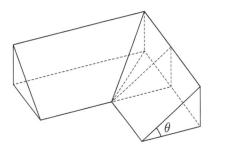

Now, $BCC'B'$ is a rectangle, $\angle B'AC' = 45°$ and $\angle AC'B'$ is a right angle, so

$$|BC| = |B'C'| = |AC'|.$$

Also,

$$|AC'| = |AC| \cos \theta,$$

so the angle at B is

$$\arctan\left(\frac{|AC|}{|BC|}\right) = \arctan\left(\frac{|AC'|}{|AC'| \cos \theta}\right) = \arctan(\sec \theta),$$

and the angle at A in the original diagram is $180°$ minus this angle.

Q283 For brevity, we shall refer to the statements (in the order given) as T1, T2, T3, F1, F2, F3 and F4.

Suppose that there are no false statements in the bag, or only one. Then F2, F3 and F4 are all false and at least two of them are in the bag: this is impossible.

Therefore, there are two or more false statements in the bag. This means that F1 and F2 are both true, and at least one of them is in the bag; so T1 is true. Thus, we have three true statements, and at least two of them are in the bag; so T2 is true. By a similar argument, T3 is true. We now have five true statements; therefore, the remaining two (F3 and F4) must both be false and must be in the bag. So, **answer**: there are two false statements in the bag.

To be quite sure that this answer is valid (and not self–contradictory), we ought to verify that, if any one of F1, F2, T1, T2, T3 is the statement which was removed, then the statements in the bag are true or false as indicated above.

Q284 Suppose that 13 equally spaced rays from the point O divide the square into 13 regions of equal area. As the rays meet the perimeter in 13 points, at least four of these points must lie on the same side of the square. So, we have a situation like this,

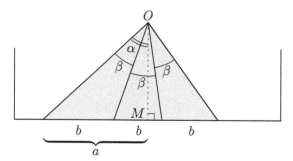

where $\beta = \left(\frac{360}{13}\right)^\circ$. We shall measure lengths on the side of the square to the right of M, and angles anticlockwise from OM; so that in the example shown, the length a is negative, as is the angle α. Note, therefore, that

$$-90^\circ < \alpha < \alpha + 3\beta < 90^\circ.$$

Now by assumption, the three shaded triangles have equal areas; and they also have equal altitudes $h = OM$; therefore, they have equal bases b, as marked in the diagram. From the four right–angled triangles in the diagram, we have

$$a = h\tan\alpha, \quad a + b = h\tan(\alpha + \beta),$$
$$a + 2b = h\tan(\alpha + 2\beta), \quad a + 3b = h\tan(\alpha + 3\beta);$$

therefore, the points (α, a), $(\alpha + \beta, a + b)$, $(\alpha + 2\beta, a + 2b)$ and $(\alpha + 3\beta, a + 3b)$ all lie on the curve $y = h\tan x$ for $-90^\circ \le x < 90^\circ$. But these four points also satisfy the linear equation

$$\frac{y - a}{b} = \frac{x - \alpha}{\beta};$$

that is, the four points are collinear; and this is impossible, as the section of the tangent graph just shown clearly does not contain four collinear points. Thus, it is impossible for an arrangement of 13 rays as described to divide the square into regions of equal area.

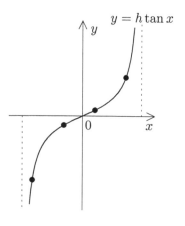

Q285 Let the height of the cone be h and its radius r (in centimetres). Let the radius of the water surface when the base is on the table be a, and let the radius of the water surface when the cone is inverted be b.

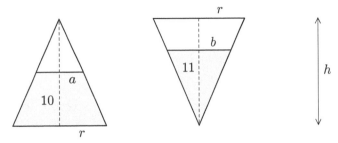

Since the volume of water in the two cases is the same, we have

$$\frac{1}{3}\pi r^2 h - \frac{1}{3}\pi a^2(h-10) = \frac{1}{3}\pi b^2(11),$$

which simplifies to

$$h - (h-10)\left(\frac{a}{r}\right)^2 = 11\left(\frac{b}{r}\right)^2.$$

But, by similar triangles, we have

$$\frac{a}{r} = \frac{h-10}{h} \quad \text{and} \quad \frac{b}{r} = \frac{11}{h};$$

substituting into the previous equation yields

$$h - \frac{(h-10)^3}{h^2} = \frac{1331}{h^2}$$

and so

$$h^3 - (h-10)^3 = 1331.$$

On simplifying the left–hand side, the h^3 terms cancel and we find the quadratic

$$30h^2 - 300h - 331 = 0$$

with positive solution

$$h = \frac{300 + \sqrt{129720}}{60} \approx 11.0028.$$

This is the height of the cone, and so the height of the empty space above the water surface is 0.0028 centimetres – which is approximately the width of one human hair!!

Q286 We can place all possible values of $n = 99a + 100b + 101c$ into groups according to the value of $k = a + b + c$. For example, group 0 corresponds to $a + b + c = 0$, giving only one possibility: $(a, b, c) = (0, 0, 0)$ and $n = 0$. Another example: group 2 corresponds to $a + b + c = 2$, giving six possible triples

$$(a, b, c) = (2, 0, 0), (1, 1, 0), (0, 2, 0), (1, 0, 1), (0, 1, 1), (0, 0, 2)$$

and five different values

$$n = 198, 199, 200, 201, 202.$$

The smallest integer in group k is $n = 99k$, given by $(a, b, c) = (k, 0, 0)$; and the largest is $n = 101k$, given by $(a, b, c) = (0, 0, k)$. Furthermore, all intermediate integers are also found in group k. To see why this is true, imagine a basket containing k balls, all labelled 99; the total of all labels is $99k$. We can, one at a time, replace a "99 ball" by a "100 ball" or a "100 ball" by a "101 ball"; each time we do this, the total increases by 1; and so we obtain every possible integer up to the point at which all k balls are labelled 101 and the total is $101k$.

To find all possible values of n, we need to take all groups collectively. Now, as we have just seen, group k consists of a sequence of consecutive integers; so does group $k+1$; and there is a gap between these groups if and only if

$$101k + 1 < 99(k+1),$$

which can be solved to give $k < 49$. Thus, group 48 goes up to $n = 4848$; group 49 starts at $n = 4851$; the intermediate values 4849 and 4850 are not in any group and are therefore not possible values of n. Moreover, there is no gap between group 49 and group 50; nor between group 50 and group 51; and so on. So, there are no further missing values of n, and the largest integer which cannot be written in the form $99a + 100b + 101c$ is 4850.

Q287 We use the AGM (arithmetic–geometric–mean) inequality, Section 3.11, which states that the average of any n positive real numbers is greater than or equal to the nth root of their product. Taking firstly the five numbers a, b, c, d, a and then the three numbers b, c, d, this gives

$$\frac{a+b+c+d+a}{5} \geq \sqrt[5]{abcda} \quad \text{and} \quad \frac{b+c+d}{3} \geq \sqrt[3]{bcd}.$$

Since $a+b+c+d = 4$, these inequalities can be written

$$4 + a \geq 5\sqrt[5]{a^2bcd} \quad \text{and} \quad 4 - a \geq 3\sqrt[3]{bcd}$$

and then multiplied to give

$$16 - a^2 \geq 15\sqrt[5]{a^2bcd}\sqrt[3]{bcd}.$$

By similar arguments, we obtain inequalities for $16 - b^2$ and $16 - c^2$ and $16 - d^2$; multiplying them and collecting surds on the right–hand side gives

$$(16 - a^2)(16 - b^2)(16 - c^2)(16 - d^2) \geq 15^4\sqrt[5]{a^2bcd\,ab^2cd\,abc^2d\,abcd^2}\sqrt[3]{bcd\,acd\,abd\,abc},$$

which simplifies to

$$(16 - a^2)(16 - b^2)(16 - c^2)(16 - d^2) \geq 15^4 a^2 b^2 c^2 d^2.$$

And now, dividing both sides by the positive number $a^2b^2c^2d^2$ solves the problem.

Q288 By symmetry, all the red triangles in the diagram are congruent. (If desired, more details are given at the end of this solution.) Therefore, we can label lengths as shown in the diagram.

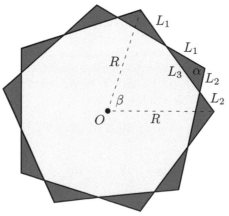

Let S_0 be the length of one side of the original polygon; let R be the distance from the centre to one vertex (that is, the circumradius of the polygon); let α be the internal angle at a vertex; and let β be the angle subtended at the centre by a side. Then by standard geometrical arguments, we have

$$\alpha = \pi - \beta, \quad P_0 = nS_0, \quad S_0 = 2R\sin(\beta/2), \quad A_0 = \tfrac{1}{2}nR^2\sin\beta.$$

The boundary of the combined figure consists of the sides L_1, L_2 taken $2n$ times, and the area of overlap is the area of the original, less n of the red triangles. If we write T for the area of one of these triangles, then

$$P = 2n(L_1 + L_2), \quad A = A_0 - nT.$$

Since the triangle has sides L_1, L_2 with included angle α, we have

$$T = \tfrac{1}{2}L_1L_2\sin\alpha = \tfrac{1}{2}L_2L_2\sin\beta.$$

Applying the Cosine Rule to this triangle and noting that $L_3 = S_0 - L_1 - L_2$ yields

$$(S_0 - L_1 - L_2)^2 = L_1^2 + L_2^2 - 2L_1L_2\cos\alpha.$$

If we expand and rearrange this, then using the identity

$$1 + \cos\alpha = 1 - \cos\beta = 2\sin^2(\beta/2),$$

we obtain (check it for yourself!)

$$2S_0(L_1 + L_2) = S_0^2 + 4L_1L_2\sin^2(\beta/2).$$

Hence,

$$\frac{P}{P_0} = \frac{2(L_1 + L_2)}{S_0} = \frac{S_0^2 + 4L_1L_2\sin^2(\beta/2)}{S_0^2} = 1 + \frac{4L_1L_2\sin^2(\beta/2)}{4R^2\sin^2(\beta/2)} = 1 + \frac{nT}{A_0}$$

and finally

$$\frac{P}{P_0} + \frac{A}{A_0} = 1 + \frac{nT}{A_0} + \frac{A_0 - nT}{A_0} = 2,$$

as claimed.

To confirm that all the red triangles are congruent, consider the following diagram, in which θ denotes the angle by which the original polygon is rotated.

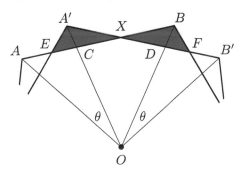

We have

$$|OA| = |OB'| = R, \quad \angle AOC = \angle B'OD \quad \text{and} \quad \angle OAC = \angle OB'D = \frac{1}{2}\alpha,$$

so $\triangle OAC$ and $\triangle OB'D$ are congruent; therefore, $|OC| = |OD|$; and $|OA'| = |OB| = R$, so by subtraction $|A'C| = |BD|$. Also, $\angle ACO = \angle B'DO$; therefore $\angle A'CX = \angle BDX$; and $\angle CA'X = \angle DBX = \frac{1}{2}\alpha$, so $\triangle CA'X$ and $\triangle DBX$ are congruent. Therefore, $|A'X| = |BX|$, $\angle A'XE = \angle BXF$ and $\angle XA'E = \angle XBF = \alpha$; and so $\triangle A'XE$ and $\triangle BXF$ are congruent, which is what we wanted to prove.

Q289 Suppose that there are y students in the class, and x of them read *Parabola*. Then

$$65\frac{1}{2}\% < \frac{x}{y} < 66\frac{1}{2}\%,$$

which can be written

$$-\frac{1}{2}\% < \frac{x}{y} - \frac{66}{100} < \frac{1}{2}\%,$$

or

$$\left| \frac{x}{y} - \frac{33}{50} \right| < \frac{1}{200}.$$

We want to find positive integers x, y satisfying this inequality, with the smallest possible value of y; since $x = 33$, $y = 50$ is an obvious solution, we can limit ourselves to solutions with $y \leq 50$. The inequality can be rewritten as

$$|50x - 33y| < \frac{y}{4},$$

and we note that the left–hand side is an integer.

For information about solving $50x - 33y = c$, where c is a given integer and x, y are required to be integers, see Section 3.2. By an easy trial and (no) error, the equation $50x - 33y = 1$ has a solution $x = 2$, $y = 3$; so the general solution of $50x - 33y = c$ is

$$x = 2c + 33t, \quad y = 3c + 50t,$$

where t is an integer. Since we are looking for $y \leq 50$, we shall choose the value of t to guarantee this.

First, note that $|c| < y/4 \leq 12\frac{1}{2}$, so $c = 0, \pm 1, \pm 2, \ldots, \pm 12$, and that $c = 0$ gives the solution $x = 33$, $y = 50$ which we know already. If $c > 0$, then $3 \leq 3c \leq 36$; this is in the range 1 to 50 already, so we take $t = 0$ and $y = 3c$. But then, we have $c < 3c/4$, which is impossible; so, this case is ruled out. Therefore, we have $c < 0$; write $c = -d$. Then $-36 \leq 3c \leq -3$, so to obtain y in the range 1 to 50 we need $t = 1$ and

$$y = -3d + 50. \tag{2.58}$$

Hence,

$$d = |c| < \frac{-3d + 50}{4},$$

which simplifies to $d \leq 7$. To obtain the smallest possible y in (2.58), we need the largest possible d; so

$$d = 7, \quad x = 19, \quad y = 29,$$

and the smallest possible number of students in the class is 29.

Check: $\frac{19}{29} = 0.6551\cdots$, which is between $65\frac{1}{2}\%$ and $66\frac{1}{2}\%$.

Q290 Let $\angle DAE = x$ and $\angle DCE = y$. Then $\angle CEB = x + y$ since the exterior angle of a triangle equals the sum of the two opposite interior angles.

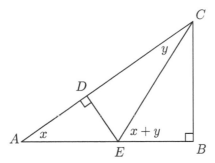

Moreover, Pythagoras' Theorem in $\triangle ADE$ gives $|AD|^2 + |DE|^2 = |AE|^2$; also $\triangle ADE$ and $\triangle ABC$ are similar since they have two equal angles, so $|AD|/|AE| = |AB|/|AC|$; therefore, $|AD||AC| = |AB||AE|$. Using these facts, we have

$$
\begin{aligned}
\cos x \cos y - \sin x \sin y &= \frac{|AD|}{|AE|}\frac{|CD|}{|CE|} - \frac{|DE|}{|AE|}\frac{|DE|}{|CE|} \\
&= \frac{|AD|(|AC| - |AD|) - |DE|^2}{|AE||CE|} \\
&= \frac{|AD||AC| - |AE|^2}{|AE||CE|} \\
&= \frac{|AB||AE| - |AE|^2}{|AE||CE|} \\
&= \frac{|AB| - |AE|}{|CE|} \\
&= \frac{|BE|}{|CE|} \\
&= \cos(x + y)
\end{aligned}
$$

as claimed.

Q291 After n minutes, each of the twins will be at some point on the north–west/south–east diagonal of the park; that is, at one of the intersections marked in red.

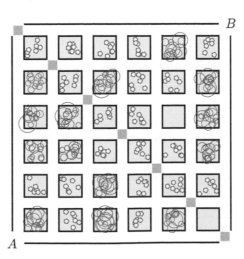

We wish to find the probability that they are at the same intersection. Now, if we set up a coordinate system such that Alex starts at $(0,0)$ and Ben starts at (n,n), then the diagonal points have coordinates $(k, n-k)$ for $k = 0, 1, 2, \ldots, n$. Alex's trip from $(0,0)$ to $(k, n-k)$ must consist of k steps in an easterly direction and $n-k$ in a northerly direction, and can be specified by a list such as $NNEN \cdots NNNE$. The number of such lists is given by the binomial coefficient $C(n, k)$. The number of ways in which Ben can reach $(k, n-k)$ is exactly the same. And the total number of possibilities for $2n$ choices of direction (n for each twin) is 2^{2n}. Since k can be any integer from 0 to n, the required probability is

$$p = \frac{1}{2^{2n}} \left(C(n,0)^2 + C(n,1)^2 + C(n,2)^2 + \cdots + C(n,n)^2 \right).$$

We can simplify the sum in brackets by considering the coefficient of x^n in the binomial expansion of

$$(x+1)^n (1+x)^n = (x+1)^{2n}$$

(see Section 3.6). A term in x^n on the left–hand side is obtained as a product of terms $x^k 1^{n-k}$ and $1^k x^{n-k}$; the coefficient is $C(n,k)C(n,k) = C(n,k)^2$. Any k from 0 to n gives a product of x^n, so the total coefficient is

$$C(n,0)^2 + C(n,1)^2 + C(n,2)^2 + \cdots + C(n,n)^2.$$

But the coefficient of x^n in $(x+1)^{2n}$ is $C(2n, n)$, which therefore equals this sum. Hence, the required probability is

$$\frac{C(2n, n)}{2^{2n}}.$$

Q292 If Ellie spoke the truth, then she and Fiona are twins, so that Fiona must also have spoken the truth. Therefore, the boy in yellow is Ben, and he spoke the truth when he said that the other boy was Alex. Since Alex is Ben's twin, he also spoke the truth, and so Ellie and Fiona are not twins. This contradicts what we discovered earlier; so the situation is impossible, and we must conclude that Ellie did not speak the truth. Since Ellie lied and the boy in red contradicted her, he must have told the truth, and since the boy in yellow is his twin, he also spoke the truth. Therefore, Alex is wearing red and Ben is wearing yellow. Moreover, this means that Fiona spoke the truth while Ellie lied, and so they cannot be twins.

Q293 It is clear from the diagram that the "boundary ellipse" should be tangent to each of the "cradle ellipses"; by symmetry, we only need to consider the tangency in the first quadrant. By calculus, the gradient of the tangent to the boundary ellipse is given by

$$\frac{2x}{a^2} + \frac{2y}{b^2} \frac{dy}{dx} = 0$$

and so

$$\frac{dy}{dx} = -\frac{b^2}{a^2} \frac{x}{y},$$

provided that $y \neq 0$. (This formula can also be determined without calculus, if you wish: see Problem 296.) Suppose that the two ellipses meet with a common tangent at (p, q), and assume initially that $q \neq 0$. Then we have

$$\frac{p^2}{a^2} + \frac{q^2}{b^2} = 1, \quad (p-1)^2 + \frac{q^2}{4} = 1, \quad -\frac{b^2}{a^2} \frac{p}{q} = -\frac{4}{1} \frac{p-1}{q}.$$

The third equation is easily solved to give

$$p = \frac{4a^2}{4a^2 - b^2}.$$

Taking b^2 times the first equation minus 4 times the second eliminates q gives

$$\frac{b^2 p^2}{a^2} - 4(p-1)^2 = b^2 - 4;$$

and we can now substitute the preceding expression for p into this. Some careful algebra results in an unexpectedly simple equation

$$b^4 - 4a^2 b^2 + 16a^2 = 0,$$

which can be treated as a quadratic polynomial in b^2 and solved to give

$$b^2 = 2a\left(a \pm \sqrt{a^2 - 4}\right).$$

We must have $a > 2$, because $a = 2$ would correspond to $q = 0$, which we have ruled out. If we take the $+$ sign in this formula, then we have $b^2 > 2a^2$ and hence

$$p = \frac{4a^2}{4a^2 - b^2} > \frac{4a^2}{2a^2} = 2,$$

which is impossible; so we must take the $-$ sign. Thus,

$$b^2 = 2a\left(a - \sqrt{a^2 - 4}\right).$$

This relation between a and b guarantees that we have an ellipse which is tangent to the "cradle" ellipses; we now need to find which of all these possibilities has the smallest area. The area of the ellipse is given by the formula $A = \pi ab$, and so we consider

$$a^2 b^2 = 2a^3\left(a - \sqrt{a^2 - 4}\right).$$

Equating the derivative of the right–hand side to zero and solving, we have

$$6a^2\left(a - \sqrt{a^2 - 4}\right) + 2a^3\left(1 - \frac{a}{\sqrt{a^2 - 4}}\right) = 0$$

so

$$8a^3 = 6a^2\sqrt{a^2 - 4} + \frac{2a^4}{\sqrt{a^2 - 4}}$$

from which it follows that

$$4a\sqrt{a^2 - 4} = 3(a^2 - 4) + a^2 = 4(a^2 - 3).$$

Therefore,

$$a^2(a^2 - 4) = (a^2 - 3)^2,$$

so $a^2 = \frac{9}{2}$, which after simplification gives $b^2 = 6$. Thomas should therefore choose the ellipse

$$\frac{x^2}{9/2} + \frac{y^2}{6} = 1,$$

which has area $A = \pi ab = 3\pi\sqrt{3}$.

Before we tell Thomas to start constructing the nursery, we'd better check to see whether he might get a better result by taking $q = 0$, which so far we have neglected. This will give $a^2 = 4$, and allowable values for b^2 will be determined by the requirement that the "boundary" ellipse must contain the "cradles": thus, the y value on the boundary ellipse in the first quadrant must be no less than the y value on the cradle. This gives

$$b^2\left(1 - \frac{x^2}{4}\right) \geq 4\left(1 - (x-1)^2\right)$$

whenever $0 < x < 2$, which simplifies to

$$b^2 \geq 16\left(1 - \frac{2}{2+x}\right).$$

If x approaches 2, then the right–hand side continually increases, approaching a value of 8. Therefore, the minimum possible value for b^2 is 8, and the minimum area for this ellipse is $4\pi\sqrt{2}$. This is larger than our previous answer, which confirms that the ellipse we found previously is the minimal solution.

Q294 For any passwords u, v using letters a, b only, we write $u \sim v$ to denote that u and v are related by a single substitution, and $u \approx v$ to denote that the passwords match; that is, they are related by a chain of substitutions.

For part (a), suppose that the "first attempt" passwords are u and v, and that they are extended by words w_1 and w_2. We want an example for which

$$u \not\approx v \quad \text{but} \quad uw_1 \approx vw_2.$$

There are many examples like this. For instance, if $u = aa$ and $v = bb$, then it is clear that u and v do not match, since they are not the same as they stand, and with only two letters, no substitutions are possible. However, appending $w_1 = b$ and $w_2 = a$ respectively gives aab and bba, which are matching passwords.

For part (b), we want

$$u \not\approx v \quad \text{but} \quad uw \approx vw;$$

unfortunately for Alex and Ben, this is not possible. To prove this, we shall show that if u, v are any words, x is any single letter and $ux \approx vx$, then $u \approx v$. Starting with $uw \approx vw$ and removing one letter at a time, this will eventually show that $u \approx v$; in other words, the "first attempt" passwords did match after all.

So, suppose that $ux \approx vx$. This means that there is a chain of single substitutions

$$ux \sim u_1x_1 \sim u_2x_2 \sim \cdots \sim u_nx_n \sim vx,$$

where $u_1, u_2, \ldots, u_n$ are words and $x_1, x_2, \ldots, x_n$ are single letters. We consider various cases.

Firstly, suppose that all the x_k are equal to x. Since any single substitution changes every letter it involves, none involves the letters x_k; every substitution occurs within the words u_k. But this means that

$$u \sim u_1 \sim u_2 \sim \cdots \sim u_n \sim v$$

and therefore $u \approx v$.

Next, suppose that none of the x_k is equal to x. Since there are only two allowable letters, they must all be equal to the other: call it y. Then, we have

$$ux \sim u_1y \sim u_2y \sim \cdots \sim u_ny \sim vx.$$

Since a substitution has clearly been made at the end of ux, the last two letters of u must be yy, and the preceding part of the word is unchanged by this substitution. Giving a similar argument involving vx, we can write

$$u = u_0 yy, \quad u_1 = u_0 xx, \quad u_n = v_0 xx, \quad v = v_0 yy.$$

Now, looking at the central part of the above chain,

$$u_1 y \sim u_2 y \sim \cdots \sim u_n y;$$

therefore by our first case, we have $u_1 \approx u_n$; that is,

$$u_0 xx \approx v_0 xx.$$

The words $u_0 x$ and $v_0 x$ are clearly shorter than u and v, so we may assume by way of induction that our main result applies to these words. Therefore, $u_0 xx \approx v_0 xx$ implies $u_0 x \approx v_0 x$, which in turn implies $u_0 \approx v_0$; hence $u_0 yy \approx v_0 yy$; that is, $u \approx v$, and the proof of this case is finished.

Finally, we consider the case when the sequence of letters $x_1, x_2, \ldots, x_n$ may consist of any arrangement of xs and ys. The sequence may then be considered as a number of xs, followed by a number of ys, followed by further xs, and so on; and the result follows from the first two cases.

Q295 First, we decide which pairs do which activities; we shall allocate individual children later. For each pair, write down the initial of the activity they *will not* be involved in. For example, if Alex and Ben do music and painting, Christine and Denise do music and reading and so on, then we write $RP \cdots$. There will be two pairs of twins who don't do music, three who don't do reading and four who don't do painting; so the number of ways to allocate activities is the same as the number of ways to write down a nine–letter word consisting of two Ms, three Rs and four Ps. The number of ways to arrange 9 letters is $9!$. However, interchanging the two Ms will give the same word, and so we must divide by $2!$; similarly, we must divide by $3!$ on account of the repeated Rs and by $4!$ on account of the repeated Ps. The number of words is therefore

$$\frac{9!}{2!\,3!\,4!} = 1260.$$

To complete the allocation, we must decide which of each pair of twins does which activity. There are 2 choices for each pair, 2^9 altogether. Therefore, the total number of options available to the teacher is

$$\frac{9!}{2!\,3!\,4!}\,2^9 = 1260 \times 512 = 645120.$$

In Problem 297, the class organisation problem becomes a little more complicated.

Q296 To find without calculus the gradient of the ellipse

$$\frac{x^2}{a^2} + \frac{y^2}{b^2} = 1$$

at the point (p, q), consider the line

$$y - q = m(x - p)$$

which has gradient m and passes through (p, q). We can eliminate y from these two equations and solve to find the x–coordinates of the points where the line meets the ellipse. In most cases, this will give $x = p$ and one other value. However, if the line is tangent to the ellipse, then we shall obtain $x = p$ only. We need to find the value of m for which this occurs. So, eliminating y and collecting terms in x gives (check for yourself!)

$$(a^2m^2 + b^2)x^2 + 2a^2m(q - mp)x + a^2[(q - mp)^2 - b^2] = 0;$$

This quadratic equation has a single root if and only if the discriminant is zero:

$$4a^4m^2(q - mp)^2 - 4(a^2m^2 + b^2)a^2[(q - mp)^2 - b^2] = 0.$$

This seems fairly nasty to solve, but if you do the algebra carefully, then you will find that many terms cancel and we end up with

$$(p^2 - a^2)m^2 - 2pqm + (q^2 - b^2) = 0. \tag{2.59}$$

Now, this looks wrong, because it seems to give two possibilities for m, whereas it is geometrically clear that there will be only one tangent to the ellipse at the given point. However, since the point (p, q) is on the ellipse, we have

$$\frac{p^2}{a^2} + \frac{q^2}{b^2} = 1,$$

which implies

$$p^2b^2 + a^2q^2 = a^2b^2 \quad \text{and so} \quad (p^2 - a^2)(q^2 - b^2) = p^2q^2.$$

Therefore, the discriminant of the quadratic in (2.59) is

$$(2pq)^2 - 4(p^2 - a^2)(q^2 - b^2) = 0$$

and there is one possibility for the gradient,

$$m = \frac{pq}{p^2 - a^2} = \frac{pqb^2}{p^2b^2 - a^2b^2} = -\frac{pqb^2}{a^2q^2} = -\frac{b^2}{a^2}\frac{p}{q},$$

as we found in Solution 293.

Secondly, we want to find the minimum possible value of

$$f(a) = 2a^4 - 2a^3\sqrt{a^2 - 4}$$

if $a \geq 2$. We can substitute $a = 2\sec\theta$ with $0 \leq \theta < \frac{\pi}{2}$: this will remove the square root term to give

$$f(a) = 32\sec^4\theta - 32\tan\theta\sec^3\theta = 32\frac{1 - \sin\theta}{\cos^4\theta}.$$

Next, eliminate the $\cos\theta$ terms by setting $t = \sin\theta$ and writing $f(a)$ as

$$f(a) = 32\frac{1 - \sin\theta}{(1 - \sin^2\theta)^2} = \frac{32}{(1 + t)(1 - t^2)},$$

To find the minimum value of $f(a)$, we want the maximum value of $(1 + t)(1 - t^2)$ for $0 \leq t < 1$. This will be a value c for which the horizontal line $t = c$ is tangent to the graph of $(1 + t)(1 - t^2)$, and for the same kind of reason as in the previous problem, we need a value of c such that the equation $(1 + t)(1 - t^2) = c$ has a double root. So we require

$$t^3 + t^2 - t + (c - 1) = (t - \alpha)^2(t - \beta)$$

for some α, β; then α will be the relevant t value (which must lie between 0 and 1), and c will be the maximum we are looking for. Expanding and equating coefficients in the above cubic equation gives

$$1 = -2\alpha - \beta, \quad -1 = 2\alpha\beta + \alpha^2, \quad c - 1 = -\alpha^2\beta.$$

Solving the first equation for β and substituting into the second gives a quadratic equation with solutions $\alpha = \frac{1}{3}, -1$; since α lies between 0 and 1, the latter possibility must be rejected. This gives $\beta = -\frac{5}{3}$ and $c = \frac{32}{27}$, and so the minimum value we seek is

$$f(a) = \frac{32}{32/27} = 27.$$

The value of a that attains this minimum is

$$a = 2\sec\theta = \sqrt{\frac{4}{1 - \sin^2\theta}} = \sqrt{\frac{4}{1 - t^2}} = \sqrt{\frac{4}{1 - \alpha^2}} = \frac{3}{\sqrt{2}},$$

as we found previously.

Q297 Step 1: choose four pairs of twins to do music: this can be done in $C(8,4)$ ways.

Step 2: suppose that k of the "non–music" pairs do painting; then $4 - k$ of the "music pairs" also do painting, and the number of ways to choose the "painting pairs" is $C(4,k)C(4,4-k)$. This leaves $4 - k$ pairs with no activity as yet: they must do reading and dancing. There remain $2k$ pairs with only one activity so far: k of them will be chosen to take the k remaining reading spots, and the other k must do dancing: there are $C(2k,k)$ ways to make this choice. This gives

$$C(4,k)C(4-k)C(2k,k)$$

options for step 2; however, k could be any number from 0 to 4, so we need to sum this expression over all values of k. Since $C(4, 4-k) = C(4,k)$, we can simplify to give the total number of options

$$C(4,0)^2C(0,0) + C(4,1)^2C(2,1) + C(4,2)^2C(4,2)$$
$$+ C(4,3)^2C(6,3) + C(4,4)^2C(8,4) = 639.$$

Step 3: choose which twin in each pair does which activity: there are 2^8 choices. Putting all this back together, we get the total number of allocations

$$C(8,4) \times 639 \times 2^8 = 11450880.$$

Q298 Interchange squares and gridlines so that the path follows the lines, as in the diagram. We may assume that the side length of the squares is 1. The path is now a lattice polygon; that is, a polygon in which every vertex has integer coordinates, and its area is given by **Pick's Theorem** (see Section 3.9):

$$A = I + \frac{B}{2} - 1,$$

where I is the number of lattice points inside the polygon and B is the number on the boundary. Since the curve visits every square in the original diagram, there are no complete squares inside the path, and therefore no lattice points inside the polygon in the new diagram: that is, $I = 0$. For the same reason, every lattice point is on the boundary, and so $B = (2n)^2$: therefore,

$$A = 2n^2 - 1.$$

Moreover, the total grey area in the new diagram is $(2n - 1)^2$: so the grey area outside the polygon is

$$(2n - 1)^2 - A = 2n^2 - 4n + 2 = 2(n - 1)^2,$$

and this is also the number of blue dots, and the number of gridline intersections in the original diagram.

As a check, the example given in the question has $n = 4$, so we predict 18 intersections outside the path, and we confirm this simply by counting the dots. This problem was inspired by the "Masyu" puzzle, which may be found at www.nikoli.co.jp/en/puzzles/masyu/.

Q299 A general formula for the sequence is $x_1 = 1$ and

$$x_{2k} = \cos(x_k), \quad x_{2k+1} = \sin(x_k)$$

for all positive integers k. We shall consider the numbers x_n in sets of $1, 2, 4, 8, \ldots$ by defining

$$S_j = \{n \in \mathbb{Z} \mid 2^j \le n < 2^{j+1}\} \quad \text{and} \quad X_j = \{x_n \mid n \in S_j\}.$$

First, we shall show that for each $j \ge 2$, the largest element in X_j is

$$x_{2^j + 2^{j-1} - 2} = \cos(\sin(\sin \cdots (\sin(\cos 1)) \cdots)),$$

where there are $j - 2$ sine terms in the left–hand side; and the smallest is

$$x_{2^j + 2^{j-1} - 1} = \sin(\sin(\sin \cdots (\sin(\cos 1)) \cdots)),$$

where there are $j - 1$ sine terms in the left–hand side. To confirm that this is true for $j = 2$, we simply calculate the four terms in X_2: we find

$$x_4 = 0.85, \quad x_5 = 0.51, \quad x_6 = 0.66, \quad x_7 = 0.74,$$

and clearly the largest is x_4 and the smallest is x_5. Now, we continue by mathematical induction (refer to Section 3.8 if you need a reminder about this important proof technique), showing that, if the claimed facts are true for some specific j, then they are also true for $j + 1$. It is clear that all values of x_n lie between 0 and 1. Therefore, when we form X_{j+1} by calculating the cosine and sine of all elements of X_j, the largest element will be either the cosine of the smallest element in X_j or the sine of the largest element in X_j. By assumption, these possibilities are

$$x_{2^{j+1} + 2^j - 2} = \cos(x_{2^j + 2^{j-1} - 1}) = \cos(\sin(\sin \cdots (\sin(\cos 1)) \cdots))$$

with $j - 1$ sine terms, and

$$x_{2^{j+1} + 2^j - 3} = \sin(x_{2^j + 2^{j-1} - 2}) = \sin(\cos(\sin \cdots (\sin(\cos 1)) \cdots))$$

with $j - 2$ sine terms, not counting the first one. To determine which of these is the larger, note that

$$x_{2^{j+1} + 2^j - 2} > \cos(\cos 1) > 0.85 \quad \text{and} \quad x_{2^{j+1} + 2^j - 3} < \sin 1 < 0.85;$$

so the largest element of X_{j+1} is $x_{2^{j+1}+2^j-2}$, as claimed. The argument for the smallest element is very similar. The choice lies between

$$x_{2^{j+1}+2^j-4} = \cos(x_{2^j+2^{j-1}-2}) > \cos 1 > 0.54$$

and

$$x_{2^{j+1}+2^j-1} = \sin(x_{2^j+2^{j-1}-1}) < \sin(\cos 1) < 0.52;$$

clearly the latter is the smaller, and is therefore the smallest element of X_{j+1}. This completes the proof.

Next, we note that the maximum element of X_j increases as j increases. This is because if we write

$$\theta = \sin(\sin \cdots (\sin(\cos 1)) \cdots)$$

with $j - 2$ sine terms, then $\sin \theta < \theta$ and so

$$x_{2^{j+1}+2^j-2} = \cos(\sin \theta) > \cos \theta = x_{2^j+2^{j-1}-2}.$$

It follows that the maximum element in any of the sets $X_0, X_1, X_2, \ldots, X_j$ is $x_{2^j+2^{j-1}-2}$. How long does it take to make this maximum greater than 0.99? We need

$$\sin(\sin \cdots (\sin(\cos 1)) \cdots) < \arccos(0.99) = 0.141539 \cdots,$$

where there are $j - 2$ sine terms on the left–hand side. So, enter 1 into a calculator, hit the "cos" button once and then the "sin" button repeatedly, keeping count of how many times you hit it, until the result satisfies this inequality. If you have a programmable calculator or equivalent software, then you will be able to automate this. Either way, you will find that 137 sine terms is not enough but 138 is. In other words, the greatest element in any of the sets $X_0, X_1, X_2, \ldots, X_{139}$ is

$$x_{2^{139}+2^{138}-2} = \cos(\sin(\sin \cdots (\sin(\cos 1)) \cdots)) = 0.989943 \cdots,$$

which is still less than 0.99, and the greatest element in X_{140} is

$$x_{2^{140}+2^{139}-2} = 0.990010 \cdots,$$

which is greater than 0.99. This looks like our answer, but we should be careful: there are many terms in X_{140} before we get to this one ($2^{139} - 2$ of them to be precise), and it is conceivable that one of them, although not the greatest element in X_{140}, might already be greater than 0.99. To resolve this question, we consider the *second* largest and *second* smallest elements in X_j. We shall show that if $j \geq 3$, then the second largest element of X_j is

$$x_{2^j+2^{j-1}+2^{j-2}-2} = \cos(\sin(\sin \cdots (\sin(\cos(\sin 1))) \cdots)))$$

with $j - 3$ sine terms in the middle, not counting the last one, and the second smallest is

$$x_{2^j+2^{j-1}+2^{j-2}-1} = \sin(\sin(\sin \cdots (\sin(\cos(\sin 1))) \cdots))) \tag{2.60}$$

starting with $j - 2$ sine terms. The proof is much the same as we did before, though, not surprisingly, we have to be a little more careful with the details. First, we check the statement by direct calculation for $j = 3, 4, 5$: this is a bit of work but is basically routine. Now suppose that the claim is true for some specific $j \geq 5$. The choice for the second largest element of X_{j+1} lies between the cosine of the second smallest element in X_j (because we know that the cosine of the smallest gives the largest element in X_{j+1}, not the second largest) and the sine of the largest. Noting that the number of sine terms in (2.60) is at least 3, the element we are looking for is either

$$x_{2^{j+1}+2^j+2^{j-1}-2} = \cos(\sin(\sin\cdots(\sin(\cos(\sin 1)))\cdots))$$
$$\geq \cos(\sin(\sin(\sin(\cos(\sin 1))))) > 0.85$$

or

$$x_{2^{j+1}+2^j-3} = \sin(\cos(\sin\cdots(\sin(\cos 1))\cdots)) < \sin 1 < 0.85\,;$$

and the former is the larger. Likewise, the choice for second smallest is between

$$x_{2^{j+1}+2^j-4} = \cos(x_{2^j+2^{j-1}-2}) > \cos 1 > 0.54$$

and

$$x_{2^{j+1}+2^j+2^{j-1}-1} = \sin(x_{2^j+2^{j-1}+2^{j-2}-1})$$
$$< \sin(\sin(\sin(\sin(\cos(\sin 1))))) < 0.53\,,$$

and the latter is the smaller. Finally, we calculate the second largest element of X_{140}: it is $x_{2^j+2^{j-1}+2^{j-2}-2}$ with $j = 140$. This is an expression with 137 sine terms in the middle,

$$x_{2^{140}+2^{139}+2^{138}-2} = \cos(\sin(\sin\cdots(\sin(\cos(\sin 1)))\cdots)) = 0.989717\cdots,$$

which is less than 0.99. Therefore, the element we found above is the only element up to X_{140} which exceeds 0.99, and it is x_n with

$$n = 2^{140} + 2^{139} - 2$$
$$= 2090694862362245919518973588060783891185662\,.$$

Comment. This may be a surprisingly large value of n, seeing that 0.99 is not very close to 1. Suppose that you tried to brute–force this question by calculating $x_2, x_3, x_4, \ldots$ sequentially until you obtained an answer greater than 0.99. With advanced numerical software, it is currently possible to perform about 2^{30} sine and cosine calculations per second; at this rate, solving the problem by brute force would take something like $2^{86} \approx 10^{26}$ years.

Q300 To determine the winner from n entrants, $n-1$ of them must lose a match. Therefore, $n - 1$ matches will be played. This answers question (a).

For (b), this shows immediately that 99 matches will be required in the first contest. Now, the best contestant will have played either 6 or 7 matches in the first round; and the second–best contestant must have been the loser in one of these matches (because nobody except the best can defeat the second–best). Therefore, there is no need to have all 99 unranked competitors in the second contest, but only these 6 or 7; there will be 5 or 6 matches; and the final winner of this contest will be the second–best arm–wrestler overall.

Now consider what happened to the third–best entrant in the first contest. He must have been beaten by either the best or second–best entrant.

- Those in the first category went on to play in the second contest, and the third–best player must have been beaten by the second–best. But since there were 6 or 7 entrants in this contest, the second–best defeated 2 others (and had a bye in the first round), or 3 others. Only these 2 or 3 are possible candidates for the third–best player.

- Now we consider those in the second category. In the first contest, the second–best player may have played 1 game only (losing to the best player in the first round); or 7 games (not receiving a bye in the first round, and eventually losing to the best player in the final); or anywhere in between. So, the number of players defeated by the second–best player in the first contest is from 0 to 6, and these are the remaining candidates for third–best arm–wrestler.

Therefore, the total number of competitors in the contest for third place is from 2 to 9; and the number of matches played will be from 1 to 8. Putting all this information together, the best three arm–wrestlers can always be found in $99 + 6 + 8 = 113$ matches; and the minimum requirement is $99 + 5 + 1 = 105$ matches.

Q301 Begin by cutting the crown at A and unfolding it into a long strip with E in the middle. Then fold it back again. The fold at E will point outwards: we denote this by O. Now given any pre–existing folds, performing another right–on–left fold will give the following result. The new folds will begin with a copy of the folds we have already, because it is just like starting with a strip of half the length and doing the same as has already been done. Then there will be an O fold, because this will always be the case for the middle fold. Then there will be the folds we have done already, but taken in reverse order; and with outward folds changed to inward and *vice versa*. The operation of performing one extra fold on an existing fold can be represented as

$$w \mapsto w O w^*,$$

where w^* means the string of symbols obtained from w by reversing the order of the symbols and swapping their identity. We know that we start with just O; folding three times gives

$$O \mapsto O(O)I$$
$$\mapsto OOI(O)OII$$
$$\mapsto OOIOOII(O)OOIIOII.$$

(The brackets have been inserted to identify the new middle element in each line, but do not have any real significance.) Now stick the crown together again at A; this will give an O fold, and so the complete circle of folds, starting at A, is

$$OOOIOOIIOOOIIOII.$$

To solve the problem, notice that the I folds always occur in pairs, with one exception of a single I between two Os. If we were to start the folding process at a different place, then this isolated I would end up in a different place; and so either it, or one of its adjacent Os, would not match the folds we have already. Therefore, it is impossible to fold the crown from a different starting point without reversing any existing folds.

Q302 We shall give a solution to this problem using ideas from *modular arithmetic*: readers who are not familiar with this topic should consult Section 3.3. So, let the digits of the secret square be $abcde$. We see immediately from the first two guesses that

$$d = 2, \quad a \neq 1, \quad b \neq 5, \quad e \neq 5.$$

The digits a, b, c, d, e consist of $1, 2, 5, 7$ and one digit x which we don't yet know (and which may be a repeat of $1, 2, 5$ or 7). We also know that $x \neq 0, 3, 6$, as these digits do not occur in the answer.

Since the secret number is a square, the sum of its digits must be a square modulo 9. Therefore,

$$15 + x \equiv 0^2, 1^2, 2^2, 3^2 \text{ or } 4^2 \equiv 0, 1, 4, 0 \text{ or } 7 \pmod 9:$$

this gives $x = 3$, which we have already ruled out, or $x = 1, 4, 7$, which remain as possibilities.

Next, the last digit of a square must be a square modulo 10; that is, $0, 1, 4, 9, 6$ or 5. So, we can rule out $e = 7$, and we have $e = 1$ or $e = 4$ (the latter only possible if $x = 4$).

Thirdly, the alternating sum $a - b + c - d + e$ is congruent to a square modulo 11, and these squares are

$$0^2, 1^2, 2^2, 3^2, 4^2, 5^2 \equiv 0, 1, 4, 9, 5, 3 \pmod{11}.$$

The alternating sum can be written as $a + b + c + d + e - 2b - 2d = 11 + x - 2b$, and so we have

$$x - 2b \equiv 0, 1, 4, 9, 5 \text{ or } 3 \pmod{11}.$$

Now suppose that $x = 7$. Then we have $e = 1$, the only possibility for b is 7, and our options are 57721 and 77521. But by direct calculation, these are not squares.

Suppose that $x = 4$. Then $b = 1, 4$ or 7 and so $x - 2b \equiv 2, 7, 1 \pmod{11}$: only the last is possible. So, our options are 47521, 57124, 57421, and none is a square.

We are left with $x = 1$, so $b = 1, 7$. Calculating $x - 2b \equiv 10, 9 \pmod{11}$ shows that the former is not possible, and the latter gives the only possibility for the secret square as $57121 = 239^2$.

Q303 We shall consider a product

$$(1 + 2x + 3x^2 + 4x^3)(\cdots) = \cdots,$$

and shall build up the second factor on the left–hand side one term at a time in such a way as to achieve what we want. Start with

$$p(x) = 1 + 2x + 3x^2 + 4x^3$$
$$q(x) = a + bx$$
$$p(x)q(x) = a + (2a + b)x + (3a + 2b)x^2 + (4a + 3b)x^3 + 4bx^4.$$

Now, we don't care what coefficients we get for the constant and the x term, but we want the x^2 coefficient to be zero. So, we cancel out the $(3a + 2b)x^2$ in the product by adding an opposite term to $q(x)$: our next attempt is

$$p(x) = 1 + 2x + 3x^2 + 4x^3$$
$$q(x) = a + bx - (3a + 2b)x^2$$
$$p(x)q(x) = a + (2a + b)x - (2a + b)x^3 - (9a + 2b)x^4 - (12a + 8b)x^5.$$

Now, this will have only square exponents provided that

$$2a + b = 0, \quad 12a + 8b = 0.$$

Unfortunately, as we have here two linear equations in two unknowns, we expect to get only the trivial solution $a = b = 0$, which does not solve our problem. (In some cases, we might be lucky and get a non–trivial solution – check for yourself that in this case we don't.) Therefore, we continue modifying $q(x)$ so as to eliminate the x^3 term:

$$p(x) = 1 + 2x + 3x^2 + 4x^3$$
$$q(x) = a + bx - (3a + 2b)x^2 + (2a + b)x^3$$
$$p(x)q(x) = a + (2a + b)x - 5ax^4 - (6a + 5b)x^5 + (8a + 4b)x^6.$$

At this stage, we do not need to eliminate the x^4 term, so we take any x^4 term in $q(x)$ and then continue to eliminate unwanted terms,

$$p(x) = 1 + 2x + 3x^2 + 4x^3$$
$$q(x) = a + bx - (3a + 2b)x^2 + (2a + b)x^3 + cx^4$$
$$p(x)q(x) = a + (2a + b)x - (5a - c)x^4 - (6a + 5b - 2c)x^5 + (8a + 4b + 3c)x^6 + 4cx^7$$

and

$$p(x) = 1 + 2x + 3x^2 + 4x^3$$
$$q(x) = a + bx - (3a + 2b)x^2 + (2a + b)x^3 + cx^4 + (6a + 5b - 2c)x^5$$
$$p(x)q(x) = a + (2a + b)x - (5a - c)x^4 + (20a + 14b - c)x^6$$
$$+ (18a + 15b - 2c)x^7 + (24a + 20b - 8c)x^8 \,.$$

At this stage, to eliminate all terms with non–square exponent, we'll need to solve three equations in three unknowns, which, as mentioned above, is unlikely to be successful. But if we continue for just one more step...

$$p(x) = 1 + 2x + 3x^2 + 4x^3$$
$$q(x) = a + bx - (3a + 2b)x^2 + (2a + b)x^3 + cx^4$$
$$+ (6a + 5b - 2c)x^5 - (20a + 14b - c)x^6$$
$$p(x)q(x) = a + (2a + b)x - (5a - c)x^4 - (22a + 13b)x^7$$
$$- (36a + 22b + 5c)x^8 - (80a + 56b - 4c)x^9$$

...then we now have only two equations in three unknowns (because we don't care what coefficient we have for x^9), and a non–zero solution is guaranteed. The equations are

$$22a + 13b = 0, \quad 36a + 22b + 5c = 0\,,$$

and, with careful working, it is not difficult to find an integer solution $a = 65$, $b = -110$, $c = 16$ (among other possibilities). Checking, we can substitute these values back into the last equation to yield

$$p(x) = 1 + 2x + 3x^2 + 4x^3$$
$$q(x) = 65 - 110x + 25x^2 + 20x^3 + 16x^4 - 192x^5 + 256x^6$$
$$p(x)q(x) = 65 + 20x - 309x^4 + 1024x^9\,,$$

and, as required, all exponents in this product are squares.

For part (b), we take any polynomial

$$p(x) = p_0 + p_1 x + p_2 x^2 + \cdots + p_n x^n$$

with integer coefficients, and we seek to show that there is a $q(x)$ with integer coefficients such that the product $p(x)q(x)$ has only terms with square exponents. Inspired by the answer to (a), we take a polynomial with unknown coefficients and degree $n^2 - n$, say

$$q(x) = q_0 + q_1 x + q_2 x^2 + \cdots + q_{n^2-n} x^{n^2-n} \,.$$

We begin by observing that this polynomial has $n^2 - n + 1$ coefficients, presently unknown. Now, the product will have terms up to x^{n^2}, so $n^2 + 1$ terms in all. This will include a constant term and terms in x, x^4, x^9, ..., x^{n^2}, and there are $n + 1$ of these. We don't care what coefficients are attached to these terms, but we want all the *other* terms, $n^2 - n$ of them, to have coefficient zero. To achieve this, we need to solve the equations

$$p_2 q_0 + p_1 q_1 + p_0 q_2 = 0$$
$$p_3 q_0 + p_2 q_1 + p_1 q_2 + p_0 q_3 = 0$$
$$p_5 q_0 + p_4 q_1 + p_3 q_2 + p_2 q_3 + p_1 q_4 + p_0 q_5 = 0$$
$$\vdots$$
$$p_n q_{n^2-n-1} + p_{n-1} q_{n^2-n} = 0\,.$$

In these equations, the p_k are the known coefficients, which are integers, the q_k are the unknowns, and the left–hand sides are the coefficients of x^2, x^3, x^5, ..., x^{n^2-1} in the product $p(x)q(x)$. Since the system involves $n^2 - n$ equations in $n^2 - n + 1$ unknowns, there is certain to be a solution with integer values for $q_0, q_1, q_2, \ldots, q_{n^2-n}$, and this solves the problem.

Comment. There is actually nothing special here about the squares, except that there are infinitely many of them. For example, we could give an almost identical proof to show that, for any $p(x)$, there exists $q(x)$ such that the exponents in the product are only primes, or only Fibonacci numbers, or... or only any infinite set of non–negative integers.

Q304 We shall solve the first problem by interpreting it in terms of graph theory – see Section 3.4 for terminology and basic results in this topic. So, draw the graph consisting of nine vertices labelled $1, 2, 3, 4, 5, 6, 7, 8, 9$, and an edge joining vertices a, b whenever the sum $a + b$ is "permissible"; that is, not a multiple of 3 or 5 or 7.

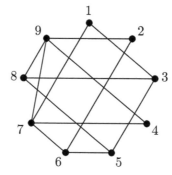

We wish to find a Hamilton cycle in this graph: that is, to select nine of the edges in such a way as to connect up all nine vertices and return to the first vertex. Since vertices $1, 2$ and 4 have degree 2, the lines shown in red below are obligatory,

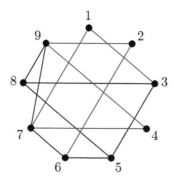

It is now easy to see that the only way to complete a Hamilton cycle is to use the path 3856. So, there is only one solution to the problem, if rotations and reflections are counted as the same solution.

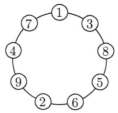

For question (b), consider nine consecutive integers and let k be the middle one, so that the least is $k - 4$ and the greatest is $k + 4$. Sums of two of these will be somewhere near $2k$.

To find a value of k which is "most likely" to fail, we look for a set of consecutive numbers in which as many as possible are multiples of 3 or 5 or 7. This can be done systematically (look up the *Chinese Remainder Theorem* if you would like to pursue this), but trial and error is probably just as easy; we find that the numbers

$$48, 49, 50, 51, 54, 55, 56, 57$$

all have forbidden factors and are therefore "impermissible sums". Choosing $2k = 52$, that is, $k = 26$, means that the sums of k with the other eight numbers in the circle are $48, 49, 50, 51, 53, 54, 55, 56$. Because only one of these sums is allowable, 26 can only have one number adjacent to it, and it is impossible to form a circle of the desired type with the numbers $22, 23, 24, 25, 26, 27, 28, 29, 30$.

Comment. As the problem depends only on divisibility by $3, 5, 7$, the existence (or not) of a solution with middle number k is effectively the same as the existence for $k + 105$. The desired arrangement is also impossible when $2k = 53 + 105$; that is, $k = 79$. These are not the only cases: an exhaustive computer search reveals that the arrangement is impossible for $k = 25, 26, 27, 28, 77, 78, 79, 80$ and possible for all other k from 1 to 105. In some of these cases, every number has at least two possible neighbours, but the neighbours cannot "fit together" in a manner which completes a Hamilton cycle.

Q305 First, we try to simplify the cube roots, beginning with the numerator. By careful but straightforward algebra, we have

$$\left(a + b\sqrt{2} + c\sqrt{3}\right)^3 = a(a^2 + 6b^2 + 9c^2) + b(3a^2 + 2b^2 + 9c^2)\sqrt{2}$$
$$+ c(3a^2 + 6b^2 + 3c^2)\sqrt{3} + (6abc)\sqrt{6},$$

and so we would like to find a, b, c such that

$$a(a^2 + 6b^2 + 9c^2) = 560, \quad b(3a^2 + 2b^2 + 9c^2) = 158,$$
$$c(3a^2 + 6b^2 + 3c^2) = 324, \quad 6abc = 90.$$

Now, a system of equations like this is generally going to be very hard to solve, but perhaps we can guess an answer? The last equation gives $abc = 15$: let's guess that a, b, c are all positive integers. In the first equation, $a^2 + 6b^2 + 9c^2$ is then a positive integer, so a is a factor of 560; but 3 is not a factor of 560 and so 3 is not a factor of a. Similarly, the second equation shows that 3 is not a factor of b and 5 is not a factor of b; the third shows that 5 is not a factor of c. Since $abc = 15$, the only remaining possibility is $a = 5$, $b = 1$, $c = 3$. Now it is important to substitute back and check this, because the whole procedure depends on a, b, c being integers, which was a guess, not a certainty. However, this is a matter of simple arithmetic, and we leave it to the reader to confirm that our guess is correct, and hence that

$$\left(5 + \sqrt{2} + 3\sqrt{3}\right)^3 = 560 + 158\sqrt{2} + 324\sqrt{3} + 90\sqrt{6}.$$

To obtain the denominator, all we need do is change the sign of b: thus,

$$\left(5 - \sqrt{2} + 3\sqrt{3}\right)^3 = 560 - 158\sqrt{2} + 324\sqrt{3} - 90\sqrt{6}.$$

Therefore, the expression is

$$\frac{5 + \sqrt{2} + 3\sqrt{3}}{5 - \sqrt{2} + 3\sqrt{3}} = \frac{\left(5 + \sqrt{2} + 3\sqrt{3}\right)\left(5 - \sqrt{2} - 3\sqrt{3}\right)}{\left(5 - \sqrt{2} + 3\sqrt{3}\right)\left(5 - \sqrt{2} - 3\sqrt{3}\right)} = \frac{\sqrt{2} + 3\sqrt{3}}{5}.$$

Q306 Substituting $y = m(x - 1) + 1$ into the equation of the ellipse and collecting terms gives the quadratic

$$(2m^2 + 1)x^2 - 4m(m - 1)x + (2m^2 - 4m - 1) = 0.$$

We could solve this by the quadratic formula, but a simpler method is to note that the two roots must add up to $4m(m - 1)/(2m^2 + 1)$. Since we know that one of the solutions is $x = 1$, we can easily find the other, and then substitute back into the equation of the line to find the corresponding y value. After simplification – please check the algebra yourself – we obtain

$$x = \frac{2m^2 - 4m - 1}{2m^2 + 1}, \quad y = \frac{-2m^2 - 2m + 1}{2m^2 + 1}, \tag{2.61}$$

which is the other point of intersection between the line and the ellipse.

For part (b), we first note that if m is rational, then the values of x, y in (a) are also rational. Moreover, the formulae give, with one exception, *all possible* rational points (x, y) satisfying the equation $x^2 + 2y^2 = 3$. For if (x, y) is a pair of rational numbers, then the line between (x, y) and $(1, 1)$ has gradient

$$m = \frac{y - 1}{x - 1},$$

which is rational, and so (x, y) is given by the above procedure with this value of m. There are two points where this argument does not quite work.

- If $(x, y) = (1, 1)$, then we have only one point on the line and m is not defined. However, it is easy to check that we do in fact get $(x, y) = (1, 1)$ by taking $m = -\frac{1}{2}$, which is the gradient of the tangent to the ellipse at $(1, 1)$.

- If $(x, y) = (1, -1)$, then the line joining the points is vertical and does not have a (finite) gradient. If you take $x = 1$, $y = -1$ in equations (2.61) and try to solve for m, then you will find that there is no solution. This is the exception referred to above – the only rational point on the ellipse which is not given by the formulae (2.61).

Since m is rational, we can write $m = u/v$, where u, v are integers. Making this substitution and multiplying numerators and denominators by v^2 gives

$$x = \frac{2u^2 - 4uv - v^2}{2u^2 + v^2}, \quad y = \frac{-2u^2 - 2uv + v^2}{2u^2 + v^2}.$$

Notice that if $u = 1$ and $v = 0$, then $x = 1$ and $y = -1$, so we have regained the missing point mentioned above. To sum up: by taking u and v to be integers, these latest formulae give all rational solutions of the equation $x^2 + 2y^2 = 3$.

Now let a, b, c be integers such that $a^2 + 2b^2 = 3c^2$. Excluding the solution $a = b = c = 0$, we can divide by c^2 to obtain

$$\left(\frac{a}{c}\right)^2 + 2\left(\frac{b}{c}\right)^2 = 3$$

– in essence, exactly the equation we have just studied! Since a/c and b/c are rational, they are given by

$$\frac{a}{c} = \frac{2u^2 - 4uv - v^2}{2u^2 + v^2}, \quad \frac{b}{c} = \frac{-2u^2 - 2uv + v^2}{2u^2 + v^2},$$

where u, v are integers. It may be that we can cancel a common factor out of both fractions (for example, if v is even, then both numerators and denominators have a factor of 2);

therefore, all possible integer solutions of $a^2 + 2b^2 = 3c^2$ having no common factors are given by taking

$$a = 2u^2 - 4uv - v^2, \quad b = -2u^2 - 2uv + v^2, \quad c = 2u^2 + v^2$$

with u, v integers, and then cancelling any common factors. The solution $a = b = c = 0$, which we ignored previously, is now given by the values $u = v = 0$.

Q307 Draw a graph (see Section 3.4) having a vertex for each coin, and an edge between two vertices whenever the corresponding coins are in the same stack. We begin with n vertices, and an edge joining each pair of distinct vertices: so, the graph is the complete graph K_n, and the number of edges is $\frac{1}{2}n(n-1)$. When the original stack has been split (over the course of one or more moves) into a number of smaller stacks, the graph representing the state of the game will consist of a number of separate smaller complete graphs. Now, splitting any stack, whether it is the original or a smaller stack, into groups of size a and b, means deleting all the edges between a set of a and a set of b vertices. There are ab such edges, and this is the score accumulated for the move. Finishing the game means deleting all the edges in the original graph, and the total final score will therefore be the initial number of edges; that is, $\frac{1}{2}n(n-1)$.

Q308 When two quantities change at equal and opposite rates, their sum remains constant. Therefore, as long as the ant moves in the way described, the sum of its distances to the two grains will equal $AP + AS$, where AP is the distance from the ant to the poison when the motion first begins, and AS is the distance from the ant to the sugar when the motion first begins. Now imagine that the ant reaches the sugar. In that case, the sum of its distances to the two grains would equal PS, the distance between the grains. Thus, we would have $PS = AP + AS$. But, unless the ant started exactly on a straight line between P and S, this would violate the triangle inequality for triangle APS. Therefore, the ant cannot reach the sugar as long as it walks in the way described.

... and this argument also shows that the only case in which the ant might reach the sugar is when it begins its journey directly between the two grains: then it will head straight towards the sugar and will eventually reach it... as long as it walks fast enough and for long enough.

The path of the ant in (a) will be a curve such that at every point on the curve, the sum of the distances to the two grains will be constant. Therefore, the ant will travel along an arc of an ellipse that has the two grains as foci.

Q309 We'll use graph theory terminology and hence will refer to the points as "vertices" and the connections between them as "edges". We'll call the sought-for configurations "a red triangle", "a blue triangle" and "a blue quadrilateral", and we'll say that two vertices joined by a red (or blue) edge are "red–neighbours" (or "blue–neighbours"). Vertices joined by two edges will be called "partners".

To prove (a), consider the vertex u. Excluding its partner v, each of the other three vertices is either a red–neighbour or a blue–neighbour of u; since there are three vertices

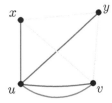

here and only two options, it must be that either two or more of the three vertices are red–neighbours of u, or two or more are blue–neighbours of u. Suppose there are two red–neighbours, and call them x, y. Then u has three red–neighbours: its partner v, and the two vertices x, y just mentioned. If any two of v, x, y are connected by a red edge, then these two, together with u, form a red triangle; if not, then v, x, y form a blue triangle. If u were to have two blue–neighbours rather than two red–neighbours, then a virtually identical argument would show that the configuration contains a blue triangle or a red triangle. This completes the proof for part (a).

Now for part (b), consider the vertex a. Excluding its partner b, each of the other six vertices is either a red–neighbour or a blue–neighbour of a; so there must be either 3 or more red–neighbours, or 4 or more blue–neighbours. Consider the former case: then (restoring b) the vertex a has 4 red–neighbours (first diagram below): either two of these are joined by a red edge, making a red triangle together with a; or all are joined by blue edges, making a blue quadrilateral.

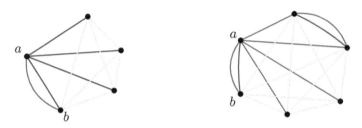

In the remaining case, a has (at least) 4 blue–neighbours among the vertices c, d and e, f and g, h. Since we have 4 blue–neighbours chosen from 3 pairs, two of them must belong to the same pair and therefore must be joined by both a red edge and a blue edge. Now restore b (second diagram above). Then the blue–neighbours of a include 5 vertices with at least one pair of vertices joined by edges of both colours. By part (a), there are two options: either these five vertices include a red triangle; or they contain a blue triangle, which together with a makes a blue quadrilateral.

We have considered every possibility, and have shown that no matter what the colouring, our configuration must contain a red triangle or a blue quadrilateral. This concludes the proof of (b).

Comment. By taking the argument for part (b) a little further, we may prove that in fact if we have a configuration of eight points in which just *three* pairs a, b and c, d and e, f have "double" connections, then the same conclusion follows: there must be either a red triangle or a blue quadrilateral. Readers may enjoy proving this for themselves.

Q310 We can write the factorisation as

$$10^{100} = \left(2^{a_1} 5^{b_1}\right) \times \left(2^{a_2} 5^{b_2}\right) \times \left(2^{a_3} 5^{b_3}\right),$$

where $a_1, \ldots, b_3$ are non–negative integers satisfying

$$a_1 + a_2 + a_3 = 100, \qquad b_1 + b_2 + b_3 = 100.$$

For part (a) of the question, there are no further restrictions on the exponents, and counting the number of possibilities for a_1, a_2, a_3 is a "dots and lines" problem of the type demonstrated in the last example of Section 3.5.2, page 253. The number of choices for a_1, a_2, a_3 is $C(102, 2) = 5151$, the number of choices for b_1, b_2, b_3 is the same, and the total number of ways to write 10^{100} as a product of three positive integers is $5151^2 = 26532801$.

When order matters, we have shown in (a) that there are 5151^2 possible factorisations. To tackle the case in which order does not matter, we need to consider when a factorisation can contain two or more equal terms. First, 10^{100} is not a cube; so there is no factorisation of the form xxx. To count factorisations in which two of the factors are the same and the other is different, choose whether the first, second or third factor is to be the different one: there are 3 options; also choose the repeated factor: it must be a number $x = 2^a 5^b$ such that $2^{2a} 5^{2b}$ is a factor of 10^{100}, so $0 \le a, b \le 50$: there are 51^2 options.

So, there are altogether 3×51^2 factorisations involving a repeated factor, and this leaves $5151^2 - 3 \times 51^2$ with no repeated factor. As we are now considering the case where order *does not* matter, the former consist of the expressions we want, counted three times each (since xxz and xzx and zxx are all regarded as the same); the latter are counted six times each: so, the number of different factorisations is

$$\frac{3 \times 51^2}{3} + \frac{5151^2 - 3 \times 51^2}{6} = 4423434.$$

Q311 By symmetry, we may assume that $a < b$. Then

$$2^a + 2^b = 2^a(1 + 2^c)$$

with $c = b - a \ge 1$. Since the first factor is a power of 2 and the second is odd, they have no common factor. So, the only way for the product to be a pth power is for each factor to be a pth power. Considering the second factor, this means that

$$1 + 2^c = n^p$$

for some integer n; it is not hard to see that n is odd and $n \ge 3$. Therefore,

$$2^c = n^p - 1 = (n-1)(n^{p-1} + n^{p-2} + \cdots + n + 1).$$

Since the product on the right-hand side equals a power of 2, each factor is a power of 2; neither factor can be $2^0 = 1$, so they are both even. But the second factor is a sum of p odd numbers, and for this to be even, p must be even.

So let $p = 2q$. Then the previous equation can be rewritten as

$$2^c = n^{2q} - 1 = (n^q - 1)(n^q + 1).$$

Therefore, $n^q - 1$ and $n^q + 1$ are powers of 2, and their difference is 2: the only possibility is that $n^q - 1 = 2$ and $n^q + 1 = 4$, so $n^q = 3$; therefore, $n = 3$ and $q = 1$. But this means $p = 2$, which is not so. We have ruled out all possibilities, and, therefore, the required equation has no solution.

Q312 Let the diameter of the semicircle be d. In the following diagram, triangles APQ and QPB are similar,

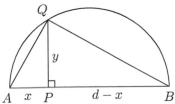

and so

$$\frac{x}{y} = \frac{y}{d-x}; \quad \text{that is,} \quad y^2 = x(d-x).$$

So, in the diagram from the question, the areas of the two triangles are

$$\frac{1}{2} \times 3\sqrt{\frac{3}{2}\left(d - \frac{3}{2}\right)} \quad \text{and} \quad \frac{1}{2} \times 2\sqrt{4(d-4)}.$$

Since these areas are equal, we have

$$9\left(\frac{3}{2}\right)\left(d - \frac{3}{2}\right) = 4(4)(d-4),$$

which simplifies to

$$27(2d - 3) = 64(d - 4)$$

and gives the diameter $d = \dfrac{35}{2}$.

Q313 Suppose that the three groups arrive at x minutes, y minutes and z minutes before the next train is due. The average waiting time will then be

$$\frac{2x + 3y + 4z}{9},$$

and we want to find the probability that $2x + 3y + 4z > 72$. The space of all possible arrival times for the three groups can be visualised as the cube specified by the inequalities

$$0 \leq x \leq 12, \quad 0 \leq y \leq 12, \quad 0 \leq z \leq 12,$$

and the probability we want is the proportion of this cube which satisfies the inequality $2x + 3y + 4z > 72$. Geometrically, this is the region within the cube and above the plane $2x + 3y + 4z = 72$. The plane intersects the edges of the cube at the four points W, X, Y, Z in the diagram.

$W = (0, 8, 12)$
$X = (0, 12, 9)$
$Y = (12, 12, 3)$
$Z = (12, 0, 12)$

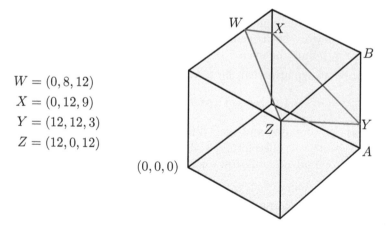

For example, to find the point Y, we note that it lies on the edge AB, where we have $x = 12$, $y = 12$: substituting into the equation of the plane gives $z = 3$. Calculations for W, X and Z are left to the reader. The region we are seeking is a truncated pyramid; the base is a right–angled triangle and the vertex is perpendicularly above the right angle. Constructing a similar pyramid on top of the truncated pyramid,

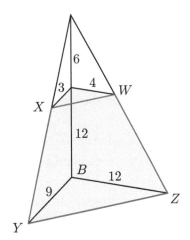

we find the volume as

$$\frac{12 \times 9 \times 18}{6} - \frac{4 \times 3 \times 6}{6} = 26 \times 12.$$

The required probability is the ratio of this volume to the whole cubical volume,

$$p = \frac{26 \times 12}{12^3} = \frac{13}{72}.$$

Q314 Begin with the well known factorisation

$$x^m - y^m = (x - y)(x^{m-1} + x^{m-2}y + \cdots + xy^{m-2} + y^{m-1}).$$

This shows that if x and y are integers, then $x - y$ is a factor of $x^m - y^m$; multiplying by integer coefficients and adding up a number of terms like this shows that $x - y$ is always a factor of $p(x) - p(y)$.

Now, the values of $p(x)$ include all the numbers $0, 1, 2, \ldots, n$, say

$$p(a_0) = 0, \quad p(a_1) = 1, \quad p(a_2) = 2, \ldots, \quad p(a_n) = n.$$

All the values a_k must be different since all the values $p(a_k)$ are different. Now from the previous paragraph, $a_1 - a_0$ is a factor of $p(a_1) - p(a_0)$; that is, $a_1 - a_0$ is a factor of 1; and since a_1, a_0 are integers, $a_1 - a_0 = \pm 1$. Treating other pairs in the same way, we have

$$a_1 - a_0 = \pm 1, \ a_2 - a_1 = \pm 1, \ldots, \ a_n - a_{n-1} = \pm 1.$$

Moreover, all the signs in these equations must be the same, or else we should have at some point $a_{k+1} - a_k = -(a_{k+2} - a_{k+1})$, so $a_k = a_{k+2}$, which is not true. We shall treat the case in which all the signs in the equations are $+$; the case in which they are all $-$ is very similar.

So, then we have $a_k = a_0 + k$ for all k, and hence

$$p(a_k) - a_k + a_0 = 0$$

for all k. This means that the polynomial $p(x) - x + a_0$ has a minimum of $n + 1$ roots. But this polynomial has degree n or less, and so the only possibility is that it is zero for all values of x. Therefore, $p(x) = x - a_0$, and this is x plus a constant, as required.

Q315 It is clear that there are many ways to choose $2n + 1$ children so that no two are in the same row or column; we have to prove that it is possible to do this in such a way that no two are twins. To make a selection in which no two children occupy the same row or column, we may proceed as follows.

- Choose a child from the same row as the teacher: this can be done in $2n$ ways.
- Specify the column of the child to be chosen from each remaining row. As there are $2n$ rows and $2n$ available columns, and the columns chosen must all be different, the number of ways to do this is $(2n)!$.

The total number of ways of making this selection is $(2n)(2n)!$.

Now count those of the above selections which also **do** contain a pair of twins. To do this,

- choose a pair of twins: this can be done in *at most* $2n^2 + 2n$ ways (possibly less, because if two twins are standing in the same row or column, then they are not an allowable choice);
- choose a child from each remaining row, without repeating columns: there are *at most* $(2n - 1)!$ ways (possibly less, because some of these $(2n - 1)!$ choices may include the teacher).

Therefore, the number of choices of children which contain a pair of twins is, at the most, $(2n^2 + 2n)(2n - 1)! = (n + 1)(2n)!$. However, as $n \geq 2$, we have $n + 1 < 2n$, and so among all selections of children with no two in the same row or column, the number including a pair of twins is less than the total number; therefore, there must be a selection which does not include a pair of twins.

Comment. If $n = 1$, then the result need not be true: for example, if four pairs of twins and a teacher are arranged as shown, then there is no selection of the required type.

$$
\begin{array}{ccc}
1 & 2 & 3 \\
4 & T & 1 \\
2 & 3 & 4
\end{array}
$$

Further comment. Note that in counting the number of selections including twins, we may have counted the same arrangement more than once. For example, suppose that the class includes twins Alex and Ben, and another pair of twins Anna and Beatriz. Then any selection containing both pairs will have been counted once where Alex and Ben are chosen first and Anna and Beatriz are "accidentally" chosen among the remaining $2n - 1$ children; and once again in the opposite case. So, the selection will have been counted twice, maybe even more. But this means that the true number of selections including twins is even less than we found above, so it does not invalidate our argument.

Q316 Let s_n be the nth number in the sequence. We shall use congruence arithmetic – see Section 3.3. It is routine to check that $s_1, s_2, \ldots, s_9$ are not multiples of 11, and that $s_9 \equiv 5$ (mod 11). Now for $n = 10, 11, \ldots, 99$, we have

$$s_n = 100s_{n-1} + n \equiv s_{n-1} + n \quad \text{(mod 11)}$$

and so

$$s_n \equiv n + (n - 1) + (n - 2) + \cdots + 10 + s_9 \quad \text{(mod 11)}.$$

Multiplying both sides of the congruence by 2, adding up an arithmetic series and using the value of s_9 from above gives

$$2s_n \equiv (n + 10)(n - 9) + 10 \quad \text{(mod 11)}.$$

We want s_n to be a multiple of 11; that is,

$$(n + 10)(n - 9) + 10 \equiv 0 \pmod{11}.$$

Expanding and subtracting $11n$ (which counts as zero modulo 11) gives

$$n^2 - 10n - 80 \equiv 0 \pmod{11};$$

completing the square and simplifying, $(n-5)^2 \equiv 105 \equiv 6 \pmod{11}$. But this is impossible, since 6 is not a square modulo 11 (see below for a reason why). So we calculate from above

$$s_{99} \equiv 99 + 98 + \cdots + 10 + s_9 \equiv 4 \pmod{11}$$

and proceed to consider appending three–digit numbers $n = 100, 101, \ldots, 999$. In this case, we have

$$s_n = 1000 s_{n-1} + n \equiv -s_{n-1} + n \pmod{11}$$

and hence

$$s_{n+1} \equiv -s_n + n + 1 \equiv s_{n-1} + 1 \pmod{11}.$$

That is, each term s_n is one more (modulo 11) than the second previous term. It is now simple to find the values

$$s_{99}, s_{100}, s_{101}, \ldots \equiv 4, 8, 5, 9, 6, 10, 7, 0, \ldots,$$

and so the smallest number in the sequence which is a multiple of 11 is

$$s_{106} = 123456789101112 \cdots 99100 \cdots 106.$$

To confirm the statement that 6 is not a square modulo 11, note that every integer is congruent modulo 11 to one of $0, \pm 1, \pm 2, \pm 3, \pm 4, \pm 5$, and the squares of these are

$$0^2, 1^2, 2^2, 3^2, 4^2, 5^2 \equiv 0, 1, 4, 9, 16, 25 \equiv 0, 1, 4, 9, 5, 3 \pmod{11},$$

and this list does not include 6.

Q317 There is no other choice of 13 squares on a 5×5 chessboard such that no two are separated by a knight's move. To prove this, observe that the diagram

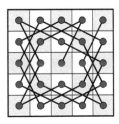

shows a path formed of knight's moves which includes all 25 squares on the board. To obtain a set of squares in which no two are separated by a knight's move, we cannot take two consecutive squares along this path. So the only way to obtain 13 squares is to take the first, third, fifth and so on of the 25 squares. Looking carefully at the diagram shows that these are precisely the black squares on the board, and there is no other way to satisfy the requirement.

Q318 We show how to do this construction inductively for n^2 digits, where the allowable digits are $0, 1, \ldots, n-1$. Clearly,

$$1100$$

does the job for $n = 2$. (It's not a circle but that would take a lot of printing space later on! You can imagine that the end is joined to the beginning so as to form a circle.) If we have already a solution for some specific n, then it must contain the pair 00 somewhere. In between these two zeros insert

$$n,\ n,\ n-1,\ n,\ n-2,\ n,\ n-3,\ \ldots,\ n,\ 0\,. \tag{2.62}$$

Then it is easy to see that every pair xy where $x = n$ or $y = n$ or both occurs in this insertion, except for $0n$. Also, every pair xy where neither x nor y is n occurs in the previous iteration, except that 00 has been split by the insertion. However, since the insertion occurs between two 0s, both these exceptions do occur after all: $0n$ where the insertion begins, and 00 where it finishes. Therefore, the combined arrangement includes all pairs of digits from 0 to n. Specifically, for $n = 10$, one possible solution is

1102212033232313044342414055454535251506656464362616077...

...6757473727170887868584838281809989797969594939291900 .

Comment. Many solutions are possible: for example, in (2.62), the digits $n-1, n-2, \ldots, 1$ could be permuted among themselves in any way.

Q319 The required greatest common divisor is 24. To prove this, we begin by noting that every one of the products is a multiple of 3, for if any product of 11 different positive integers is not a multiple of 3, then the sum of the numbers is at least

$$1 + 2 + 4 + 5 + 7 + 8 + 10 + 11 + 13 + 14 + 16 = 91$$

and cannot be 82. Similarly, every product must be a multiple of 8, for if not, then the numbers include at most two even numbers, and their sum is at least

$$1 + 2 + 3 + 4 + 5 + 7 + 9 + 11 + 13 + 15 + 17 = 87.$$

Therefore, the greatest common divisor of all products is at least 24. However, we can find cases in which the product does not have 16 as a factor, for example,

$$1 + 2 + 3 + 5 + 6 + 7 + 9 + 10 + 11 + 13 + 15 = 82;$$

so 16 is not a factor of *all* the products under consideration, and is therefore not a factor of the greatest common divisor. Likewise, we can find cases where the product does not have 9 as a factor, for example,

$$1 + 2 + 3 + 4 + 5 + 7 + 8 + 10 + 11 + 14 + 17 = 82,$$

and where the product does not have 5 as a factor, for example,

$$1 + 2 + 3 + 4 + 6 + 7 + 8 + 9 + 11 + 12 + 19 = 82,$$

and, finally, where the product does not have any prime factor greater than 5, for example,

$$1 + 2 + 3 + 4 + 5 + 6 + 8 + 9 + 10 + 16 + 18 = 82.$$

Therefore, the greatest common divisor cannot be greater than 24, so it is exactly 24.

Q320 Such a path must consist of m moves to the right and n moves up, together with **either** an extra move right and one left, **or** an extra move up and one down. In the first case, the left move cannot have a right move immediately before or after it as this would mean some point was visited twice; the left cannot occur before the first right as it would take the path off the grid, and it cannot occur after the last right as it would have come from off the grid. A similar argument applies to the case where there is an extra up and a down move; therefore, a path such as we are seeking must consist of one of the following mutually exclusive possibilities:

- $m+1$ single moves right, $n-2$ single moves up, 1 sequence of three moves up–left–up;
- $m-2$ single moves right, $n+1$ single moves up, 1 sequence of three moves right–down–right.

To count the first type, we begin by arranging $m+2$ letters R and $n-2$ letters U: this can be done in $C(m+n, n-2)$ ways. Then, we replace one of the letters R by ULU: this can be done for any of the $m+2$ letters R except the first or last, so there are m options. A similar argument for the second case gives the total number of paths as

$$m \binom{m+n}{n-2} + n \binom{m+n}{m-2}.$$

Q321 Let N be an 8–digit number formed from the digits $1, 2, 3, 4, 5, 6, 7, 8$, and write N as a concatenation of two–digit numbers $n_1 n_2 n_3 n_4$. Since

$$10^2 \equiv -1 \pmod{101},$$

we have

$$N = 10^6 n_1 + 10^4 n_2 + 10^2 n_3 + n_4$$
$$\equiv -n_1 + n_2 - n_3 + n_4 \pmod{101}.$$

So, N is a multiple of 101 if and only if $R = n_1 - n_2 + n_3 - n_4$ is a multiple of 101.

Since we want the largest value of N, we try taking $n_1 = 87$ and $n_2 = 65$; that is, $N = 8765dddd$. If we can find solutions in this case, then they cannot be exceeded by any other solutions. This gives

$$R = 22 + n_3 - n_4,$$

which must be a multiple of 101. Since n_3 and n_4 consist of the digits $1, 2, 3, 4$, we have

$$22 + 12 - 43 \leq R \leq 22 + 43 - 12,$$

that is, $-9 \leq R \leq 53$. As R is a multiple of 101, it must be zero, and we have $n_4 = n_3 + 22$. Since the digits of n_3, n_4 are $1, 2, 3, 4$ and we want the largest possible n_3, we have $n_3 = 21$ and $n_4 = 43$. So, the required number is $N = 87652143$.

Q322 Suppose that a square s^2 has digits $d_1 d_2 \cdots d_n d_1 d_2 \cdots d_n$, and let x be the number with digits $d_1 d_2 \cdots d_n$. Then we have

$$s^2 = x + 10^n x = (10^n + 1)x.$$

Let a^2 be the largest square factor of $10^n + 1$; then we can write

$$10^n + 1 = a^2 b \quad \text{and so} \quad s^2 = a^2 b x,$$

where b is an integer with no square factor other than 1. For the right–hand side to be a square, x must have the same prime factors as b, once each; and any additional prime factors of x must occur an even number of times. That is, we have

$$x = bc^2$$

for some integer c. Now since x has n digits,

$$b \le x < 10^n < 10^n + 1 = a^2 b,$$

and so a must be greater than 1. With computing assistance, we can find the factorisations of 11, 101, 1001, 10001 and so on; it turns out that the first number in this sequence to have a square factor greater than 1 is

$$10^{11} + 1 = 100000000001 = 11^2 \times 826446281,$$

and so we have

$$n = 11, \quad a = 11, \quad b = 826446281.$$

Using again the fact that x has n digits (not beginning with zero), we have $10^{n-1} \le x < 10^n$ and, hence,

$$\frac{10^{10}}{826446281} \le c^2 < \frac{10^{11}}{826446281},$$

that is, $13 \le c^2 \le 120$; to obtain the smallest possible value for x, we take $c^2 = 16$, which yields $x = 13223140496$. So, the smallest square consisting of two identical digit–strings is

$$1322314049613223140496,$$

which you can confirm is equal to 36363636364^2.

Q323 Recall that we are investigating sets S of integers having the following property for all real quadratics $f(x) = ax^2 + bx + c$:

"if $f(x)$ is an integer for all x in S, then $f(x)$ is an integer for all integers x".

Since we are looking at Integer values of Quadratics, we shall call such a set an *IQ–set*. We already know that $S = \{0, 1, 2\}$ is an IQ–set, and our first task is to find all possible IQ–sets of three integers. We approach the problem by stages.

- Step 1: if $S = \{0, 1, s\}$ with $s > 2$, then S is not an IQ–set. **Proof.** The quadratic

$$f(x) = \frac{x^2 - x}{s(s-1)}$$

  is an integer for $x = 0, 1, s$, as is easily checked; however,

$$f(2) = \frac{2}{s(s-1)}$$

  is not an integer, because the denominator has a factor $s > 2$.

- Step 2: if $S = \{0, s, t\}$ with $1 < s < t$, then S is not an IQ–set. **Proof.** Similar to the above: consider

$$f(x) = \frac{x^2 - sx}{t(t-s)}$$

  and check that while $f(0)$, $f(s)$ and $f(t)$ are integers, $f(1)$ is not.

- Step 3: it follows from steps 1 and 2 that the only three–element IQ–set in which the smallest element is 0 is our original example $\{0, 1, 2\}$. To finish the problem, we show that any other IQ–set of three elements is closely related to this one.

- Step 4. Let $S = \{s, t, u\}$ and $T = \{s-k, t-k, u-k\}$, where k is any integer. Then T is an IQ–set if and only if S is. **Proof.** Suppose that S is an IQ–set; we aim to show that T is an IQ–set. Let $f(x)$ be any real quadratic such that $f(s-k)$, $f(t-k)$ and $f(u-k)$ are integers; we have to prove that $f(x)$ is an integer for all integers x. The assumption that S is an IQ–set means that the stated property applies to *any* quadratic: so we may apply it to the quadratic $g(x) = f(x-k)$. This is an integer for $x = s, t, u$; so $g(x)$ is an integer for all integers x; so $g(x+k) = f(x)$ is an integer for all integers x. Thus, T is an IQ–set. We also need to show that if T is an IQ–set, then so is S; the argument for this is virtually identical and will be omitted.

We've done all the hard work, and it is now relatively easy to find all three–element IQ–sets. Let $S = \{s, t, u\}$ with $s < t < u$. Then

S is an IQ–set

$\Leftrightarrow \quad T = \{0, t-s, u-s\}$ is an IQ–set $\qquad$ [step 4, taking $k = s$]

$\Leftrightarrow \quad t - s = 1, u - s = 2$ $\qquad$ [step 3]

$\Leftrightarrow \quad S = \{s, s+1, s+2\}.$

That is, a three–element set has the property we seek if and only if it consists of three consecutive integers.

It is not hard to see that if a set contains three consecutive integers and more besides, then this does not affect the IQ–property. However, $S = \{0, 1, 3, 5\}$ is also an IQ–set, despite the fact that it does not contain three consecutive integers. To prove this, consider the quadratic $f(x) = ax^2 + bx + c$, and suppose that $f(0)$, $f(1)$, $f(3)$, $f(5)$ are all integers. That is,

$$c, \quad a+b+c, \quad 9a+3b+c, \quad 25a+5b+c$$

are integers. It follows that $a + b = (a+b+c) - c$ is the difference of two integers and hence is itself an integer; and likewise that

$$2a = (25a + 5b + c) - 3(9a + 3b + c) + 4(a + b + c) - 2c$$

is an integer. Since we now know that $2a$ and $a+b$ and c are integers, we can follow word for word the final part of Solution 182 to show that $f(x)$ is always an integer if x is an integer.

Q324 By carefully studying the solution of Problem 169, we can see that it will still apply in the case of 120 residents, *unless* all of the inequalities in the solution become equalities. (We are being deliberately vague here, so as not to spoil the earlier puzzle for any readers who have not yet attempted it.) So we will need one apartment having exactly 15 residents, and 15 other apartments having a total of 105 residents. One way of ensuring that "intolerable" apartments persist indefinitely is to devise a scenario in which the numbers of occupants in various apartments are the same after a redistribution as they were before. So, suppose that all but 16 apartments are absolutely uninhabitable and always remain empty; and that the sixteen liveable apartments house

$$15, 14, 13, \ldots, 2, 1, 0$$

residents respectively. Then the residents of the intolerable apartment redistribute themselves to the other 15 of these 16 apartments, one to each, so that the numbers of residents become

$$0, 15, 14, \ldots, 3, 2, 1.$$

Since these numbers are exactly the same as they were before, they will also be the same after the next redistribution; and so on indefinitely. Hence, there will always be an apartment with 15 or more residents.

Q325 Define $a_k = p_1^2 p_2^2 \cdots p_{k-1}^2 p_k$, where $p_1, p_2, \ldots, p_n$ are the primes in increasing order, and let S be the set of all a_k with $k \geq 1$. Explicitly,

$$S = \{\, 2,\ 2^2 \times 3,\ 2^2 \times 3^2 \times 5,\ 2^2 \times 3^2 \times 5^2 \times 7, \ldots \,\}$$
$$= \{\, 2,\ 12,\ 180,\ 6300,\ 485100,\ 69369300, \ldots \,\}.$$

For any k, we have the following important properties: p_k is a factor of a_k, but p_k^2 is not a factor of a_k; also, a_k is a factor of all larger elements of S, and the quotient is a multiple of p_k. So, take any subset

$$A = \{\, a_{m_1},\ a_{m_2}, \ldots,\ a_{m_n} \,\}$$

of S, where we assume, as we may, that m_1 is the smallest of all the subscripts. The sum of these elements is

$$N = a_{m_1} + a_{m_2} + \cdots + a_{m_n} = a_{m_1}\left(1 + \frac{a_{m_2}}{a_{m_1}} + \cdots + \frac{a_{m_n}}{a_{m_1}}\right).$$

By our previous remarks, all the quotients a_{m_k}/a_{m_1} for $k > 1$ are integers, and are multiples of p_{m_1}. So, the expression in brackets is 1 plus a sum of multiples of p_{m_1}, and therefore is not a multiple of p_{m_1}. (Note that this is still true even if there are no terms in the brackets after the initial 1.) Hence, the only factor p_{m_1} in the prime factorisation of N is the single p_{m_1} in a_{m_1}. In the prime factorisation of a bth power, every prime must have an exponent which is a multiple of b; but that of N has p_{m_1} with exponent 1, and so N cannot be a bth power with $b > 1$.

Q326 Finding an arrangement of the numbers $1, 2, \ldots, 64$ on a chessboard in which the minimum difference between adjacent numbers is 15 is to a large extent a matter of trial and error. However, intelligent trial and error is always better than mindless trial and error! – so we discuss some ideas before giving an answer. We begin by placing the following numbers in a row of eight squares:

1	17	2	18	3	19	4	20

The rationale for doing this is that we have a row where differences between adjacent numbers are 15 or 16, while using only a selection from the smallest numbers available: in this way, we may hope to use larger numbers in the next row, and thereby avoid having any differences less than 15 between the two rows. Cataloguing similar rows of the same form to use all numbers up to 32, and then beginning again at 33, we have

1	17	2	18	3	19	4	20
5	21	6	22	7	23	8	24
9	25	10	26	11	27	12	28
13	29	14	30	15	31	16	32
33	49	34	50	35	51	36	52
37	53	38	54	39	55	40	56
41	57	42	58	43	59	44	60
45	61	46	62	47	63	48	64

Now it is obvious that this arrangement includes many differences less than 15 and therefore does not solve our problem; but perhaps we can succeed by using the same rows in a different order? It is clear that the row beginning with 1 cannot be next to those beginning with 5, 9 or 13; a careful examination shows that it *can* be next to those beginning with 33, 37, 41 or 43. Considering other rows in the same way, we find that the possible adjacencies between rows are given by the following diagram.

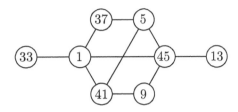

We need to find in this graph a path connecting all eight vertices without repetition (known as a *Hamilton path* – see Section 3.4.2), and there is little difficulty in finding that the only possibility is 33–1–37–5–41–9–45–13, or its reverse. Therefore, the following placement of numbers meets our requirements.

33	49	34	50	35	51	36	52
1	17	2	18	3	19	4	20
37	53	38	54	39	55	40	56
5	21	6	22	7	23	8	24
41	57	42	58	43	59	44	60
9	25	10	26	11	27	12	28
45	61	46	62	47	63	48	64
13	29	14	30	15	31	16	32

To tackle part (b), suppose that we have placed the sixty–four numbers in such a way that the minimum difference between adjacent numbers is 16 or more: that is, adjacent numbers never differ by 15 or less. As in Solution 87, page 92, divide the chessboard into sixteen 2×2 "cells". No two of the sixteen numbers $1, 2, \ldots, 16$ can occur in the same cell, as this would give two adjacent squares with a difference of 15 or less. So each must be in a different cell. Label the cells $1, 2, \ldots, 16$ according to which of these numbers it contains: so, cell k will contain the number k, and three further numbers from $17, 18, \ldots, 64$. Which cell contains the number 17? It cannot be cell k for $k = 2, 3, \ldots, 16$, as there would then be adjacent squares with difference $17 - k < 16$; so it must be cell 1. The number 18 cannot be in cells $3, 4, \ldots, 16$ for a similar reason; nor can it be in cell 1, as we now know that that cell also contains 17; so it is in cell 2. Continuing in this way for the numbers $19, 20, \ldots, 32$, and then similarly for $33, 34, \ldots, 48$ and then $49, 50, \ldots, 64$, we see that the numbers in cell k are

$$C_k = \{\, k,\ k + 16,\ k + 32,\ k + 48 \,\}.$$

Now consider any of the four cells *not* meeting the boundary of the chessboard, say, cell m. The number m in this cell is on the same side of the cell as either $m + 32$ or $m + 48$ (perhaps both). In either case, m and the other number are not on the boundary of the chessboard, and hence are both adjacent to two numbers x, y in an adjoining cell.

(a) Suppose the numbers referred to are m and $m + 32$. Then x and y cannot be from 1 to $m + 15$, as all of these numbers are within 15 of m; they cannot be $m + 16$, as that is in C_m; they cannot be from $m + 17$ to $m + 47$, as these are within 15 of $m + 32$; and they cannot be $m + 48$ as that is in C_m. Therefore, both x and y belong to

$$\{ m + 49, \ m + 50, \ldots, \ 64 \},$$

a set of at most 15 consecutive integers; so x and y differ by less than 16, and this possibility must be ruled out.

(b) If the numbers referred to are m and $m + 48$, then for similar reasons, x and y cannot be from 1 to $m + 15$, nor $m + 16$, nor $m + 32$, nor from $m + 33$ to 64; so they both belong to

$$\{ m + 17, \ m + 18, \ldots, \ m + 31 \},$$

and this again must be ruled out.

There are no options left, and we conclude that placing the numbers with minimum difference 16 or more between adjacent squares is impossible.

Q327 The smallest value of s is $11^2 = 121$. To see this, first note that the smallest possible sum of eleven different positive *odd numbers* is

$$1 + 3 + 5 + \cdots + 21 = 121;$$

so if $s < 121$ and positive integers $x_1, x_2, x_3, \ldots, x_{11}$ add up to s, then at least one of the x_k must be even. So in this case, every product under consideration is even, and the GCD (greatest common divisor) of all the products must be 2, if not more.

This shows that the required GCD cannot be 1 if the sum s is less than 121; to show that the GCD is equal to 1 when $s = 121$, we follow the ideas of Solution 319. Consider the following three sums of eleven different positive integers:

$$1 + 3 + 5 + 7 + 9 + 11 + 13 + 15 + 17 + 19 + 21 = 121;$$
$$1 + 2 + 4 + 5 + 7 + 8 + 10 + 11 + 13 + 14 + 46 = 121;$$
$$1 + 2 + 3 + 4 + 6 + 8 + 12 + 16 + 18 + 24 + 27 = 121.$$

In the first case, none of the summands has 2 as a factor; so 2 is not a factor of their product, and is not a factor of the GCD of all products. For similar reasons, the second sum shows that 3 is not a factor of the GCD. In the third case, none of the summands has any prime factor greater than 3; so their product has no prime factor greater than 3, and the GCD cannot have such a factor either. We have shown that in the case $s = 121$, the GCD of all products has no prime factor at all, and therefore is equal to 1.

If we consider a sum $s > 121$, then it need not be true that the GCD of all products is still 1. Indeed, let s be an even number greater than 121. Any eleven integers with sum s must include at least one even number (because the sum of eleven odd numbers is odd); therefore, by the same argument as we used in the first paragraph of this solution, the GCD of all products of eleven integers adding up to s must be at least 2.

Q328 We count the admissible arrangements according to the maximum number of dots in any one line.

- To choose five dots all in the same line, first select one of the 16 lines, and then select 5 dots from the 8 in that line. The number of choices is $16C(8, 5)$.

- For an arrangement with a maximum of four dots in the same line, choose the line and then 4 of its 8 dots, giving $16C(8,4)$ options as above; then choose the fifth dot from the 56 *not* on this line.

- Similarly, if the maximum number of dots in the one line is three, then we choose the line, choose 3 dots from it and 2 from the remaining 56. However, in this case, any arrangement with three dots in each of two lines, as, for example, shown by the red dots,

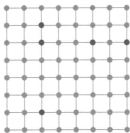

will have been counted twice and therefore must be subtracted once. To make such an arrangement, first select the dot common to the horizontal and vertical line, 64 ways; then choose 2 dots from the remaining 7 in each of the lines intersecting at that dot, $C(7,2)^2$ ways. By the principle of inclusion/exclusion (see Section 3.5.3), the final number of options in this case is $16C(8,3)C(56,2) - 64C(7,2)^2$.

So, the total number of choices of five dots from the 8×8 grid in which at least three lie in the same horizontal or vertical line is

$$16C(8,5) + 16C(8,4) \times 56 + 16C(8,3)C(56,2) - 64C(7,2)^2 = 1415232.$$

Q329 First we look at the leftmost "dotted curve", here shown as a sequence of blue dots.

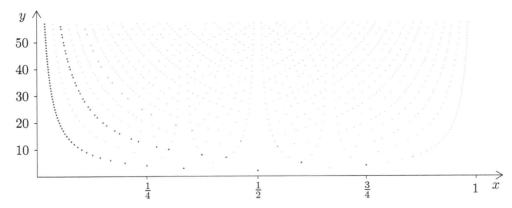

These are the points on the graph corresponding to fractions x with numerator $p = 1$. In this case we have

$$x = \frac{1}{q} \quad \text{and} \quad y = f(x) = f\left(\frac{1}{q}\right) = q = \frac{1}{x}.$$

That is, all of these points lie on the hyperbola $y = 1/x$, which explains why they give the appearance of a smooth curve. The "curves" above this one are formed by fractions x with numerators $2, 3, 4, \ldots$. Likewise, the points on the rightmost "dotted curve" correspond to $p = q - 1$; we have

$$x = \frac{q-1}{q}, \quad y = f\left(\frac{q-1}{q}\right) = q = \frac{1}{1-x},$$

and these are therefore points on the curve $y = 1/(1 - x)$, a hyperbola with asymptote $x = 1$. The two "dotted curves" forming a "tunnel" above the point $x = \frac{1}{2}$ (and other tunnels elsewhere) can be explained similarly: we invite readers to fill in the details.

One further observation: notice that, for example, the third "curve" on the left (red in the diagram) appears to have gaps in it. This is the curve corresponding to numerator 3, that is, $x = 3/q$. However, if q is a multiple of 3, then we will have to cancel the 3 before computing $f(x)$. For instance, if $q = 12$, then $x = \frac{3}{12}$ does not have denominator 12; rather, we write $x = \frac{1}{4}$ and the denominator is 4. Therefore, this point is "missing" from the $p = 3$ curve and belongs to the $p = 1$ curve instead; the same applies to every third point. For similar reasons, the $p = 6$ curve (orange) includes only the first and fifth out of every six consecutive points, missing the second, third, fourth and sixth; and likewise for other numerators.

CHAPTER 3

Some Useful Problem-solving Techniques

I N THE PRESENT CHAPTER, we discuss briefly a number of techniques which may be useful for tackling our collection of *Parabolic Problems*. No attempt has been made to give a detailed exposition of these topics, and anyone seeking proofs of the statements made (which we would wholeheartedly recommend!) should consult standard texts. We have limited ourselves in each topic to a concise explanation and one or two useful examples. In most cases, many further examples may be found in our problem solutions.

3.1 GREATEST COMMON DIVISORS

The greatest common divisor of two integers a and b is the largest integer which is a factor of both a and b; it is also sometimes known as the "highest common factor" of a and b. The greatest common divisor (GCD for short) of a and b makes sense as long as a and b are not both zero. In finding GCDs, it makes no difference whether we are considering positive or negative numbers, so for simplicity we can stick to positive integers.

The GCD of a and b can be found by factorising the two numbers, but if they are large, then this will involve a lot of work. A much more efficient method is provided by the **Euclidean Algorithm**. Divide a by b to give an integer quotient with remainder r_1; then divide b by r_1 to give remainder r_2; then r_1 by r_2 to give remainder r_3; and so on. Eventually, we will arrive at a division which goes exactly, with no remainder; and then $\gcd(a, b)$ is the last divisor, or the last non–zero remainder. In symbols: if we calculate

$$a = q_1 b + r_1$$
$$b = q_2 r_1 + r_2$$
$$r_1 = q_3 r_2 + r_3$$
$$r_2 = q_4 r_3 + r_4$$
$$\vdots$$
$$r_{n-2} = q_n r_{n-1} + r_n$$
$$r_{n-1} = q_{n+1} r_n,$$

then $\gcd(a, b) = r_n$.

Example. Find the greatest common divisor of 4916 and 1284.

DOI: 10.1201/9781003396413-3

Solution. Performing the Euclidean Algorithm calculations beginning with 4916 and 1284, we obtain

$$4916 = 3 \times 1284 + 1064$$
$$1284 = 1 \times 1064 + 220$$
$$1064 = 4 \times 220 + 184$$
$$220 = 1 \times 184 + 36$$
$$184 = 5 \times 36 + 4$$
$$36 = 9 \times 4 + 0.$$

Therefore, $\gcd(4916, 1284) = 4$. It follows that $\gcd(-4916, 1284)$ is also equal to 4.

As well as giving a superior method of calculating GCDs, the Euclidean Algorithm has an important by–product. It enables us with very little calculation to write the GCD as an **integer linear combination** of the original numbers. That is, by using the Euclidean Algorithm applied to a and b, we can find integers x and y such that

$$ax + by = \gcd(a, b).$$

The procedure is to use the steps we have already calculated for the Euclidean Algorithm in reverse, making each remainder the subject of its equation, and then to substitute it into the previous equation.

Example. Find integers x and y such that $4916x + 1284y = 4$.

Comment. The point is that we want x and y to be *integers*: it would be very easy to find many fractions x and y which satisfy the equation.

Solution. We use the Euclidean Algorithm above, and write each remainder as the subject of its equation:

$$4 = 184 - 5 \times 36$$
$$36 = 220 - 1 \times 184$$
$$184 = 1064 - 4 \times 220$$
$$220 = 1284 - 1 \times 1064$$
$$1064 = 4916 - 3 \times 1284.$$

Of course, this step does not actually involve any calculation whatsoever! Now we substitute each equation into the previous one, collecting terms at each stage before proceeding:

$$4 = 184 - 5 \times 36$$
$$= 184 - 5 \times (220 - 1 \times 184)$$
$$= -5 \times 220 + 6 \times 184$$
$$= -5 \times 220 + 6 \times (1064 - 4 \times 220)$$
$$= 6 \times 1064 - 29 \times 220$$
$$= 6 \times 1064 - 29 \times (1284 - 1 \times 1064)$$
$$= -29 \times 1284 + 35 \times 1064$$
$$= -29 \times 1284 + 35 \times (4916 - 3 \times 1284)$$
$$= 35 \times 4916 - 134 \times 1284.$$

Comparing the equation we wanted with the one we have just found,

$$4916x + 1284y = 4 \quad \text{and} \quad 35 \times 4916 - 134 \times 1284 = 4,$$

we may read off a solution

$$x = 35, \quad y = -134.$$

We have found one pair of possible values for x and y. In fact, there will be infinitely many possible solutions of this equation: we pursue this topic in Section 3.2.

3.2   SOLVING LINEAR DIOPHANTINE EQUATIONS

Given integers a, b and c, consider the problem of finding integers x, y such that

$$ax + by = c.$$

With the restriction that the solutions be integers, this problem is referred to as a linear **Diophantine equation**. We wish firstly to determine whether or not such integers exist; if there are solutions, we then want to find them all. In some problems, we shall have additional requirements, for example, to find the solution in which x and y are both positive and x has the smallest possible positive value.

The first step is to determine the greatest common divisor of the left–hand–side coefficients a and b: this can be done by using the Euclidean Algorithm, as shown in Section 3.1. If $g = \gcd(a, b)$ is not a factor of c, it is not hard to see that the equation has no solution. It is also true, though notably less obvious, that if g is a factor of c, then there are solutions. To find one solution, we use the Euclidean Algorithm to obtain a solution of $ax + by = g$, and then "scale up" to find a solution of the required equation.

Example. Find an integer solution of the linear Diophantine equation

$$4916x + 1284y = 28.$$

Solution. In Section 3.1, we found that $\gcd(4916, 1284) = 4$; since 4 is a factor of 28, our present equation has a solution. We also saw that $x = 35$, $y = -134$ is a solution of $4916x + 1284y = 4$ by making use of the equation

$$35 \times 4916 - 134 \times 1284 = 4.$$

To obtain an equation in which the right–hand side is 28, we multiply both sides by 7, doing so in such a way as not to change the coefficients 4916 and 1284 on the left–hand side:

$$245 \times 4916 - 938 \times 1284 = 28.$$

Therefore, $4916x + 1284y = 28$ has a solution $x = 245$, $y = -938$.

Once we have found a single solution, the following result shows us how we may write down all solutions.

Theorem 3.1. The Bézout Property. *Let a, b and c be integers such that $\gcd(a, b) = 1$, and suppose that the equation*

$$ax + by = c$$

has an integer solution $x = x_0$, $y = y_0$. Then all solutions of this equation are given by

$$x = x_0 + bt, \quad y = y_0 - at$$

where t is an integer.

Example. Find all integer solutions of the equation $10x - 17y = 251$.

Solution. We apply the Euclidean Algorithm to 17 and 10, and then "run it backwards":

$$17 = 1 \times 10 + 7 \qquad\qquad 1 = 7 - 2 \times (10 - 7)$$
$$10 = 1 \times 7 + 3 \qquad\qquad\quad = -2 \times 10 + 3 \times (17 - 10)$$
$$7 = 2 \times 3 + 1 \qquad\qquad\quad = 3 \times 17 - 5 \times 10.$$

Multiplying both sides by 251 and comparing with our desired equation, we have

$$-1255 \times 10 + 753 \times 17 = 251, \quad 10x - 17y = 251,$$

and so a solution is $x = -1255$, $y = -753$. From the Euclidean Algorithm, $\gcd(10, 17) = 1$, so by Theorem 3.1, the complete solution is

$$x = -1255 - 17t, \quad y = -753 - 10t,$$

where t is an integer.

Comments. With a bit of practice, a good deal of the writing in the Euclidean Algorithm can be abbreviated. In the above working, we have not written down the last line of the Euclidean Algorithm, as it is obvious that the division by 1 will go exactly. We have not written the step in which we make each remainder the subject of its equation but have done it in our heads. And we have left out a number of intermediate steps in the "backwards" calculation.

Indeed, sometimes it is not necessary to use the Euclidean Algorithm at all! In the present example, it is fairly obvious that 10 and 17 have no common factor but 1; and if you are sufficiently familiar with your multiplication tables, then you may be able to spot by trial and error that the equation $10x - 17y = 251$ has a solution $x = 20$, $y = -3$. Therefore, another way to write the solution, involving (almost) no working, is

$$x = 20 - 17t, \quad y = -3 - 10t. \tag{3.1}$$

Example. Find all solutions of the Diophantine equation $4916x + 1284y = 28$.

Solution. We cannot use Theorem 3.1 directly since the coefficients of x and y have GCD 4, not 1. However, all we need do is cancel 4 from both sides to give the equivalent equation

$$1229x + 321y = 7.$$

This equation has a solution just as before, $x = 245$, $y = -938$, and now the coefficients have no common factor, so the complete solution is

$$x = 245 + 321t, \quad y = -938 - 1229t \tag{3.2}$$

where t is an integer.

To conclude, we give two examples involving extra conditions.

Example. Find the solution of $10x - 17y = 251$ in which both x and y are positive integers and x is as small as possible.

Solution. From the alternative form given in equation (3.1), the solutions may be written as

$$x = 20 - 17t, \quad y = -3 - 10t,$$

where t is an integer. These expressions are both positive if and only if

$$t < \frac{20}{17} = 1 + \frac{3}{17} \quad \text{and} \quad t < -\frac{3}{10} = -1 + \frac{7}{10};$$

since t is an integer, these statements are true simultaneously if and only if $t \leq -1$. To obtain the smallest possible x value, we need the largest possible t value, so we take $t = -1$, giving

$$x = 37, \quad y = 7$$

as the desired solution.

Example. How many integer solutions has the equation $4916x + 1284y = 28$ in which both x and y are less than 2024?

Solution. Using the expressions found in (3.2), we require

$$245 + 321t < 2024 \quad \text{and} \quad -938 - 1229t < 2024;$$

that is,

$$t < \frac{1779}{321} = 5 + \frac{174}{321} \quad \text{and} \quad t > -\frac{2962}{1229} = -3 + \frac{725}{1229}.$$

Since t is once again an integer, this is the same as $-2 \leq t \leq 5$, and so there are eight solutions in integers less than 2024.

3.3 MODULAR ARITHMETIC

Definition 3.1. *Let m be an integer. Integers a and b are* **congruent modulo** *m, written*

$$a \equiv b \pmod{m},$$

if m is a factor of the difference $a - b$.

We can think of *modular arithmetic* as a system of arithmetic in which only the integers $\{0, 1, 2, \ldots, m - 1\}$ are necessary (although it is sometimes convenient to refer to integers outside this set), and where any multiple of m "counts as zero". For example, it may be helpful to verbalise the congruence

$$10 + 7 \equiv 5 \pmod{12} \tag{3.3}$$

as "$10 + 7 = 17 = 12 + 5$, but 12 counts as 0 so this is the same as 5". This is very similar to how we might work out that "7 hours after 10 o'clock is 5 o'clock", and for this reason, modular arithmetic is sometimes referred to as *clock arithmetic*. Of course, it is also very easy to confirm that (3.3) is true by going back to the definition and substituting $a = 10+7$, $b = 5$, $m = 12$.

It is also possible to subtract and multiply in modular arithmetic, and to combine the basic operations into more complex expressions. Working an example modulo 13, we could calculate $2 - 6 \times 7 = -40 = -3 \times 13 - 1$, and "13 counts as 0", so

$$2 - 6 \times 7 \equiv -1 \pmod{13}.$$

As mentioned above, it is sometimes convenient not to insist upon obtaining a final answer from 0 to $m - 1$; but if we wish to do so, then we would write this last example as

$$2 - 6 \times 7 \equiv 12 \pmod{13}.$$

Arguably, the previous answer is "simpler".

Looking carefully at the definition, it should be clear that a congruence statement is just another way of expressing divisibility, and in that sense, is "nothing new". However, modular arithmetic is very useful because congruences share many of the important properties of equality.

Theorem 3.2. Properties of congruences. *Let m be an integer.*

- *Congruence modulo m shares the three fundamental properties of equality. Specifically, for any integers a, b and c, we have*
  - $a \equiv a \pmod{m}$;
  - *if $a \equiv b \pmod{m}$, then $b \equiv a \pmod{m}$;*
  - *if $a \equiv b \pmod{m}$ and $b \equiv c \pmod{m}$, then $a \equiv c \pmod{m}$.*

- *If we have two congruences to the same modulus, then we may add, subtract and multiply the congruences just as we do for equations. Specifically, if $a \equiv b \pmod{m}$ and $c \equiv d \pmod{m}$, then*
  - $a + c \equiv b + d \pmod{m}$;
  - $a - c \equiv b - d \pmod{m}$;
  - $ac \equiv bd \pmod{m}$.

  Moreover, if $a \equiv b \pmod{m}$ and n is a non–negative integer, then
  - $a^n \equiv b^n \pmod{m}$.

Simplification in modular arithmetic. We can simplify an integer a modulo m, for suitable values of m, by easy calculations based on the digits of a.

- Any positive integer is congruent modulo 9 to the sum of its digits.
- Any positive integer is congruent modulo 10 to its last digit.
- Any positive integer is congruent modulo 11 to the *alternating sum* of its digits: that is, the last digit, minus the second last, plus the third last, minus the fourth last, and so on.

There are similar tests for simplification modulo 2, 3, 4, 5, and (with a greater level of sophistication) modulo other integers. We leave these for readers to investigate for themselves.

Example. Consider $a = 2024$. By purely mental calculation, we have

$$2024 \equiv 8 \pmod{9}$$
$$2024 \equiv 4 \pmod{10}$$
$$2024 \equiv 4 - 2 + 0 - 2 \equiv 0 \pmod{11}.$$

Observe that $a \equiv 0 \pmod{m}$ is just another way of saying that a is a multiple of m. So we have just shown, without recourse to division, that 2024 is a multiple of 11. It is also possible to extend these ideas to negative integers: the above calculations give easily

$$-2024 \equiv -8 \equiv 1 \pmod{9}$$
$$-2024 \equiv -4 \equiv 6 \pmod{10}$$
$$-2024 \equiv 0 \pmod{11}.$$

Powers. Modular arithmetic can be helpful in puzzles involving squares (or higher powers), since for any m, there are only a limited number of possible values for a square modulo m.

Example. Find all squares modulo 11. **Solution.** Any integer is congruent to one of the numbers $0, 1, 2, \ldots, 10$ modulo 11. So for any square a^2, we have

$$a^2 \equiv 0^2,\, 1^2,\, 2^2,\, 3^2,\, 4^2,\, 5^2,\, 6^2,\, 7^2,\, 8^2,\, 9^2,\, 10^2 \pmod{11}$$
$$\equiv 0,\, 1,\, 4,\, 9,\, 5,\, 3,\, 3,\, 5,\, 9,\, 4,\, 1 \pmod{11}$$

respectively. In summary: any square is congruent to $0, 1, 3, 4, 5$ or 9 modulo 11. Note, by the way, that since $6 \equiv -5 \pmod{11}$, we have automatically $6^2 \equiv 5^2 \pmod{11}$, and we could have reduced the previous working by half.

Another example. Show that if a, b, c are integers such that $a^3 + b^3 + c^3$ is a multiple of 7, then at least one of a, b, c is a multiple of 7. **Solution.** The non–zero cubes modulo 7 are

$$1^3 \equiv 1, \quad 2^3 \equiv 1, \quad 3^3 \equiv 6, \quad 4^3 \equiv 1, \quad 5^3 \equiv 6, \quad 6^3 \equiv 6 \pmod{7},$$

that is, 1 and 6 only, or, more simply, 1 and -1 only. The sum of three such numbers cannot be 0 modulo 7.

We conclude with a theorem which is often of great utility for simplifying powers in modular arithmetic. Don't let the adjective "little" diminish the importance of this theorem! The name is traditional, and presumably arose in order to distinguish the result from "Fermat's Last Theorem".

Theorem 3.3. Fermat's "Little" Theorem. *If p is prime, a is an integer and a is not a multiple of p, then*

$$a^{p-1} \equiv 1 \pmod{p}.$$

One of the consequences of Fermat's Little Theorem is that if k is a factor of $p - 1$ (or more generally, if k and $p - 1$ have a common factor greater than 1), then there will be only a limited number of kth powers modulo p. For example, take $p = 13$, a prime, and $k = 4$, a factor of 12. Suppose that x is a non–zero fourth power modulo 13. Then for some a, we have that

$$x \equiv a^4 \pmod{13} \qquad \text{implies} \qquad x^3 \equiv a^{12} \equiv 1 \pmod{13}.$$

Now, the equation $x^3 \equiv 1 \pmod{13}$ can have at most three solutions, and so there are at most 3 non–zero fourth powers modulo 13. As a contrasting example, 5 and 12 have no common factor, and you may check that a fifth power may have any of the thirteen possible remainders modulo 13.

A word of caution regarding the argument of the previous paragraph: the statement that $x^3 \equiv 1 \pmod{13}$ can have at most three solutions is true because 13 is prime. Similar claims may not be true for other moduli: for instance, the equation $x^3 \equiv 1 \pmod{8}$ has four solutions $x \equiv 1, 3, 5, 7 \pmod{8}$.

3.4 GRAPH THEORY

A **graph** or **network** can be thought of as a collection of dots, together with lines connecting some of the dots. (This has nothing to do with the graph of a function $y = f(x)$, an entirely different topic.)

The terminology in use is that a graph consists of a set of **vertices**, some pairs of which are joined by **edges**. We write uv for an edge joining vertices u and v, and we say that u and v are the **endpoints** of this edge.

The graph G with vertices $\{a, b, c, d, e, f, g\}$ and edges $\{ab, bc, bd, be, cf, cg\}$ is represented in two ways by the drawings below. When drawing these graph representations, the edges can be straight or bendy, and the vertices can be drawn anywhere (except on top of each other): the only things that matter are the number of vertices, and the way in which they are connected by edges. Sometimes, we include vertex labels, and, sometimes, we just draw dots without vertex labels, as in the graph above.

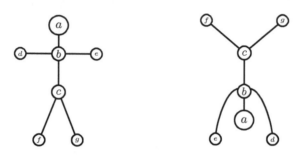

The **degree** of a vertex v, written $\deg v$, is the number of edges that have v as an endpoint. In terms of graph drawings, $\deg v$ is the number of lines that touch v. Although we will not be using such graphs here, we mention for the sake of completeness that a **loop** on v – that is, an edge which has v for both its endpoints – counts twice towards the degree of v.

Example. In the graph G above, the "hand" and "head" vertices a, d, e, f, g each have degree 1, whereas $\deg b = 4$ and $\deg c = 3$.

The following lemma expresses a simple but exceedingly important identity.

Theorem 3.4. The Handshaking Lemma. *In any graph, the sum of the degrees of all vertices is equal to twice the number of edges.*

Proof. A proof of this lemma is virtually given by its title. Think of the vertices as people, and suppose that any two "people" u, v shake hands exactly when uv is an edge in the graph. The number of times that u shakes hands is $\deg u$. Each "handshake" uv is counted twice in this way, once by u and once by v, so the sum of degrees is twice the number of handshakes, as stated.

Example. The graph in the previous example has 6 edges; the sum of the degrees of all vertices is $1 + 4 + 3 + 1 + 1 + 1 + 1 = 12$, which is twice the number of edges, as predicted by the Handshaking Lemma.

Problem. Prove that there is no graph with 5 vertices of degrees $5, 4, 3, 2, 1$.

Solution. The sum of the degrees in such a graph would be $5 + 4 + 3 + 2 + 1 = 15$, and so by the Handshaking Lemma, the number of edges would be $\frac{15}{2}$, which is not possible.

A graph G is **connected** if we can get from any vertex u in G to any other vertex v in G by stepping along edges, from one vertex to another, starting from u and ending at v.

Example. Below are a connected graph and a disconnected graph.

3.4.1 Planar Graphs and Euler's Formula

A graph is **planar** if it can be represented by dots and lines drawn in the plane in such a way that none of the edge lines cross each other. Such a drawing of a planar graph is called a **planar representation** of the graph.

Example. The graph G drawn below to the left is planar since we can re–draw it as the planar representation on the right, where no two lines cross each other.

G also G

Planar representations of a planar graph have **regions**. These are areas that are enclosed by edge lines, including the outer region around the graph. The **degree** of a region r, written as $\deg r$, is the number of edges that border r.

Example. The planar representation of the graph G from the previous example has four regions: the three inner triangular regions, drawn below in different colours, and the white outer region. The degree of each of the four regions r is $\deg r = 3$.

Below is a "regions version" of the Handshaking Lemma. The proof of this lemma is essentially the same as that of the previous version.

Theorem 3.5. The Handshaking Lemma for regions. *In any planar representation of a planar graph G, the sum of the degrees of all regions is equal to twice the number of edges.*

Example. The sum of region degrees for the graph in the previous example is $4 \times 3 = 12$, which is twice the number of edges, namely 6, as claimed by the Handshaking Lemma for regions.

Comment. By the two versions of the Handshaking Lemma, the sum of the vertex degrees and the sum of the region degrees are the same for any planar representation of a planar graph. This means that the sum of region degrees is the same, regardless of which planar representation we might draw.

We are now able to present a classical and very useful formula.

Theorem 3.6. Euler's Formula. *If a planar connected graph G has v vertices and e edges, and if some planar representation of G has r regions, then*

$$v - e + r = 2.$$

Comment. We noted above that the *sum* of region degrees is the same, regardless of which planar representation we might draw. By Euler's Formula, we see that the *number* of regions is constant as well.

Example. The graph G depicted below on the left has $v = 6$ vertices, $e = 8$ edges and $r = 4$ regions labeled w, x, y, z, with region degrees $3, 4, 3, 6$, respectively. We confirm that $v - e + r = 2$, as claimed by Euler's Formula. The drawing on the right is a different planar representation of the same graph G, and the degrees of the regions are now different: $3, 5, 3, 5$. However, the number of regions is the same, and Euler's Formula still holds.

The Handshaking Lemmas and Euler's Formula can be very effective when used together, as the following example demonstrates.

Example. Five towns are to be connected by roads so that there is a direct road from each town to each other town. If the towns happened to lie in a circle and the roads were straight, then the towns and roads would lie as in the picture to the right. However, the town planners would like to design the roads so that no two roads cross each other. Is this possible?

Solution. We can define a graph G to have the towns as vertices and the roads as edges. The problem is then to determine whether or not G is planar. Suppose that G is planar, and consider some planar representation of G. Let r be the number of regions, and observe that each region must have degree at least 3. By the Handshaking Lemma for regions and Euler's Formula,

$$2e = \sum_{r \in R} \deg r \geq 3r \quad \text{and} \quad v - e + r = 2.$$

Therefore, $2e \geq 3(2 - v + e)$, which simplifies to $e \leq 3v - 6$. But since $e = 10$ and $v = 5$, this is not true. Therefore, G is not planar, and no such road system is possible.

Directed graphs. The solutions to some of the problems in our collection require the use of *directed graphs*. These can be conceptualised as graphs in which each edge has a specific direction, generally shown on diagrams by means of an arrow, such as in the diagram below.

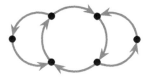

Note that while in an undirected graph an edge between vertices u and v can be written either as uv or vu, we have to be more careful with directed graphs: in this case, uv will denote an edge (arrow) from u to v, which is not the same as vu, an edge from v to u. Directed graphs have many interesting and important properties: readers who work through our *Parabolic Problems* will discover a few for themselves.

3.4.2 Hamilton Paths and Cycles

Definition 3.2. *A* **Hamilton path** *in a graph is a path which uses the edges of the graph to "travel" from vertex to vertex, taking in every vertex once and only once. A* **Hamilton cycle** *travels from some vertex back to the same vertex, taking in every other vertex once and only once.*

Example. On a dodecahedron, select any vertex; then travel along the edges in such a way as to return to the start after visiting every other vertex exactly once.

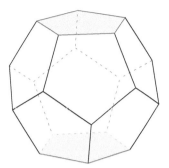 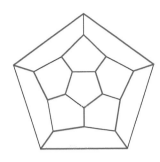

It's quite feasible to find a path on the three–dimensional drawing of a dodecahedron (above left); but, probably, even easier to do so if the dodecahedron is "flattened out" to give an equivalent graph (above right, with one of many possible Hamilton cycles shown in red). The puzzle was devised in 1857 by the Irish mathematician William Rowan Hamilton, and was later marketed under the name "A Voyage round the World", each vertex being labelled with the name of a prominent city. It failed to achieve any significant commercial success, but (perhaps more importantly) resulted in Hamilton's name being attached to this essential concept of graph theory.

It's very easy to see how Hamilton's puzzle, especially in its commercial realisation, relates to the idea of travelling along various permissible routes. However, some engaging and important problems with no obvious relation to travel can be solved by expressing the essential features of the problem in a graph, and then seeking a Hamilton path or cycle in the graph. One recent application is to genome sequencing: a paper by Shepherd Chikomana and Xiaoxue Hu (available online) employs graph–theoretic ideas to study how fragmentary nucleotides can be reassembled into a DNA molecule. We give a more recreational "application".

Example. A steeple contains four bells of different pitches: we shall label them as $1, 2, 3, 4$. It is desired to ring all four successively; then again, in a different order; and so on, until all 24 possible orders have been performed; and then to continue from the beginning. Owing to the physical limitations on bell–ringing, the only allowable difference between successive groups of four is for two bells occupying adjacent places to be swapped. For instance,

$$1, 2, 3, 4; \quad 2, 1, 3, 4; \quad 2, 3, 1, 4$$

is an allowable start, but

$$1, 2, 3, 4; \quad 1, 3, 4, 2; \quad 3, 1, 2, 4$$

is not. It is not hard to find a possible sequence of 24 changes by trial and error, or indeed to develop a systematic procedure which obviates the need for trial and error.

Now suppose that we introduce an additional rule: the odd–numbered bells 1 and 3 may not swap unless they are in the middle two positions. Is it now possible to ring all 24 changes and continue from the beginning?

Solution. We construct a graph having 24 vertices labelled with the 24 possible orders of bells; two vertices will be adjacent if and only if the corresponding orders can be rung successively. Thus, 1234 will be adjacent to 2134, 1324 and 1243. Each vertex will be adjacent to three others, unless the first two or last two digits are odd: 1324 is adjacent to 1234 and to 1342, but not to 3124. A little work gives the whole graph as shown,

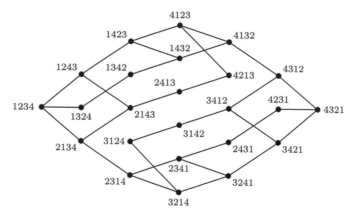

and in order to obtain an allowable sequence of 24 groups of four, we need to find a Hamilton cycle in this graph. Now just looking at the graph, we see immediately that it divides into two symmetric halves: for a Hamilton cycle, we will have to get from 2134, visiting all vertices in the "upper left" part of the graph once each, to 4132, and then from 4312 through the "lower right" section to 2314. In the upper left half, we will have to include the paths 1234–1324–1342–1432 and 2143–2413–4213–4123 (or their reverses) in our path; and having realised this, it is not hard to find various possibilities by inspection. One peal satisfying our requirements is

$$2, 1, 3, 4; \quad 2, 1, 4, 3; \quad 2, 4, 1, 3; \quad 4, 2, 1, 3; \quad 4, 1, 2, 3; \quad 1, 4, 2, 3;$$
$$1, 2, 4, 3; \quad 1, 2, 3, 4; \quad 1, 3, 2, 4; \quad 1, 3, 4, 2; \quad 1, 4, 3, 2; \quad 4, 1, 3, 2;$$
$$4, 3, 1, 2; \quad 3, 4, 1, 2; \quad 3, 1, 4, 2; \quad 3, 1, 2, 4; \quad 3, 2, 1, 4; \quad 3, 2, 4, 1;$$
$$3, 4, 2, 1; \quad 4, 3, 2, 1; \quad 4, 2, 3, 1; \quad 2, 4, 3, 1; \quad 2, 3, 4, 1; \quad 2, 3, 1, 4.$$

Since the end of this sequence joins up with the beginning in an allowable way, the peal can be rung again from the beginning, and so on indefinitely.

In general, finding a Hamilton path or cycle in a given graph can be quite difficult – essentially a matter of trial and error, though frequently assisted by observations such as we made in the preceding example. There are not many simple results which can guarantee the existence of a Hamilton cycle: we mention one of the most famous. Essentially, the proof of this theorem is one of our *Parabolic Problems* – we leave readers to discover which one, and to attempt the proof. The word **simple** in the statement means that no vertex in the graph has an edge "looping back" to itself, and that no two vertices are directly connected by two or more edges.

Theorem 3.7. Dirac's Theorem. *Let G be a simple graph with n vertices, where $n \geq 3$. If every vertex of G has degree at least $n/2$, then G has a Hamilton cycle.*

3.4.3 Ramsey Theory

Ramsey theory is a fascinating area of graph theory: often intensely difficult, its most basic questions are nevertheless reasonably accessible. We can paraphrase the main idea by saying that we seek to show that any random or disordered system, if large enough, must contain some small glimpses of order.

First, we need some simple terminology: the **complete graph** on n vertices, denoted K_n, is the graph which consists of n vertices, with one edge joining each pair of distinct vertices. By way of example, the diagram illustrates the complete graphs K_n for $n = 3, 4, 5, 6$.

We can now give a simple problem in Ramsey theory.

Example. The complete graph K_6 has each of its 15 edges coloured red or blue. Prove that however this colouring is done, there will always be either a red triangle or a blue triangle. (To say this more precisely: either a copy of K_3 with all edges coloured red, or a copy of K_3 with all its edges blue.)

Solution. Consider any vertex v in K_6. It is adjacent to every other vertex and therefore has 5 incident edges. Since every edge bears one of two colours, there must be at least 3 edges incident on v which have the same colour: by symmetry, we may assume that the colour is red. Consider the three vertices w_1, w_2, w_3 adjacent to v along these red edges. If any two of them are joined by a red edge, then those two, together with v, form a red triangle. If, on the other hand, no two of w_1, w_2, w_3 are joined by red edges, then all are joined by blue edges, and they form a blue triangle. So, our graph necessarily contains either a red or a blue copy of K_3.

Ramsey numbers. If we reconsider the previous problem, starting with K_5 instead of K_6 and once again colouring each edge either red or blue, we find that it is *not necessarily true* that there is either a red or a blue triangle: this can be proved quite simply by giving an example of a colouring of K_5 with no single–colour K_3. So, for this problem, 6 is the "magic number" – the smallest number of vertices which *guarantees* that a red–blue colouring of the edges in the complete graph will *always* include a red or a blue K_3. We write $R(3,3) = 6$, where the two 3s refer to the two possible colours of K_3 sought in the question. Similarly, the Ramsey number $R(s,t)$ is defined to be the smallest number of vertices for which a red–blue colouring of the edges in the complete graph *must* include a red K_s or a blue K_t.

Part of the reason for the difficulty of Ramsey theory problems is that, even for relatively small cases, the number of possible red–blue colourings of edges is so large that a solution by trial and error is, essentially, impossible. If we consider K_6, the number of colourings is already 2^{15}. To solve the above problem by brute force would require checking every single one of these 32768 configurations for a monochromatic triangle: perhaps manageable by computer, but certainly not by hand. If we wanted to do the same for 18 vertices, we would need to examine 2^{153} cases: some idea of the magnitude of this task can be formed by calculating that if we were able to process, let's say, a billion billion billion cases per second, the job would still take longer than the estimated age of the universe. Yet by using more intelligent methods than brute force, it has been found possible to prove that $R(4,4) = 18$. If we go one step further and ask for $R(5,5)$, its exact value is still unknown, though it has been proved to lie between 43 and 48.

Some of our *Parabolic Problems* are, in effect, questions about Ramsey theory. We have also included a puzzle where the complete graphs are modified by allowing two vertices to be joined not only by one edge either red or blue, but by two edges, one of each colour.

3.5   BASIC COMBINATORICS

The mathematical topic of combinatorics is sometimes referred to as "clever counting". It is the science of counting in situations where it is difficult or impossible to line up objects in a row, point at them and say "one, two, three, ... ".

3.5.1   Powers, Permutations and Combinations

Suppose that we have a collection of n different objects and that we wish to make r choices from these objects. The number of ways of doing so depends on whether or not it is possible to choose the same object more than once; and on whether or not the order of the choices matters.

- If "repetition is allowed" (that is, it is possible to choose the same item more than once) and order is important, then the number of ways to make our selection is n^r.

- If repetition is not allowed (that is, each item is chosen once or not at all) and order is important, the selection is referred to as a **permutation** of r objects chosen from n; the number of possible permutations is denoted $P(n,r)$ and is given by

$$P(n,r) = n(n-1)(n-2)\cdots(n-(r-1)).$$

We can also write this as

$$P(n,r) = \frac{n!}{(n-r)!},$$

where $m!$ (read "m factorial") is the product of all positive integers from 1 to m, that is,

$$m! = m(m-1)(m-2)\cdots \times 2 \times 1.$$

- If repetition is not allowed and order is not important, the selection is referred to as a **combination** of r objects chosen from n; the number of possible combinations is

$$C(n,r) = \frac{n(n-1)(n-2)\cdots(n-(r-1))}{r!} = \frac{n!}{r!\,(n-r)!}.$$

- If repetition is allowed and order is not important, the number of possible selections is $C(n+r-1,r)$.

Comments. The combination numbers $C(n,r)$ are often written with an entirely different notation,

$$C(n,r) = \binom{n}{r}.$$

We shall use both notations. They are sometimes also written as C^n_r or C^r_n (with the n and r the other way around), but we shall avoid these. Likewise, $P(n,r)$ is sometimes written P^n_r. The combination numbers are also often referred to as **binomial coefficients**, as they have an important connection with the Binomial Theorem and Pascal's Triangle: see Section 3.6.

We shall not prove the first three items above. We shall later, however, discuss how to prove the last, as it can be demonstrated by means of a very flexible and ingenious argument which can be adapted to a wide variety of problems.

Examples. While sometimes a problem may explicitly specify whether or not repetition is allowed and whether or not order is important, frequently you will have to work this out for yourself.

- Each of 5 children is to choose for themselves one of 11 different toys from a catalogue. Each toy is available in unlimited quantities. How many possible outcomes are there? **Solution**. There are $n = 11$ items, and $r = 5$ choices are to be made. The same item may be selected more than once (because there are unlimited quantities of each), and the order is important (because a child may not want to get the toy chosen by someone else). So, the number of outcomes is $11^5 = 161051$.

- A collection of 11 different toys has been donated to a playgroup, and 5 children are allowed to choose one each for themselves. How many possible outcomes now? **Solution**. Since there is now only one of each toy, the same item cannot be chosen more than once. Order is important for the same reason as above. So, the number of outcomes is $P(11, 5) = 55440$.

- A catalogue lists 11 different toys, each available in unlimited quantities. Each of 5 children is to choose one toy to go into the playgroup's shared toybox. How many different assortments of toys are possible? **Solution**. Once again, there are 11 items and 5 of them are to be chosen. Repetition is allowed, as more than one of the same toy could be chosen. Order in this case is *not* important since the toys are to go into a shared toybox: for example, a choice of toys A, A, A, B, C would give the same collection as C, A, A, B, A. So, the number of possible collections is $C(15, 5) = 3003$. **Comment**. Note the difference between this and our first example! One lesson to be learned here is that it is essential to read a problem very closely, and carefully consider the issues involved.

The above examples illustrate problems in which, once we have identified the appropriate conditions, the answer follows in one step. Frequently, however, solving a puzzle requires a number of successive steps.

Example. How many 11–digit positive integers are there in which 0 is never used, there are 3 *different* odd digits, and there are 8 even digits? **Solution**. Note that it is not enough to choose the digits: we must also specify where they go. It is recommended to set out each step carefully and separately.

1. Choose the places for the odd digits. (Note that we are choosing places, not digits, in this step!) There are 11 places and we have to choose 3 of them. We cannot choose the same place more than once – for example, you cannot put 3 odd digits in places $1, 1, 2$. The order of choosing places is *not* important: since we have not yet chosen what the odd digits will be, saying that they are to go in places $3, 4, 5$ is the same as saying that they are to go in places $4, 3, 5$. Therefore, the selection of places is a combination, and the number of options is $C(11, 3)$.

2. Choose the odd digits. There are 5 of them and we have to pick 3. Repetition is not allowed – this is explicitly stated in the question. Order is important, as having the same digits in a different order will give a different number. So, the number of choices is $P(5, 3)$.

3. Choose the even digits. There are 4 of them, we must choose 8, repetition is possible, order is important, so the number of possibilities is 4^8.

The total number of 11–digit integers subject to the stated conditions is

$$C(11, 3) \times P(5, 3) \times 4^8.$$

Another example. What happens if we modify the previous problem by allowing the number to contain 0? **Solution**. The catch here is that we may not use 0 for the first digit – for example, 000123 is not really a six–digit number. So, we'll have to adapt the above ideas and count cases separately, depending on whether the first digit is even or odd.

- If the first digit is odd, it can be chosen in 5 ways. There are then $C(10, 2)$ choices of places for the other odd digits; $P(4, 2)$ choices for the identity of these digits; and 5^8 choices for the even digits.

- If the first digit is even, it can be chosen in 4 ways, and there are $C(10, 3)$ choices of places for odd digits; $P(5, 3)$ choices of odd digits; 5^7 choices for the remaining even digits.

Since we have two alternatives here, we have to add the number of options for each. So the number of integers in this problem is

$$\left(5 \times C(10, 2) \times P(4, 2) \times 5^8\right) + \left(4 \times C(10, 3) \times P(5, 3) \times 5^7\right).$$

Possibly an easier approach: count all the 11–digit "numbers" which may have 0 as the first digit, then subtract all those which actually do have 0 as the first digit. Counting the latter is the same as counting 10–digit integers under similar conditions to those stated above: we leave it up to the reader to see that this method gives the number of possibilities as

$$C(11, 3) \times P(5, 3) \times 5^8 - C(10, 3) \times P(5, 3) \times 5^7.$$

This looks different from our previous answer, but by some careful algebra (or by direct calculation) you may check that it works out to be the same.

3.5.2   The "Dots and Lines" Method

Problem. Suppose that we have a bag containing 14 identical dice – ordinary six–sided dice – and we shake them out onto a table and observe the numbers facing upwards. There are various possible outcomes, for example all 1s; or all 6s; or two 1s, six 2s, no 3s, five 4s, a 5 and no 6s; and many more. How many different outcomes are possible altogether?

Solution. To get our thoughts straight, after shaking out the dice, let's arrange them into a row: all the 1s first, then the 2s and so on. We could visualise this as a row of 14 dots, all identical (because the 14 dice are all identical), and we could draw lines to separate those showing 1 from those showing 2; those showing 2 from those showing 3; and so on. The last example given above would be diagrammed as

$$\bullet\ \bullet\ |\ \bullet\ \bullet\ \bullet\ \bullet\ \bullet\ \bullet\ |\ |\ \bullet\ \bullet\ \bullet\ \bullet\ \bullet\ |\ \bullet\ |$$

Note that since there are 6 possible outcomes for each die, we need 5 lines. So, the number of possible overall outcomes is the same as the number of diagrams such as that we have just drawn: that is, the number of arrangements of 14 dots and 5 lines in a row. To count these, we argue as follows.

1. There are $14 + 5 = 19$ locations, and we have to choose 14 of these for the dots. Note that we are choosing *locations* – not dots, not lines!

2. Two dots cannot be placed in the same location, so repetition is not possible.

3. The order of our choices is not important since, for example, having dots in locations $1, 2, \ldots, 13, 14$ would be the same as having them in locations $14, 13, \ldots, 2, 1$.

Therefore, the number of possible outcomes from rolling 14 identical dice is $C(19, 14) = 11628$.

Alternative solution. We can express the problem in the following terms: we have $n = 6$ "items" (the numbers available on the dice) and we have to choose $r = 14$ of them (the numbers which appear on the dice after they are shaken from the bag). Repetition is allowed since it is possible for the same number to come up more than once. (Actually in this case, it is not just possible, but certain.) Order is not important since the dice are all identical and therefore, for instance, $1, 1, 1, \ldots, 1, 1, 2$ is the same as $1, 2, 1, \ldots, 1, 1, 1$. So from Section 3.5.1, the number of possible outcomes is $C(n + r - 1, r) = C(19, 14)$, as above.

In fact, the "dots and lines" method used in our first argument can be applied to the problem of choosing r items from n, repetition allowed, order not important, and this will prove the $C(n + r - 1, r)$ formula that we gave in Section 3.5.1. However, it is often good to think in terms of dots and lines, as the method can frequently be adapted to deal with more complex problems.

Example. We have 100 identical balls which are to be placed into buckets numbered $1, 2, 3, 4$. The number of balls in bucket 1 must be a multiple of 5 plus one extra ball; in bucket 2 it must be a multiple of 5 plus 2 extra balls; and similarly for buckets 3 and 4. In how many ways can this be done? **Solution.** We set aside 1 ball to go in the first bucket, 2 balls for the second, 3 for the third and 4 for the fourth, 10 in all. We split the remaining 90 into 18 groups of 5, and represent each group by a dot. The 18 dots have to be arranged in a row together with 3 lines to separate them into the 4 buckets. So, the number of possibilities is $C(21, 3) = 1330$. To find a specific arrangement, the number of balls in the first bucket will be 5 for each dot before the first line, plus one of our reserved dots; and likewise for the other buckets: this guarantees that the required conditions will hold. For instance, the arrangement of dots and lines

$$\bullet \; \bullet \; \bullet \; \bullet \; \bullet \; \bullet \; \bullet \mid \bullet \; \bullet \; \bullet \mid \mid \bullet \; \bullet \; \bullet \; \bullet \; \bullet \; \bullet \; \bullet \; \bullet$$

would mean that the numbers of balls in the four buckets should be $36, 17, 3$ and 44.

Another example. How many "words" of 5 letters can be made from the English alphabet, if the letters must occur in alphabetical order? A "word" does not have to be a "real" dictionary word but can be any string of five letters in alphabetical order, for example, $DHHSU$. **Solution.** There are 26 letters in the English alphabet and we have to choose 5 of them. Repetition is allowed. The order of our choices is *not* important, as it will be fixed by the requirement of alphabetical order – for example, choosing S, H, U, D, H would give the same word as that mentioned in the question. So, we can depict a choice of (for example) a D, two Hs, an S and a U as

$$\mid \mid \mid \bullet \mid \mid \mid \mid \bullet \; \bullet \mid \mid \mid \mid \mid \mid \mid \mid \mid \mid \mid \bullet \mid \mid \bullet \mid \mid \mid \mid.$$

There are $5 + 25$ locations, we must choose 5 of them for the dots, and the number of possible words is $C(30, 5) = 142506$.

A final example. How many solutions are there of the equation

$$x_1 + x_2 + x_3 + x_4 + x_5 + x_6 + x_7 = 13,$$

if each x_k is to be a non–negative integer? **Solution.** One example of a solution is

$$x_1 = 0, \; x_2 = 0, \; x_3 = 4, \; x_4 = 1, \; x_5 = 0, \; x_6 = 7, \; x_7 = 1,$$

which can be represented as a "dots and lines" diagram

$$\mid \mid \bullet \; \bullet \; \bullet \; \bullet \mid \bullet \mid \mid \bullet \; \bullet \; \bullet \; \bullet \; \bullet \; \bullet \; \bullet \mid \bullet$$

consisting of 13 dots and 6 lines. The number of solutions to the equation is the number of such arrangements, which is $C(19, 6) = 27132$.

3.5.3   The Principle of Inclusion/Exclusion

The principle of inclusion/exclusion is a counting technique which is frequently useful when we want to count items satisfying at least one of various properties, and where "overlap" between the properties is possible.

Example. Among the $C(19, 6)$ solutions of the equation

$$x_1 + x_2 + x_3 + x_4 + x_5 + x_6 + x_7 = 13, \tag{3.4}$$

which we studied at the end of Section 3.5.2, how many have the property that either $x_1 \geq 4$ or $x_2 \geq 5$? As in the previous problem, each x_k is to be a non–negative integer.

Solution. We shall again address this problem by visualising solutions of (3.4) as arrangements of 13 dots and 6 lines. To determine the number of solutions in which $x_1 \geq 4$, we set aside 4 dots and arrange the remaining 9 in a row together with the 6 lines: the number of ways of doing this is $C(15, 6)$. We then place the reserved dots into "section 1" of the arrangement, which guarantees that $x_1 \geq 4$. Thus, the number of solutions is $C(15, 6)$. A very similar argument shows that the number of solutions in which $x_2 \geq 5$ is $C(14, 6)$, so at this stage we have $C(15, 6) + C(14, 6)$ solutions. However – and this is the point of the inclusion/exclusion method, – there are some solutions in which both $x_1 \geq 4$ and $x_2 \geq 5$ simultaneously; these will have been counted twice, and, therefore, we need to determine their number and subtract them once. To count these solutions, set aside 9 dots; arrange the remaining 4 dots in a row with 6 lines; place 4 of the reserved dots into the first and 5 into the second section. The number of solutions in this case is $C(10, 6)$. **Answer.** The number of solutions in non–negative integers of

$$x_1 + x_2 + x_3 + x_4 + x_5 + x_6 + x_7 = 13$$

in which either $x_1 \geq 4$ or $x_2 \geq 5$ is

$$C(15, 6) + C(14, 6) - C(10, 6) = 7798.$$

To cope with more intricate situations than this, we shall need to introduce some notation. For any two sets A and B, their **union** $A \cup B$ is the set of all elements which are in A or B, or both. Their **intersection** $A \cap B$ is the set of all elements which are in both A and B simultaneously. We will often consider a **universal set** $\mathcal{U}$ consisting of all items which are relevant for some specific problem; then the **complement** of A is the set A^c consisting of all elements of $\mathcal{U}$ which are not in A. The number of elements or **cardinality** of a finite set A is denoted by $|A|$.

For instance, in the previous problem we might have chosen $\mathcal{U}$ to be the set of all solutions of $x_1 + x_2 + \cdots + x_7 = 13$ in non–negative integers; A_1 to be the set of all solutions in which $x_1 \geq 4$; and A_2 the set of solutions with $x_2 \geq 5$. Then our counting technique is expressed by the result

$$|A_1 \cup A_2| = |A_1| + |A_2| - |A_1 \cap A_2|.$$

This is, in fact, the basic version of the method of inclusion/exclusion: extended versions apply to the union of more than two sets.

Theorem 3.8. *The Principle of Inclusion/Exclusion. If $A_1, A_2, \ldots, A_n$ are finite sets, then the number of elements in their union is found by adding the number in each set; subtracting the number in each intersection of two sets; adding the number in each intersection of three* *sets; and so on. In symbols, we may write*

$$|A_1 \cup A_2 \cup \cdots \cup A_n| = |A_1| + |A_2| + \cdots$$
$$- |A_1 \cap A_2| - |A_1 \cap A_3| - \cdots$$
$$+ |A_1 \cap A_2 \cap A_3| + \cdots$$
$$- \cdots$$
$$\pm |A_1 \cap A_2 \cap \cdots \cap A_n|,$$

where the $\pm$ sign in the last row will be $+$ if n is odd, $-$ if n is even.

Example. Referring again to non–negative integer solutions of $x_1 + x_2 + \cdots + x_7 = 13$, how many solutions are there in which $x_k < 4$ for each k?

Solution. Let $\mathcal{U}$ be the set of all solutions of the equation, and for each k, let A_k be the set of solutions in which $x_k \geq 4$. We wish to count the solutions which are in *none* of the A_k; in other words, all the solutions in $\mathcal{U}$ *except for* those in one or more of the A_k: that is, we want

$$|A_1^c \cap A_2^c \cap \cdots \cap A_7^c| = |\mathcal{U}| - |A_1 \cup A_2 \cup \cdots \cup A_7|. \tag{3.5}$$

We already know that $|\mathcal{U}| = C(19, 6)$. Therefore, we need to find the cardinality of a union of sets; there are solutions which are in two or more of the A_k, that is, "overlap" is possible: this is an indication that inclusion/exclusion may be an appropriate technique.

We calculated in the previous example that $|A_1| = C(15, 6)$ and gave an example of finding the cardinality of an intersection; applying similar ideas will show in this case that $|A_1 \cap A_2| = C(11, 6)$ and $|A_1 \cap A_2 \cap A_3| = C(7, 6)$. The intersection of four or more sets will consist of solutions in which $x_k \geq 4$ for four or more terms; but this is impossible, since the sum of all terms is to be 13. So by the principle of inclusion/exclusion, we have

$$|A_1 \cup A_2 \cup \cdots \cup A_7| = |A_1| + \cdots - |A_1 \cap A_2| - \cdots + |A_1 \cap A_2 \cap A_3| + \cdots.$$

Now, there are seven terms like $|A_1|$, all of which have the same cardinality. The number of terms like $|A_1 \cap A_2|$ is $C(7, 2)$, because there are 7 sets A_k and we have to choose 2 of them to obtain a term like this; similarly, there are $C(7, 3)$ terms like $|A_1 \cap A_2 \cap A_3|$. Remembering from equation (3.5) that all of this has to be subtracted from $|\mathcal{U}|$, the number of solutions we seek is

$$C(19, 6) - 7C(15, 6) + C(7, 2)C(11, 6) - C(7, 3)C(7, 6).$$

For our final example, we introduce a slight complication.

Example. How many different words can be made by arranging all eight letters of PARABOLA, if each word must include two adjacent As?

Solution. For $k = 1, 2, \ldots, 7$, let S_k be the set of words made from the letters of PARABOLA, such that A occurs in places k and $k + 1$. The number of words we need is

$$|S_1 \cup S_2 \cup \cdots \cup S_7|.$$

It is possible for a word to be in two of these sets – for example, AAABLOPR is in both S_1 and S_2, – so we use inclusion/exclusion. To choose a word in S_1, we note that we have two As whose location is specified; and six more letters ABLOPR, which can be arranged in 6! ways. The same is true for all S_k. For a word in $S_1 \cap S_2$, the first three letters are all As and we have to arrange the others: 5! ways. Note, however, that the same is *not* true for all two–way intersections: for instance, words in $S_1 \cap S_7$ must have As in places $1, 2, 7, 8$, which is impossible. In fact, there are six equal cardinalities

$$|S_1 \cap S_2| = |S_2 \cap S_3| = \cdots = |S_6 \cap S_7| = 5!,$$

and all others are zero. It is impossible for any arrangement of PARABOLA to be in three or more of the S_k, so the required number of words is

$$7 \times 6! - 6 \times 5! = 4320.$$

Exercise. Give an alternative approach to this problem, using the dots and lines method to show that the number of arrangements of PARABOLA we *do not* want is $C(6,3)5!$; and check that this is consistent with the number we just found.

3.5.4 Combinatorial Proofs

A beautifully elegant technique for proving algebraic equalities relies on methods of combinatorics. The idea is to invent a counting problem – this may take a good deal of imagination, and is generally the difficult part of the process – then solve it in two ways, finding as one answer the left–hand side of the desired equality, and as another, the right–hand side. Even if the two expressions obtained appear very different, they are both the answer to the same question, and therefore they must be equal.

Since we are appealing to counting techniques, this method is most likely to be successful when the expressions in question involve powers, factorials and binomial coefficients, which, as we have seen, frequently appear in counting problems.

Example. Show that binomial coefficients satisfy the identity

$$r\binom{n}{r} = n\binom{n-1}{r-1}$$

whenever $1 \leq r \leq n$.

Solution. A club with n members is to elect a committee of r members. One member of the committee is to be chosen as president; the others are merely regarded as "ordinary members". In how many ways can these choices be made? Firstly: choose the committee by selecting r of the n members, no repetition, order not important; then choose one of the r committee members as president. The number of choices is

$$r\binom{n}{r}. \tag{3.6}$$

Alternatively, choose the president first: this person could be any one of the n club members. Then choose the $r-1$ ordinary committee members from the $n-1$ remaining club members. The number of choices is

$$n\binom{n-1}{r-1}. \tag{3.7}$$

Since (3.6) and (3.7) are answers to the same problem, they must be equal, and this is what we wished to prove.

Comment. It is also possible to prove the above identity by using the factorial formula for binomial coefficients, page 250, and doing a little algebra. This is (arguably) easier than a combinatorial proof, but (unarguably, in our opinion) less interesting!

Example. Prove that

$$m^n - \binom{m}{1}(m-1)^n + \binom{m}{2}(m-2)^n - \cdots + (-1)^{m-1}\binom{m}{m-1}1^n = 0$$

whenever $m > n$.

Solution. A club with n members has m committees. Each member belongs to exactly one committee. Each committee must have at least one member, but other than this, there are no restrictions on the number of members in each committee. In how many ways can this club be organised into committees?

In choosing one of the m committees for each of the n members, repetition is possible (more than one person may belong to the same committee) and order is important (interchanging the committees chosen for two people would result in a different situation): so the number of choices is m^n. However, this includes choices where some of the committees have no members: we shall use the principle of inclusion/exclusion to count these choices and subtract them from the total. Number the committees from 1 to m, and let S_k be the set of choices in which committee k is empty. Then

- to count S_1, we have to choose from $m - 1$ committees for each of the n members: there are $(m - 1)^n$ ways to do this, so $|S_1| = (m - 1)^n$; the same is true for each of the $\binom{m}{1}$ terms $|S_k|$;

- to count $S_1 \cap S_2$, we have to choose from $m - 2$ committees for each of the n members, so $|S_1 \cap S_2| = (m - 2)^n$; there are $\binom{m}{2}$ terms like this;

- and so on.

Using inclusion/exclusion, the number of arrangements in which every committee has at least one member is

$$|S_1^c \cap S_2^c \cap \cdots \cap S_m^c| = |\mathcal{U}| - |S_1| - \cdots + |S_1 \cap S_2| + \cdots$$

$$= m^n - \binom{m}{1}(m - 1)^n + \binom{m}{2}(m - 2)^n - \cdots$$

$$+ (-1)^{m-1}\binom{m}{m - 1}1^n + (-1)^m\binom{m}{m}0^n;$$

since, clearly, the last term can be omitted, this expression is just the left–hand side of our desired identity. But now consider what happens when $m > n$. Then the number of committees exceeds the number of members; as each member is on one committee only, for each committee to have one or more members is *impossible*; and so the number of arrangements is zero! Hence, the expression just found must equal zero whenever $m > n$, and this is what we wanted to prove.

A final example. In Problem 93, you were asked to find two quite different–looking expressions for N, the answer to a counting problem. So this was, in effect, a combinatorial proof that the two expressions are equal. By generalising the ideas in the solution, you can prove that

$$\sum_{k=m}^{n-m}\binom{k}{m}\binom{n - k}{m} = \binom{n + 1}{2m + 1}$$

whenever $m, n \geq 0$ and $2m \leq n$. Here the *summation notation* on the left–hand side denotes the sum of all the values of the expression shown, when k takes all values from m to $n - m$.

3.5.5   The Pigeonhole Principle

Suppose that a number of pigeons are resting in some pigeonholes. In its simplest form, the **Pigeonhole Principle** states that if there are more pigeons than pigeonholes, then at least two pigeons will share a pigeonhole:

Strictly speaking, this is not a counting technique like the other topics in this section; rather, it is a proof technique which makes use of counting methods. In its general form, the Principle can be stated as follows.

Theorem 3.9. The Pigeonhole Principle. *If there are more than kn pigeons occupying n pigeon holes, then some pigeonhole will contain more than k pigeons.*

Example. Show that, if $n+1$ distinct numbers are chosen from the numbers $1, 2, \ldots, 2n$, then at least two of the chosen numbers are coprime (that is, they have no common factor except for 1).

Solution. Define the "pigeons" to be the $n+1$ chosen numbers and the "pigeonholes" to be the sets $\{1, 2\}$, $\{3, 4\}, \ldots,$ $\{2n-1, 2n\}$. Since there are $n+1$ pigeons and only n pigeonholes, at least one of the pigeonholes must contain both its pigeons; that is, we can find two of the chosen numbers in the same set. These two numbers are consecutive and therefore coprime.

Example. Show that, if 19 points are chosen from a 3×3 square, then there must be three points of the 19 which form a triangle with area $\frac{1}{2}$ or less. (We allow the possibility of three collinear points, forming a "triangle" of area zero.)

Solution. Sometimes, we express the Pigeonhole Principle in terms of "objects" contained in "boxes". So, suppose that we have 19 points in a 3×3 square; let the objects be these 19 points, and let the boxes be the 9 regions obtained by dividing the 3×3 square into nine 1×1 subsquares.

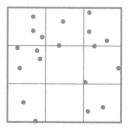

Since there are 9 boxes and more than 2×9 objects, there must be a box containing more than 2 of the given points, that is, a 1×1 square containing at least three points. We invite readers to finish the problem by showing that three points inside or on the boundary of a 1×1 square cannot form a triangle of area greater than $\frac{1}{2}$.

Example. Prove that any sequence $a_1, a_2, \ldots, a_{n^2+1}$ of $n^2 + 1$ distinct integers has a subsequence of $n + 1$ numbers that is either increasing or decreasing.

By way of illustration, take $n = 3$ and consider the sequence of $n^2 + 1$ distinct integers

$$5\ 7\ 9\ 4\ 6\ 3\ 8\ 0\ 1\ 2.$$

This sequence has no increasing subsequence of size $n+1 = 4$, but it does have a decreasing subsequence of size $n+1 = 4$, for instance $7\ 6\ 3\ 2$.

Solution. Suppose that no increasing subsequence contains $n + 1$ numbers. Let ℓ_i be the longest length of an increasing subsequence starting in a_i. These numbers $\ell_1, \ell_2, \ldots, \ell_{n^2+1}$ will be the pigeons. Note that ℓ_i can only be one of the numbers $1, 2, \ldots, n$: these numbers will be the pigeonholes. By the Pigeonhole Principle, some pigeonhole contains at least $n+1$

pigeons; that is, we can find at least $n + 1$ of the ℓ_i values that are identical. If $\ell_i = \ell_j$ for some $i < j$, then $a_i \not< a_j$, so $a_i > a_j$. We therefore have a decreasing subsequence of at least $n + 1$ numbers, and this is what we wanted to prove.

This result is known as the **Erdős–Szekeres Theorem**, after the two great Hungarian mathematicians Paul Erdős and George Szekeres who first proved it. George took up a position at the University of New South Wales in 1963, and was one of the founders of *Parabola*. He contributed many articles and problems to the magazine over a period of 40 years, finally "retiring" at the age of 93.

3.6 THE BINOMIAL THEOREM

Theorem 3.10. The Binomial Theorem. *If n is a non–negative integer, then*

$$(x + y)^n = \sum_{k=0}^{n} \binom{n}{k} x^{n-k} y^k = x^n + \binom{n}{1} x^{n-1} y + \binom{n}{2} x^{n-2} y^2 + \cdots + y^n.$$

Here, the factors $\binom{n}{k}$ are the binomial coefficients defined in Section 3.5.1. Note that the first and last coefficients are $\binom{n}{0} = \binom{n}{n} = 1$.

Pascal's Triangle. Values of binomial coefficients can be calculated by formulae given in Section 3.5.1, or read off Pascal's Triangle. The beginning of the triangle looks like this.

$$
\begin{array}{ccccccccccccc}
 & & & & & & 1 & & & & & & \\
 & & & & & 1 & & 1 & & & & & \\
 & & & & 1 & & 2 & & 1 & & & & \\
 & & & 1 & & 3 & & 3 & & 1 & & & \\
 & & 1 & & 4 & & 6 & & 4 & & 1 & & \\
 & 1 & & 5 & & 10 & & 10 & & 5 & & 1 & \\
1 & & 6 & & 15 & & 20 & & 15 & & 6 & & 1 \\
\end{array}
$$

$$1 \quad 7 \quad 21 \quad 35 \quad 35 \quad 21 \quad 7 \quad 1$$

These rows, and further rows as far as desired, are obtained by writing down a 1 at the end of each row, and calculating each "internal" element as the sum of the two above it in the previous row. The binomial coefficient $\binom{n}{k}$ can be read off as the kth element of row n in Pascal's Triangle, where both elements and rows are counted starting from 0: for instance, $\binom{7}{5} = C(7,5) = 21$.

Various identities involving binomial coefficients can be proved by substituting suitable values for x and y, or by algebraically manipulating sums in various ways.

Example. By substituting $x = y = 1$, we have

$$\binom{n}{0} + \binom{n}{1} + \binom{n}{2} + \cdots + \binom{n}{n-2} + \binom{n}{n-1} + \binom{n}{n} = 2^n.$$

That is, the numbers in row n of Pascal's Triangle add up to 2^n.

Another example. It is algebraically obvious that

$$(1 + x)^{m+n} = (1 + x)^m (1 + x)^n. \tag{3.8}$$

Let's use the Binomial Theorem to determine the coefficient of x^r on each side. We'll assume that $r \leq m \leq n$: this is not actually necessary, but simplifies some of the details in the following argument. First, according to the Binomial Theorem, the left–hand side is

$$(1 + x)^{m+n} = \sum_{r=0}^{m+n} \binom{m+n}{r} x^r,$$

and here we may simply read off the coefficient of x^r: it is $\binom{m+n}{r}$. Now we can also apply the theorem to each of the two factors on the right–hand side,

$$(1 + x)^m = \sum_{j=0}^{m} \binom{m}{j} x^j \quad \text{and} \quad (1 + x)^n = \sum_{k=0}^{n} \binom{n}{k} x^k.$$

When we multiply these expressions together, we shall obtain a sum of terms like

$$\binom{m}{j} \binom{n}{k} x^{j+k}; \tag{3.9}$$

we want to collect all these in which $j + k = r$. Since the powers of x in both expressions go up to x^r at least, we shall obtain suitable products for any value of j from 0 to r, with the corresponding value $k = r - j$. Therefore, the coefficient of x^r on the right–hand side is the sum of the coefficient in (3.9) for j from 0 to r, that is,

$$\sum_{j=0}^{r} \binom{m}{j} \binom{n}{r-j}.$$

But because of the equality (3.8), the coefficients we have just found on the left– and right–hand sides must be equal. That is,

$$\binom{m+n}{r} = \sum_{j=0}^{r} \binom{m}{j} \binom{n}{r-j},$$

provided that $r \leq m \leq n$. If you prefer to avoid the summation notation, this can be written as

$$\binom{m+n}{r} = \binom{m}{0} \binom{n}{r} + \binom{m}{1} \binom{n}{r-1} + \binom{m}{2} \binom{n}{r-2} + \cdots + \binom{m}{r} \binom{n}{0}.$$

A particular instance would be

$$\binom{13}{3} = \binom{5}{0} \binom{8}{3} + \binom{5}{1} \binom{8}{2} + \binom{5}{2} \binom{8}{1} + \binom{5}{3} \binom{8}{0},$$

which you may easily confirm by evaluating the binomial coefficients.

For **one more example**, see Solution 291 on page 205.

3.7   SOME TRIGONOMETRIC FORMULAE

There are vast numbers of formulae involving the trigonometric functions. We restrict ourselves to a few which appear in our solutions to the puzzles in this book, or which might be found helpful in alternative approaches. Readers are referred to any of the many relevant textbooks for proofs, and for more sophisticated results.

Basic formulae. For any real numbers x, y, the cosine and sine functions satisfy the formulae

$$\cos^2 x + \sin^2 x = 1$$
$$\cos(x + y) = \cos x \cos y - \sin x \sin y$$
$$\sin(x + y) = \sin x \cos y + \cos x \sin y .$$

Proof. We won't prove these formulae here, as they can be found in many standard texts. But if you would like to try for yourself, then you could follow the method suggested in Problem 290.

Other trigonometric functions. For appropriate values of x we define the tangent, secant, cosecant and cotangent functions

$$\tan x = \frac{\sin x}{\cos x}, \quad \sec x = \frac{1}{\cos x}, \quad \csc x = \frac{1}{\sin x}, \quad \cot x = \frac{1}{\tan x}.$$

Sometimes $\operatorname{cosec} x$ is written instead of $\csc x$, but we find it more elegant to have three–letter names for all of these functions!

Derived formulae. If x, y are real numbers, then

$$1 + \tan^2 x = \sec^2 x$$
$$\tan(x + y) = \frac{\tan x + \tan y}{1 - \tan x \tan y}$$
$$\cos 2x = \cos^2 x - \sin^2 x = 2\cos^2 x - 1 = 1 - 2\sin^2 x$$
$$\sin 2x = 2 \sin x \cos x$$
$$\cos x \cos y = \frac{\cos(x - y) + \cos(x + y)}{2}$$
$$\sin x \sin y = \frac{\cos(x - y) - \cos(x + y)}{2}$$
$$\sin x \cos y = \frac{\sin(x - y) + \sin(x + y)}{2}$$
$$\cos x + \cos y = 2 \cos\left(\frac{x + y}{2}\right) \cos\left(\frac{x - y}{2}\right)$$
$$\sin x + \sin y = 2 \sin\left(\frac{x + y}{2}\right) \cos\left(\frac{x - y}{2}\right) .$$

The last five of these are frequently referred to as the "products–to–sums" and "sums–to–products" identities. We decidedly *do not* recommend memorising them, but rather, being aware that they exist and being able to look them up when needed.

Triangle theorems. Suppose that a triangle has angles A, B, C and that the sides opposite these angles have lengths a, b, c respectively. Then we have the **Sine Rule**

$$\frac{a}{\sin A} = \frac{b}{\sin B} = \frac{c}{\sin C},$$

and the **Cosine Rule**

$$c^2 = a^2 + b^2 - 2ab \cos C.$$

The area T of the triangle is given by

$$T = \frac{1}{2} bc \sin A,$$

or by similar formulae involving the other angles. **Heron's Formula** (named for the Greek mathematician Heron, who lived in Alexandria in the first century C.E.) gives the area in terms of the side lengths a, b, c and the semi–perimeter $s = \frac{1}{2}(a + b + c)$ as

$$T = \sqrt{s(s - a)(s - b)(s - c)}.$$

3.8   PROOF BY MATHEMATICAL INDUCTION

Mathematical induction is an important method of proving statements concerning the positive integers (and related situations). For a statement involving a variable n, if we prove

- that the statement is true when $n = 1$; and
- that if the statement is true for any *specific* integer $n \geq 1$, then it is also true for $n + 1$;

then it follows that the statement is true for all integers $n \geq 1$. The first step is referred to as the **basis** of the induction, the second as the **inductive step**.

Example. Let x be a real number, $0 < x < 1$. Then for any positive integer n, we have

$$(1 - x)^n \geq 1 - nx.$$

Proof. When $n = 1$, the desired statement is $1 - x \geq 1 - x$. This is certainly true, and we have proved the basis step.

Now suppose that the inequality is true for some specific integer $n \geq 1$. That is,

$$(1 - x)^n \geq 1 - nx.$$

Multiplying both sides by $1 - x$, which is a positive number, we obtain

$$(1 - x)^{n+1} \geq (1 - x)(1 - nx).$$

Expanding the right–hand side,

$$(1 - x)^{n+1} \geq 1 - (n + 1)x + nx^2;$$

but since $nx^2 \geq 0$, this implies

$$(1 - x)^{n+1} \geq 1 - (n + 1)x.$$

We have shown that if the desired inequality is true for some specific integer n, it is also true for $n + 1$, and this completes the inductive step. Since we earlier confirmed that the basis case is true, we conclude by mathematical induction that the inequality is true for all integers $n \geq 1$. This completes the proof.

To make the method of induction work, we need some fairly simple connection between the required statement for one value of n and that for the previous value. Sometimes, the clearest connection involves not just the previous instance but the previous two instances. In such a case, it may be possible to use a variant method of induction. For a statement involving a variable n, suppose we can prove that

- the statement is true for $n = 1$ and for $n = 2$;
- for any integer $n \geq 1$, if the statement is true for n and for $n + 1$, then it is also true for $n + 2$.

Then, again, the statement will be true for all integers $n \geq 1$.

Example. The Fibonacci sequence $F_0, F_1, F_2, \ldots$ is defined by the conditions

$$F_0 = 0, \quad F_1 = 1 \quad \text{and} \quad F_{n+2} = F_{n+1} + F_n \text{ for } n \geq 0 :$$

that is, it starts with $0, 1$, and every number thereafter is the sum of the previous two. The beginning of the sequence is

$$0, \ 1, \ 1, \ 2, \ 3, \ 5, \ 8, \ 13, \ 21, \ 34, \ 55, \ 89, \ 144, \ 233, \ 377, \ 610, \ 987.$$

Let m be a non–negative integer. Then for every integer $n \geq 0$, the integer $F_{n+m} - F_n F_{m+1}$ is a multiple of F_m.

Proof. When $n = 0$, the claim is that F_m is a multiple of F_m, which is clearly true. When $n = 1$, the claim is that $F_{m+1} - F_{m+1}$ is a multiple of F_m, which is true because 0 is a multiple of any integer. We have established that the basis cases are true.

Now, suppose that the result is true for some specific integer n, and also for $n + 1$. That is, we are assuming that each of

$$F_{n+m} - F_n F_{m+1} \quad \text{and} \quad F_{n+1+m} - F_{n+1} F_{m+1} \tag{3.10}$$

is a multiple of F_m; we wish to show that

$$F_{n+2+m} - F_{n+2} F_{m+1} \tag{3.11}$$

is also a multiple of F_m. To do this, note from the recurrence defining the Fibonacci numbers that

$$F_{n+2} = F_{n+1} + F_n \quad \text{and} \quad F_{n+2+m} = F_{n+1+m} + F_{n+m}.$$

Using these, (3.11) can be written as

$$(F_{n+1+m} + F_{n+m}) - (F_{n+1} + F_n)F_{m+1} = (F_{n+1+m} - F_{n+1} F_{m+1}) + (F_{n+m} - F_n F_{m+1}).$$

The two bracketed expressions on the right–hand side are precisely the expressions in (3.10), and each is a multiple of F_m; therefore, so is their sum. We have proved that (3.11) is a multiple of F_m, and this completes the inductive step.

Therefore, by mathematical induction, $F_{n+m} - F_n F_{m+1}$ is a multiple of F_m for every $m \geq 0$.

Comments. In this example, the initial value for n is 0, not 1 as we have suggested previously. It will be seen that this makes no significant difference to the procedure. Sometimes, the truth of a statement will depend not on the previous one or two instances, but on three or more. The method of induction can be adapted to this kind of situation too: we leave this to the reader.

The result just proved may seem unmotivated and of no great interest; we have included it because it is the key step in proving a fascinating fact about Fibonacci numbers, which may also be demonstrated by induction. This time, it will be a "one–step" induction, that is, our original fundamental form of mathematical induction.

Example. Let m and n be positive integers. If m is a factor of n, then F_m is a factor of F_n.

Proof. We re–express the statement as

"if m and k are positive integers, then F_m is a factor of F_{km}"

and prove it by induction on the variable k. For the basis: if $k = 1$, the claim is that F_m is a factor of F_m, which is certainly true.

For the inductive step, suppose that the statement is true for a certain integer $k \geq 1$, that is,

$$F_m \text{ is a factor of } F_{km}.$$

We shall use this, together with the result of the previous example, to show that

$$F_m \text{ is a factor of } F_{(k+1)m}. \tag{3.12}$$

To do this, we write

$$F_{(k+1)m} = F_{km+m} = (F_{km+m} - F_{km}F_{m+1}) + F_{km}F_{m+1}.$$

Now, the bracketed term on the right–hand side is a multiple of F_m, as is seen by choosing $n = km$ in our previous example; the final term is a multiple of F_{km} and hence, by our inductive assumption, of F_m; and so the sum is a multiple of F_m. This establishes (3.12), and by induction, the proof is complete.

It's always difficult to drag oneself away from the topic of Fibonacci numbers, but we observe that what we have just proved leads fairly quickly to another remarkable result, which readers may care to tackle for themselves:

if F_n is prime, then n is prime, except for the case $F_4 = 3$.

In fact, the principle of induction can be extended even further: we may show that each instance of our required statement follows not from one or two previous instances, or indeed any fixed number of instances, but from *all* previous instances. If we can prove

- that a statement is true when $n = 1$;
- that for any specific integer $n \geq 1$, if the statement is true for all values $1, 2, 3, \ldots, n$, then it is true for $n + 1$;

then we may deduce that the statement is true for all $n \geq 1$. This is sometimes known as "extended" or "strong" induction.

Example. Every positive rational number less than 1 can be written as a sum of distinct unit fractions.

Comment. By a "unit fraction", we mean a fraction with numerator 1. Making the above statement more explicit, we seek to prove that for every positive integer m the following is true: for each integer $n > m$, there exist positive integers $d_1, d_2, \ldots, d_s$, all different, such that

$$\frac{m}{n} = \frac{1}{d_1} + \frac{1}{d_2} + \cdots + \frac{1}{d_s}.$$

Proof by induction on m. Let $m = 1$ and $n > 1$; choose $s = 1$ and $d_1 = n$. Then

$$\frac{m}{n} = \frac{1}{d_1},$$

so the statement is true. This proves the basis of the induction.

Now let m be a specific integer, $m \geq 2$, and suppose that the result is true for fractions with numerators $1, 2, \ldots, m - 1$. We must deduce that it is true for numerator m. Let n be an integer with $n > m$; since $0 < \frac{m}{n} < 1$, there exists $q \geq 2$ such that

$$\frac{1}{q} \leq \frac{m}{n} < \frac{1}{q - 1}, \tag{3.13}$$

and we can write $\frac{m}{n}$ as a unit fraction plus a remainder,

$$\frac{m}{n} = \frac{1}{q} + \frac{qm - n}{nq}. \tag{3.14}$$

Now, the inequalities (3.13) imply

$$0 \leq qm - n < m.$$

First, consider the case $qm - n = 0$. Then (3.14) shows that $\frac{m}{n}$ is itself a unit fraction, and we have the desired conclusion with $s = 1$ and $d_1 = q$. If, on the other hand, $qm - n$ is not zero, then it is one of the integers $1, 2, \ldots, m - 1$, and we may apply the inductive hypothesis. Since $qm - n < m < n < nq$, there exist positive integers $d_1, d_2, \ldots, d_s$, all of them different, such that

$$\frac{qm - n}{nq} = \frac{1}{d_1} + \frac{1}{d_2} + \cdots + \frac{1}{d_s}.$$

Because

$$\frac{qm - n}{nq} < \frac{n}{nq} = \frac{1}{q},$$

every one of the denominators on the right–hand satisfies $d_k > q$; therefore, all the denominators on the right–hand side of

$$\frac{m}{n} = \frac{1}{q} + \frac{1}{d_1} + \frac{1}{d_2} + \cdots + \frac{1}{d_s}$$

are different, and the result is true for fractions with numerator m. By extended induction, the theorem is proved.

Comments. Actually, rational numbers greater than 1 can also be written as sums of distinct unit fractions. This is a bit harder to prove. The result (though probably not the proof) was known to ancient Egyptian mathematicians and scribes around 2000 B.C.

Since there are huge numbers of resources concerning mathematical induction in textbooks and online, we shall limit ourselves to the above examples. We conclude by mentioning some well–known exercises, and some important results, that can be proved by induction: we encourage readers to investigate further for themselves.

- Many, many summation formulae, for instance,

$$1^2 + 2^2 + 3^2 + \cdots + n^2 = \frac{n(n + 1)(2n + 1)}{6};$$

$$1^3 + 2^3 + 3^3 + \cdots + n^3 = (1 + 2 + 3 + \cdots + n)^2;$$

$$\sin\theta + \sin 2\theta + \sin 3\theta + \cdots + \sin n\theta = \frac{\sin\theta + \sin n\theta - \sin(n + 1)\theta}{2 - 2\cos\theta};$$

$$\frac{1}{1 \times 2} + \frac{1}{2 \times 3} + \frac{1}{3 \times 4} + \cdots + \frac{1}{n(n + 1)} = \frac{n}{n + 1}.$$

- Further facts about the Fibonacci sequence, other specific sequences, and sequences in general.
- Inequalities involving powers and factorials, for example, $3^n < n!$ whenever $n \geq 7$.
- Any positive integer can be factorised into primes.
- Important facts about *formal languages*, which have applications to computer languages and information technology.

3.9 PICK'S THEOREM

A **lattice point** in the (x, y) plane is a point whose coordinates are both integers. A **lattice polygon** is a polygon whose vertices are all lattice points. Such a polygon is referred to as **simple** if its edges never cross over each other and it contains no "holes". The diagram shows three examples of lattice polygons; only the leftmost is simple.

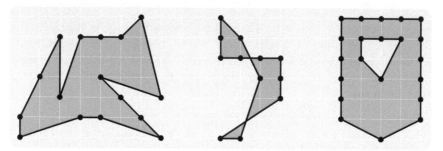

There is a delightfully easy way of finding the area of a simple lattice polygon!

Theorem 3.11. Pick's Theorem. *The area A of a simple lattice polygon is given by*

$$A = I + \frac{B}{2} - 1,$$

where I is the number of lattice points inside the polygon and B is the number on its boundary.

Examples. In the following diagram,

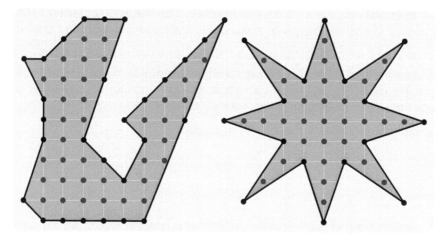

the polygon on the left has $I = 36$, $B = 28$ and area 49. That on the right has $I = 29$, $B = 16$ and area 36.

3.10 ROOTS AND COEFFICIENTS OF POLYNOMIALS

Consider a monic polynomial of degree n – **monic** means that the coefficient of the highest–degree term is 1 – say,

$$f(x) = x^n + a_{n-1}x^{n-1} + a_{n-2}x^{n-2} + \cdots + a_1 x + a_0.$$

A **root** of this polynomial is any number c such that $f(c) = 0$. A polynomial of degree n will always have precisely n roots, provided that repeated roots are counted repeatedly, and that both real and complex roots are counted[1]. The coefficients and the roots of a polynomial are related by the following important equations.

Theorem 3.12. Roots and coefficients of polynomials. *Suppose that the monic polynomial*

$$f(x) = x^n + a_{n-1}x^{n-1} + a_{n-2}x^{n-2} + \cdots + a_1 x + a_0$$

has roots $c_1, c_2, c_3, \ldots, c_n$. *Then*

$$c_1 + c_2 + c_3 + \cdots + c_n = -a_{n-1}$$
$$c_1 c_2 + c_1 c_3 + \cdots + c_{n-1} c_n = a_{n-2}$$
$$c_1 c_2 c_3 \cdots c_n = (-1)^n a_0;$$

and in general, the sum of all possible products of the roots $c_1, c_2, c_3, \ldots, c_n$, *taken k at a time, is equal to* $(-1)^k$ *times the coefficient* a_{n-k}.

Since the roots of $f(x)$ are the same as the roots of $af(x)$, where a is a non–zero constant, it is easy to adapt this result to the case of a non–monic polynomial. The quadratic case is particularly well known: if the quadratic polynomial $ax^2 + bx + c$ has roots α and β, then

$$\alpha + \beta = -\frac{b}{a} \quad \text{and} \quad \alpha\beta = \frac{c}{a}.$$

Example. We have the factorisation

$$f(x) = x^4 + 5x^3 - x^2 - 17x + 12 = (x-1)^2(x+3)(x+4),$$

and so $f(x)$ has the four roots (including repetitions) $c_1 = 1$, $c_2 = 1$, $c_3 = -3$, $c_4 = -4$. Confirming the relations cited in the theorem,

$$c_1 + c_2 + c_3 + c_4 = -5$$
$$c_1 c_2 + c_1 c_3 + c_1 c_4 + c_2 c_3 + c_2 c_4 + c_3 c_4 = -1$$
$$c_1 c_2 c_3 + c_1 c_2 c_4 + c_1 c_3 c_4 + c_2 c_3 c_4 = -(-17)$$
$$c_1 c_2 c_3 c_4 = 12.$$

The connection between roots and coefficients is very helpful for problems like the following.

Problem. If the roots of the cubic

$$x^3 + 20x^2 + 9x - 59$$

are α, β, γ, then find the monic cubic having roots $\alpha^2, \beta^2, \gamma^2$.

Solution. From Theorem 3.12, we have

$$\alpha + \beta + \gamma = -20, \quad \alpha\beta + \beta\gamma + \gamma\alpha = 9, \quad \alpha\beta\gamma = 59;$$

the coefficients of the desired polynomial are

$$-(\alpha^2 + \beta^2 + \gamma^2), \quad \alpha^2\beta^2 + \beta^2\gamma^2 + \gamma^2\alpha^2, \quad -\alpha^2\beta^2\gamma^2;$$

---

[1] For readers who have not yet met complex numbers: they are not needed for any of the problems in this book.

and we wish to evaluate the latter by using the former. There is in fact a definite procedure for doing this, but for a small–degree example, one may as well use trial and error. Firstly, we note the expansion

$$(\alpha + \beta + \gamma)^2 = (\alpha^2 + \beta^2 + \gamma^2) + 2(\alpha\beta + \beta\gamma + \gamma\alpha),$$

which yields

$$\alpha^2 + \beta^2 + \gamma^2 = (\alpha + \beta + \gamma)^2 - 2(\alpha\beta + \beta\gamma + \gamma\alpha) = (-20)^2 - 2(9) = 382.$$

Similarly,

$$(\alpha\beta + \beta\gamma + \gamma\alpha)^2 = (\alpha^2\beta^2 + \beta^2\gamma^2 + \gamma^2\alpha^2) + 2(\alpha\beta^2\gamma + \beta\gamma^2\alpha + \gamma\alpha^2\beta)$$
$$= (\alpha^2\beta^2 + \beta^2\gamma^2 + \gamma^2\alpha^2) + 2(\alpha\beta\gamma)(\alpha + \beta + \gamma)$$

and so

$$\alpha^2\beta^2 + \beta^2\gamma^2 + \gamma^2\alpha^2 = (9)^2 - 2(59)(-20) = 2441.$$

The last coefficient is the easiest:

$$\alpha^2\beta^2\gamma^2 = (\alpha\beta\gamma)^2 = 59^2 = 3481.$$

Therefore, the polynomial with roots α^2, β^2, γ^2 is

$$x^3 - 382x^2 + 2441x - 3481.$$

Another important pair of results connects the roots of a polynomial with its division properties.

Theorem 3.13. The Remainder Theorem. *Let $f(x)$ be a polynomial. If we divide $f(x)$ by $x - a$ to obtain a quotient $q(x)$ and a remainder r, that is,*

$$f(x) = q(x)(x - a) + r,$$

then the remainder is $r = f(a)$.

Corollary 3.14. The Factor Theorem. *For any polynomial $f(x)$ and any number a, we have*

$$x - a \text{ is a factor of } f(x) \qquad \Longleftrightarrow \qquad f(a) = 0.$$

Example. Find the remainder when the polynomial

$$f(x) = x^2(x^2 - 1)(x^2 - 4)(x^2 - 9) \cdots (x^2 - 2023^2)$$

is divided by $x - 2024$.

Solution. By the Remainder Theorem, the remainder is just $f(2024)$. To evaluate this, note that

$$f(x) = x^2(x + 1)(x - 1)(x + 2)(x - 2) \cdots (x + 2023)(x - 2023)$$

and so

$$f(2024) = 2024^2 \times 2025 \times 2023 \times 2026 \times 2022 \times \cdots \times 4047 \times 1.$$

The product includes the integers from 1 to 4047 once each, with 2024 also occurring one extra time. Therefore, the remainder is

$$f(2024) = 2024 \times 4047!.$$

3.11 INEQUALITIES

Inequalities which may appear very intricate can sometimes be proved using a few basic techniques.

Theorem 3.15. The Arithmetic–Geometric Mean Inequality. *For any positive numbers* x, y, *we have*

$$\frac{x+y}{2} \geq \sqrt{xy};$$

and the only case in which the left–hand side and right–hand side are equal is when $x = y$. *More generally, for any* n *positive numbers* $x_1, x_2, \ldots, x_n$, *we have*

$$\frac{x_1 + x_2 + \cdots + x_n}{n} \geq (x_1 x_2 \cdots x_n)^{1/n},$$

with equality if and only if all the x_k *are equal.*

The reason for the name of this theorem is that the left–hand sides of the inequalities are known as the **arithmetic means** of the numbers involved, the right–hand sides as their **geometric means**. We remark that the second part of this result may be proved by a remarkably ingenious variant of the method of mathematical induction. It is easy to see that the result is true when $n = 1$, and not very hard to show that it is true for $n = 2$. We may then prove that, if the inequality is true for some specific positive integer n, then it is also true for $2n$; moreover, if it is true for n, then it is true for $n - 1$. We leave readers to convince themselves that doing this proves the result for all positive integers n; and to look up the details of the last two steps.

Theorem 3.16. The Cauchy–Schwarz Inequality. *For any real numbers* $x_1, x_2, \ldots, x_n$ *and* $y_1, y_2, \ldots, y_n$, *we have*

$$(x_1 y_1 + x_2 y_2 + \cdots + x_n y_n)^2 \leq (x_1^2 + x_2^2 + \cdots + x_n^2)(y_1^2 + y_2^2 + \cdots + y_n^2).$$

Equality holds if and only if there exists a constant a *such that* $y_k = a x_k$ *for all* k, *or a constant* b *such that* $x_k = b y_k$ *for all* k.

Frequently, the key to proving an inequality is to make a clever choice of the variables appearing in one of these basic results.

Example. Suppose that $a_1, a_2, \ldots, a_n$ and $b_1, b_2, \ldots, b_n$ are positive numbers and that $a_1 + a_2 + \cdots + a_n \leq b_1 + b_2 + \cdots + b_n$. Prove that

$$b_1 + b_2 + \cdots + b_n \leq \frac{b_1^2}{a_1} + \frac{b_2^2}{a_2} + \cdots + \frac{b_n^2}{a_n}.$$

Solution. Apply the Cauchy–Schwarz Inequality to

$$x_k = \frac{b_k}{\sqrt{a_k}} \quad \text{and} \quad y_k = \sqrt{a_k}.$$

Then $x_k y_k = b_k$ and, hence,

$$(b_1 + b_2 + \cdots + b_n)^2 \leq \left(\frac{b_1^2}{a_1} + \frac{b_2^2}{a_2} + \cdots + \frac{b_n^2}{a_n} \right)(a_1 + a_2 + \cdots + a_n).$$

Since $a_1 + a_2 + \cdots + a_n \leq b_1 + b_2 + \cdots + b_n$, we have

$$(b_1 + b_2 + \cdots + b_n)^2 \leq \left(\frac{b_1^2}{a_1} + \frac{b_2^2}{a_2} + \cdots + \frac{b_n^2}{a_n} \right)(b_1 + b_2 + \cdots + b_n),$$

and the result follows.

CHAPTER 4

Problem Sources

O VER ITS SIXTY YEARS OF EXISTENCE, *Parabola* has succeeded in publishing many thought–provoking, ingenious and entertaining problems and puzzles. While the majority have been provided by the problem editors, the journal has also been fortunate in receiving numerous contributions from interested readers. Records from the early years of *Parabola* are scanty, but it is likely that a majority of the problems published in these years were devised by Charles Cox (chief editor and problems editor 1964–1991) and George Szekeres (editor 1964–2004). Many geometry problems in these years were very likely contributed by Esther Szekeres. It is possible that some problems were obtained from uncredited sources; it is possible that some were widely circulated problems, though the present authors have tried to avoid including problems which are exceptionally well known. We have attempted below to give credit to all people and organisations who have contributed work appearing in our selection of *Parabolic Problems*: we apologise in advance for the inevitable omissions.

South Australian School Mathematics Competition	1.
Sin Keong Tong	24, 280, 285, 286, 300, 310, 313, 318, 321.
Wisconsin Mathematical, Engineering and Scientific Talent Search	42, 44, 78, 81.
International Mathematical Olympiad http://www.imo-official.org/problems.aspx	47, 59, 65, 70, 71, 72, 76.
British Mathematical Olympiad	62, 63.
Eőtvős competition (Hungary)	73, 74.
Leningrad Mathematical Olympiads 1987-1991 (Contests in Mathematics Series; Vol. 1), 0th Edition, by Dmitry Fomin and Alexey Kirichenko, MathPro Press, 1994	169.
Walter Vannini	198.
Akash Pardeshi	253.
Shared under the CC-BY-SA license http://math.stackexchange.com/questions/2773177/	264.
Shared under the CC-BY-SA license http://math.stackexchange.com/questions/256178/	307.
Shiva Oswal	273.
James Franklin	282, 289.
Toyesh Prakash Sharma	287, 305.
Jason Zimba	288, 308.
Soham Dutta	304.

DOI: 10.1201/9781003396413-4

CHAPTER 5

List of Problems by Topic

T HE PRESENT AUTHORS feel that one of the particularly enjoyable features of problems appearing in *Parabola* has always been the variety of topics presented, in that no attempt has been made to collect together puzzles in similar areas. This (lack of) organisation has been maintained in *Parabolic Problems* for the same reason; and also in order to preserve the characteristic style of *Parabola's* first sixty years. Recognising, however, that some readers may especially want to tackle problems of a specific type, in this section we have categorised problems by subject area. Some have been listed in more than one category, often reflecting the variety of mathematical ideas we have employed in their solution.

Our intention is that the classification of a problem should indicate the principal aspects of the problem or its solution, and not any secondary work which may be required in solving the problem. For instance, a question whose solution employs straightforward use of routine algebra will not for that reason alone be listed in the "algebra" category.

Assigning a problem to a specific category inevitably runs the risk of providing a "spoiler" for the solution: therefore, readers wishing to solve our *Parabolic Problems* entirely by their own efforts should consult this list only with caution.

Algebra. Puzzles in which an apparently difficult algebraic problem can be resolved by an insightful idea. Also, algebraic aspects of functions and expressions which may not be commonly met in school syllabi.

Problems 1, 16, 33, 39, 41, 62, 70, 76, 75, 83, 89, 111, 112, 117, 126, 130, 137, 154, 171, 172, 176, 180, 182, 184, 199, 200, 202, 205, 208, 211, 215, 218, 219, 220, 221, 234, 236, 249, 252, 254, 263, 266, 269, 270, 271, 276, 303, 305, 314, 323.

Arithmetic. Puzzles that require the solver to find numbers satisfying specified conditions. This category overlaps with number theory, but normally applies to problems in which general properties of numbers are not needed.

Problems 2, 5, 24, 27, 29, 77, 78, 127, 128, 132, 144, 160, 172, 198, 203, 204, 214, 217, 221, 232, 233, 237, 273, 277, 280, 281, 289, 299, 302, 310.

Calculus. Problems that can be solved by calculus techniques.

Problems 6, 149, 253.

Combinatorics. Problems that can be solved by combinatorial counting techniques such as those described in Section 3.5.

Problems 14, 28, 34, 35, 38, 36, 48, 51, 54, 57, 60, 67, 68, 73, 84, 87, 93, 90, 94, 95, 100, 103, 104, 105, 107, 114, 115, 116, 123, 125, 133, 136, 135, 141, 147, 152, 165, 166, 169, 193, 201, 206, 208, 207, 216, 227, 228, 231, 246, 258, 260, 262, 276, 284, 291, 294, 295, 297, 300, 313, 315, 318, 320, 324, 326, 328.

DOI: 10.1201/9781003396413-5

Don't be afraid of large numbers. It's often entertaining when a problem which, to all appearances, involves only "ordinary sized" numbers turns out to have a very large answer – just so long as the solution involves plenty of ingenuity, and does not rely solely on extensive calculations. This category also includes certain problems in which the final answer is surprisingly *small*.

Problems 86, 90, 137, 195, 198, 220, 232, 299, 316, 322.

Games and puzzles. Human beings have always amused themselves with games. Many such entertainments can be analysed with the aid of a little mathematical thinking; conversely, games and puzzles will often give rise to challenging mathematical problems.

Problems 24, 26, 34, 40, 132, 142, 167, 168, 174, 203, 246, 248, 265, 272, 275, 317.

Geometry. "Classical" Euclidean geometry as well as coordinate geometry.

Problems 1, 4, 6, 7, 9, 12, 11, 15, 17, 20, 21, 23, 32, 34, 37, 41, 42, 44, 43, 47, 50, 52, 53, 55, 56, 58, 61, 64, 68, 69, 72, 79, 82, 88, 91, 99, 103, 106, 110, 113, 118, 119, 120, 121, 122, 124, 133, 134, 140, 139, 143, 146, 147, 148, 151, 153, 156, 161, 162, 166, 170, 177, 179, 181, 183, 186, 185, 187, 188, 189, 192, 197, 202, 209, 210, 212, 213, 216, 222, 226, 229, 230, 236, 245, 247, 250, 251, 253, 257, 261, 278, 279, 282, 284, 285, 288, 290, 293, 298, 301, 306, 308, 312.

Graph theory. See Section 3.4 for the basics of graph theory. While, superficially, this field of mathematics only studies networks consisting of points and lines, it may happen that problems without any evident connection to such topics can be solved by an astute application of graph–theoretic ideas.

Problems 13, 18, 26, 36, 40, 63, 68, 80, 131, 163, 166, 238, 240, 243, 244, 304, 307, 309.

Induction. Problems that can be solved by the method of induction.

Problems 6, 18, 43, 65, 70, 73, 83, 94, 100, 102, 154, 171, 193, 238, 294, 299.

Inequalities. Frequently an ingenious approach can make the most intractable–seeming inequalities relatively straightforward.

Problems 1, 6, 21, 27, 40, 127, 144, 146, 153, 157, 158, 173, 180, 186, 188, 190, 239, 241, 245, 253, 272, 287.

Logic. Problems involving truth/falsity of given statements, or deductions which can be made from certain circumstances.

Problems 2, 8, 14, 48, 66, 129, 132, 134, 135, 138, 141, 150, 163, 174, 191, 223, 225, 246, 268, 272, 283, 292, 294, 301, 302.

No calculus. Problems that can can be solved easily by calculus techniques but which are more fun and challenging to solve without those techniques.

Problems 222, 224, 267, 293, 296.

Number theory. Generally speaking, problems involving properties of the integers, including divisibility, primes, squares and powers. Also, questions involving rationality and irrationality.

Problems 3, 5, 7, 10, 19, 22, 23, 26, 30, 31, 45, 46, 49, 52, 59, 65, 66, 71, 73, 81, 85, 86, 92, 96, 97, 98, 101, 102, 108, 109, 128, 135, 142, 145, 155, 159, 164, 175, 178, 182, 194, 195, 196, 197, 216, 229, 235, 255, 256, 258, 262, 259, 264, 274, 286, 306, 306, 311, 316, 319, 321, 322, 323, 325, 327, 329.

Sequences. Problems involving sequences – that is, finite or infinite lists of numbers. Some of these questions involve the well–known Fibonacci sequence.

Problems 28, 41, 74, 76, 113, 242, 270.

9 781032 483191